International Symposium on Entrance Channel Effect on the Reaction Mechanism in Heavy Ion Collisions 2013

Journal of Physics: Conference Series Volume 515

Messina, Italy
6 – 8 November 2013

Editors:

G. Giardina
A. Nasirov
G. Mandaglio

ISBN: 978-1-63439-296-9
ISSN: 1742-6588

Printed from e-media with permission by:

Curran Associates, Inc.
57 Morehouse Lane
Red Hook, NY 12571

Some format issues inherent in the e-media version may also appear in this print version.

Printed by Curran Associates, Inc. (2014)

For permission requests, please contact the Institute of Physics
at the address below.

Institute of Physics
Dirac House, Temple Back
Bristol BS1 6BE UK

Phone: 44 1 17 929 7481
Fax: 44 1 17 920 0979

techtracking@iop.org

Additional copies of this publication are available from:

Curran Associates, Inc.
57 Morehouse Lane
Red Hook, NY 12571 USA
Phone: 845-758-0400
Fax: 845-758-2634
Email: curran@proceedings.com
Web: www.proceedings.com

TABLE OF CONTENTS

Preface

The aim of the Symposium has been to widen and detail the discussion of problems arising in front of experimental and theoretical groups, and to find overlap between different approaches and methods which are devoted to the studying dynamics of nuclear reactions. Therefore, the reaction product yields are determined by various processes in competition. The main topics of the Symposium have been devoted to the following well sounded problems of nuclear reactions:

- The synthesis of superheavy elements and the study of exotic nuclei far from the valley of the beta stability. The production mechanism of the observed new elements and isotopes.
- The study of transfer reactions as a way to understand mechanism of evolution of from the deep-inelastic collisions to fusion regime.
- The study of non-equilibrium stage of the reaction mechanism and distribution of the excitation energy between binary reaction products including spontaneous fission products are still important to have a correct presentation about the whole reaction mechanism.
- The similarities and difference between fusion–fission and quasifission products. Unambiguity in estimation of the realistic fusion cross sections by the experimental and theoretical methods.
- Angular anisotropy of the complete and incomplete fusion reaction products.
- The effect of the nuclear shell structure in formation of the mass symmetric and asymmetric fission products.
- The investigation of the role of angular momentum, mass asymmetry and orientation angles of the symmetry axes of colliding nuclei in the entrance channel in formation of the evaporation residues, mass and angular distribution of the fusion–fission and quasifission products.
- Multi-fragmentation and symmetry energy.

The new experimental and theoretical investigations on these and related topics allow researchers to improve knowledge about nucleus–nucleus interaction dynamics and to make conclusions about perspectives in the study of the landscape of islands superheavy elements and exotic nuclei far from the valley of the beta stability.

Giorgio Giardina
Avazbek K Nasirov
Giuseppe Mandaglio
Editors of the Proceedings of ECHIC2013

Journal of Physics: Conference Series **515** (2014) 011002 doi:10.1088/1742-6596/515/1/011002

Committees

International Symposium

Entrance Channel Effect on the Reaction Mechanism in Heavy Ion Collisions

Messina (Italy), 6–8 November, 2013

Editors:

Giorgio Giardina (University of Messina), Avazbek K Nasirov (JINR of Dubna) and Giuseppe Mandaglio (CSFNSM of Catania)

Chairman: Giorgio Giardina (Messina)
Co-Chairman: Avazbek Nasirov (JINR-Dubna)
Co-Chairman: Giuseppe Mandaglio (Messina)
Co-Chairman: Katsuhisa Nishio (JAEA-Tokai)

Scientific Secretaries: Francesca Curciarello and Veronica De Leo (Messina)

Local Organizing Committee

- L Auditore
- F Curciarello
- V De Leo
- G Fazio
- G Giardina
- G Mandaglio
- M Trimarchi
- A Trifirò.

Journal of Physics: Conference Series **515** (2014) 011002 doi:10.1088/1742-6596/515/1/011002

International Advisory Committee

- N V Antonenko (JINR, Russia)
- Birger B Back (ANL, USA)
- Bao-An Li (TAMU-Commerce, USA)
- L Calabretta (LNS, Italy)
- R K Choudhury (BARC – India)
- L Corradi (LNL-Padova)
- G Giardina (Messina, Italy)
- D Hinde (Camberra, Australia)
- S Hofmann (GSI, Germany)
- M G Itkis (JINR, Dubna)
- E Kozulin (JINR, Dubna)
- Yu-Gang Ma (Shangai, China)
- Jie Meng (Beijing, China)
- A K Nasirov (JINR, Russia)
- J B Natowitz (TAMU, USA)
- K Nishio (JAEA, Japan)
- Yu Ts Oganessian (JINR, Russia)
- A Pagano (INFN-CT, Italy)
- K Pomorski (Lublin, Poland)
- A Sobiczewski (Warsaw, Poland)

Organizing Institutions:
University of Messina and Fondazione Bonino-Pulejo (Messina)

Topics:
Deep inelastic collisions and nucleon transfer reactions
Mass asymmetric and more symmetric entrance channels
Competition between quasifission and complete fusion
Fast fission and fusion–fission
Investigation on nuclear fragment production
Synthesis of superheavy elements

Sponsored by: University of Messina, Fondazione Bonino-Pulejo (Messina) and INFN Sezione di Catania.

Website: http://newcleo.unime.it/ECHIC2013

International Symposium

Entrance Channel Effect on the Reaction Mechanism in Heavy Ion Collisions

Messina (Italy) November 6-8, 2013

Chairman: Giorgio Giardina (Messina)
Co-Chairman: Avazbek Nasirov (JINR-Dubna)
Co-Chairman: Giuseppe Mandaglio (Messina)
Co-Chairman: Katsuhisa Nishio (JAEA-Tokai)

Organized by:
University of Messina
Fondazione Bonino-Pulejo (Messina)

Scientific Secretaries: Francesca Curciarello and Veronica De Leo (Messina)

Topics:
Deep inelastic collisions and nucleon transfer reactions
Mass asymmetric and more symmetric entrance channels
Competition between quasifission and complete fusion
Fast fission and Fusion-fission
Investigation on nuclear fragment production
Synthesis of superheavy elements

International Advisory Committee:

N.V. Antonenko (JINR, Russia)
Birger B. Back (ANL, USA)
Bao-An Li (TAMU-Commerce, USA)
L. Calabretta (LNS, Italy)
R. K. Choudhury (BARC – India)
L. Corradi (LNL-Padova)
G. Giardina (Messina, Italy)
D. Hinde (Camberra, Australia)
S. Hofmann (GSI, Germany)
M.G. Itkis (JINR, Dubna)
E. Kozulin (JINR, Dubna)
Yu-Gang Ma (Shangai, China)
Jie Meng (Beijing, China)
A.K. Nasirov (JINR, Russia)
J. B. Natowitz (TAMU, USA)
K. Nishio (JAEA, Japan)
Yu. Ts. Oganessian (JINR, Russia)
A. Pagano (INFN-CT, Italy)
K. Pomorski (Lublin, Poland)
F. Riggi (Catania, Italy)
A. Sobiczewski (Warsaw, Poland)

Local Organizing Committee: L. Auditore, F. Curciarello, V. De Leo, G. Fazio, G. Giardina, G. Mandaglio, M. Trimarchi, A. Trifirò.

Symposium Proceedings will be published in
Journal of Physics: Conference Series

Contact: loc@unime.it

Website: http://newcleo.unime.it/Events/ECHIC2013

Peer review statement

All papers published in this volume of *Journal of Physics: Conference Series* have been peer reviewed through processes administered by the proceedings Editors. Reviews were conducted by expert referees to the professional and scientific standards expected of a proceedings journal published by IOP Publishing.

Nuclear reactions at near-barrier energies with quantum diffusion approach

V.V.Sargsyan[1,2], G. Scamps[3], G.G.Adamian[1], N.V.Antonenko[1], D. Lacroix[4], W.Scheid[5] and H.Q.Zhang[6]

[1]Joint Institute for Nuclear Research, 141980 Dubna, Russia
[2]International Center for Advanced Studies, Yerevan State University, 0025 Yerevan, Armenia
[3]GANIL, 14076 Chen Cedex, France
[4]Institut de Physique Nucléaire, Université Paris-Sud, F-91406 Orsay Cedex, France
[5]Institut für Theoretische Physik der Justus–Liebig–Universität, D–35392 Giessen, Germany
[6]China Institute of Atomic Energy, Post Office Box 275, Beijing 102413, China

E-mail: adamian@theor.jinr.ru

Abstract. The role of neutron transfer in the fusion (capture) reactions is discussed.

1. Introduction

The nuclear deformation and neutron-transfer process have been identified as playing a major role in the magnitude of the sub-barrier fusion (capture) cross sections [1]. There are a several experimental evidences which confirm the importance of nuclear deformation on the fusion. The influence of nuclear deformation is straightforward. If the target nucleus is prolate in the ground state, the Coulomb field on its tips is lower than on its sides, that then increases the capture or fusion probability at energies below the barrier corresponding to the spherical nuclei. The role of neutron transfer reactions is less clear. The importance of neutron transfer with positive Q-values on nuclear fusion (capture) originates from the fact that neutrons are insensitive to the Coulomb barrier and therefore they can start being transferred at larger separations before the projectile is captured by target-nucleus. Therefore, it is generally thought that the sub-barrier fusion cross section will increase because of the neutron transfer.

2. Quantum diffusion approach for capture

In the quantum diffusion approach [2, 3, 4, 5, 6] the capture of the projectile by the target-nucleus is described with a single relevant collective variable: the relative distance R between the colliding nuclei. This approach takes into consideration the fluctuation and dissipation effects in collisions of heavy ions which model the coupling of the relative motion with various channels (for example, the non-collective single-particle excitations, low-lying collective dynamical modes of the target and projectile). The nuclear static deformation effects are taken into account through the dependence of the nucleus-nucleus potential on the deformations and mutual orientations of the colliding nuclei. We have to mention that many quantum-mechanical and non-Markovian effects accompanying the passage through the potential barrier are taken into consideration in our formalism [2, 3].

The capture cross section is a sum of partial capture cross sections [2, 3]

$$\begin{aligned}\sigma_{cap}(E_{\rm c.m.}) &= \sum_J \sigma_{\rm cap}(E_{\rm c.m.}, J) = \\ &= \frac{\pi\hbar^2}{2\mu E_{\rm c.m.}} \sum_J (2J+1) \int_0^{\pi/2} d\theta_1 \sin(\theta_1) \int_0^{\pi/2} d\theta_2 \sin(\theta_2) P_{\rm cap}(E_{\rm c.m.}, J, \theta_1, \theta_2), \quad (1)\end{aligned}$$

where $\mu = m_0 A_1 A_2/(A_1 + A_2)$ is the reduced mass (m_0 is the nucleon mass), and the summation is over the possible values of angular momentum J at a given bombarding energy $E_{\rm c.m.}$. Knowing the potential of the interacting nuclei for each orientation with the angles $\theta_i (i = 1, 2)$, one can obtain the partial capture probability $P_{\rm cap}$ which is defined by the passing probability of the potential barrier in the relative distance R coordinate at a given J. The value of $P_{\rm cap}$ is obtained by integrating the propagator G from the initial state (R_0, P_0) at time $t = 0$ to the final state (R, P) at time t:

$$P_{\rm cap} = \lim_{t\to\infty} \int_{-\infty}^{r_{\rm in}} dR \int_{-\infty}^{\infty} dP \; G(R, P, t|R_0, P_0, 0) = \lim_{t\to\infty} \frac{1}{2} {\rm erfc} \left[\frac{-r_{\rm in} + \overline{R(t)}}{\sqrt{\Sigma_{RR}(t)}} \right]. \quad (2)$$

The second line in (2) is obtained by using the propagator $G = \pi^{-1}|\det \mathbf{\Sigma}^{-1}|^{1/2} \exp(-\mathbf{q}^T\mathbf{\Sigma}^{-1}\mathbf{q})$ ($\mathbf{q}^T = [q_R, q_P]$, $q_R(t) = R - \overline{R(t)}$, $q_P(t) = P - \overline{P(t)}$, $\overline{R(t=0)} = R_0$, $\overline{P(t=0)} = P_0$, $\Sigma_{kk'}(t) = 2\overline{q_k(t)q_{k'}(t)}$, $\Sigma_{kk'}(t=0) = 0$, $\{k, k'\} = \{R, P\}$, P is a momentum) calculated for an inverted oscillator which approximates the nucleus-nucleus potential V in the variable R. The frequency ω of this oscillator with an internal turning point $r_{\rm in}$ is defined from the condition of equality of the classical actions of approximated and realistic potential barriers of the same hight at given J. This approximation is well justified for the reactions and energy range, which are here considered.

We assume that the sub-barrier capture mainly depends on the optimal one-neutron ($Q_{1n} > Q_{2n}$) or two-neutron ($Q_{2n} > Q_{1n}$) transfer with the positive Q-value. Our assumption is that, just before the projectile is captured by the target-nucleus (just before the crossing of the Coulomb barrier) which is a slow process, the transfer occurs and can lead to the population of the first excited collective state in the recipient nucleus [7] (the donor nucleus remains in the ground state). So, the motion to the N/Z equilibrium starts in the system before the capture because it is energetically favorable in the dinuclear system in the vicinity of the Coulomb barrier. For the reactions under consideration, the average change of mass asymmetry is connected to the one- or two-neutron transfer ($1n$- or $2n$-transfer). Since after the transfer the mass numbers, the isotopic composition and the deformation parameters of the interacting nuclei, and, correspondingly, the height $V_b = V(R_b)$ and shape of the Coulomb barrier are changed, one can expect an enhancement or suppression of the capture. If after the neutron transfer the deformations of interacting nuclei increase (decrease), the capture probability increases (decreases). When the isotopic dependence of the nucleus-nucleus potential is weak and after the transfer the deformations of interacting nuclei do not change, there is no effect of the neutron transfer on the capture. In comparison with Ref. [8], we assume that the negative transfer $Q-$values do not play visible role in the capture process. Our scenario was verified in the description of many reactions [3, 4, 5, 6].

3. Results of calculations

Because the capture cross section is equal to the complete fusion cross section for the reactions treated, the quantum diffusion approach for the capture is applied to study the complete fusion. All calculated results are obtained with the same set of parameters as in Ref. [2]. Realistic friction coefficient in the relative distance coordinate $\hbar\lambda$=2 MeV is used. Its value is close to that calculated within the mean-field approaches [9, 10]. For the nuclear part of the nucleus-nucleus potential, the double-folding formalism with the Skyrme-type density-dependent effective nucleon-nucleon interaction is used [2, 3]. The parameters of the nucleus-nucleus interaction potential $V(R)$ are adjusted to describe the experimental data at energies above the Coulomb barrier corresponding to spherical nuclei. The absolute values of the experimental quadrupole deformation parameters β_2 of even-even deformed nuclei in the ground state and of the first excited collective states of nuclei are taken from Ref. [11]. For the nuclei deformed in the ground state, the β_2 in the first excited collective state is similar to the β_2 in the ground state. For the quadruple deformation parameter of an odd nucleus, we choose the maximal value from the deformation parameters of neighboring even-even nuclei (for example, $\beta_2(^{231}\mathrm{Th})=\beta_2(^{233}\mathrm{Th})=\beta_2(^{232}\mathrm{Th})=0.261$). For the double magic and neighboring nuclei, we take $\beta_2 = 0$ in the ground state. Since there are uncertainties in the definition of the values of β_2 in light-mass nuclei, one can extract the ground-state quadrupole deformation parameters of these nuclei from a comparison of the calculated capture cross sections with the existing experimental data. By describing the reactions ^{12}C+^{208}Pb, ^{18}O+^{208}Pb, ^{32}S+^{90}Zr, ^{58}Ni + ^{58}Ni, and ^{64}Ni + ^{64}Ni, where there are no neutron transfer channels with positive Q-values, we extract the ground-state quadrupole deformation parameters β_2=-0.3, 0.1, 0.312, 0.05, and 0.087, for the nuclei ^{12}C, ^{18}O, ^{32}S, ^{58}Ni, and ^{64}Ni, respectively, which are used in our calculations.

3.1. Role of neutron transfer in capture process at sub-barrier energies

After the neutron transfer in the reaction $^{40}\mathrm{Ca}(\beta_2 = 0) + {}^{48}\mathrm{Ca}(\beta_2 = 0) \rightarrow {}^{42}\mathrm{Ca}(\beta_2 = 0.247) + {}^{46}\mathrm{Ca}(\beta_2 = 0)$ [$Q_{2n} = 2.6$ MeV] or $^{40}\mathrm{Ca}(\beta_2 = 0) + {}^{116}\mathrm{Sn}(\beta_2 = 0.112) \rightarrow {}^{42}\mathrm{Ca}(\beta_2 = 0.247) + {}^{114}\mathrm{Sn}(\beta_2 = 0.121)$ [$Q_{2n} = 2.8$ MeV], or $^{40}\mathrm{Ca}(\beta_2 = 0) + {}^{124}\mathrm{Sn}(\beta_2 = 0.095) \rightarrow {}^{42}\mathrm{Ca}(\beta_2 = 0.247) + {}^{122}\mathrm{Sn}(\beta_2 = 0.1)$ [$Q_{2n} = 5.4$ MeV], the deformation of the nuclei increases and the mass asymmetry of the system decreases, and, thus, the value of the Coulomb barrier decreases and the capture cross section becomes larger (Fig. 1). In Fig. 2, we observe the same behavior in the reactions $^{58}\mathrm{Ni}(\beta_2 = 0.05) + {}^{64}\mathrm{Ni}(\beta_2 = 0.087) \rightarrow {}^{60}\mathrm{Ni}(\beta_2 = 0.207) + {}^{62}\mathrm{Ni}(\beta_2 \approx 0.1)$ [$Q_{2n} = 3.9$ MeV] and $^{64}\mathrm{Ni}(\beta_2 = 0.087) + {}^{132}\mathrm{Sn}(\beta_2 = 0) \rightarrow {}^{66}\mathrm{Ni}(\beta_2 = 0.158) + {}^{130}\mathrm{Sn}(\beta_2 = 0)$ [$Q_{2n} = 2.5$ MeV]. One can see a good agreement between the calculated results and the experimental data [12, 13, 14, 15, 16]. So, the observed capture enhancement at sub-barrier energies in the reactions mentioned above is related to the two-neutron transfer channel. One can see that at energies above and near the Coulomb barrier the cross sections with and without two-neutron transfer are almost similar.

One can find reactions with a positive Q-values of the two-neutron transfer where the transfer weakly influences or even suppresses the capture process. This happens if after the transfer the deformations of the nuclei do not change much or even decrease. For instance, in the reactions $^{60}\mathrm{Ni}(\beta_2 \approx 0.1) + {}^{100}\mathrm{Mo}(\beta_2 = 0.231) \rightarrow {}^{62}\mathrm{Ni}(\beta_2 = 0.198) + {}^{98}\mathrm{Mo}(\beta_2 = 0.168)$ [$Q_{2n} = 4.2$ MeV], $^{64}\mathrm{Ni}(\beta_2 \approx 0.087) + {}^{100}\mathrm{Mo}(\beta_2 = 0.231) \rightarrow {}^{66}\mathrm{Ni}(\beta_2 = 0.158) + {}^{98}\mathrm{Mo}(\beta_2 = 0.168)$ [$Q_{2n} = 0.94$ MeV], and $^{60}\mathrm{Ni}(\beta_2 \approx 0.1) + {}^{150}\mathrm{Nd}(\beta_2 = 0.285) \rightarrow {}^{62}\mathrm{Ni}(\beta_2 = 0.198) + {}^{148}\mathrm{Nd}(\beta_2 = 0.204)$ [$Q_{2n} = 6$ MeV] we expect a weak dependence of the capture cross section on the neutron transfer (Fig. 3). There is the experimental evidence [17] of such an effect for the ^{60}Ni + ^{100}Mo reaction. So, the two-neutron transfer channel with large positive Q_{2n}-value weakly influences the fusion (capture) cross section. The reduced capture cross sections in the reactions ^{60}Ni + ^{100}Mo,^{150}Nd are close to each other in contrast to those in the reactions ^{64}Ni + ^{58}Ni,^{132}Sn. The ^{60}Ni + ^{150}Nd reaction has even a small suppression due to the neutron transfer.

Figures 4 and 5 show the capture excitation function for the reactions ^{32}S+Pd,Ru as a function of the bombarding energy. One can see a relatively good agreement between the calculated results and the experimental data [18]. The Q_{2n}-values for the $2n$-transfer processes are positive for all reactions with ^{32}S. After the $2n$-transfer (before the capture) in the reactions $^{32}\mathrm{S}(\beta_2 = 0.312) + {}^{106}\mathrm{Pd}(\beta_2 = 0.229) \rightarrow {}^{34}\mathrm{S}(\beta_2 = 0.252) + {}^{104}\mathrm{Pd}(\beta_2 = 0.209)$,

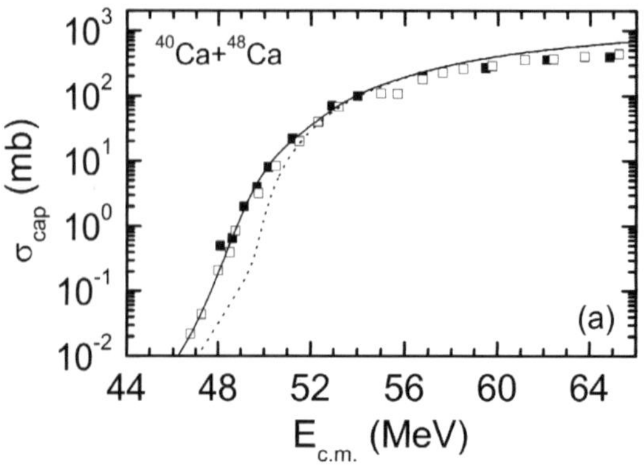

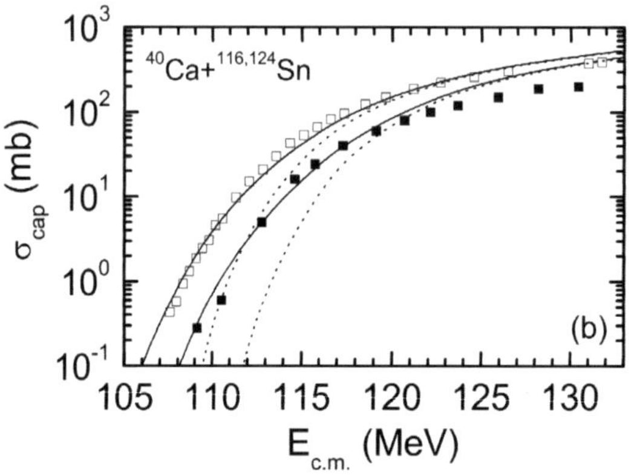

Figure 1. The calculated (lines) and experimental (symbols) [12, 13, 14] capture cross sections vs $E_{\rm c.m.}$ for the reactions ^{40}Ca+^{48}Ca (a) and ^{40}Ca+^{116,124}Sn (b). The calculated capture cross sections without taking into account the neutron pair transfer are shown by dotted lines.

^{32}S(β_2 = 0.312)+^{104}Pd(β_2 = 0.209)→^{34}S(β_2 = 0.252)+^{102}Pd(β_2 = 0.196) or ^{32}S(β_2 = 0.312)+^{104}Ru(β_2 = 0.271)→^{34}S(β_2 = 0.252)+^{102}Ru(β_2 = 0.24), ^{32}S(β_2 = 0.312)+^{102}Ru(β_2 = 0.24)→^{34}S(β_2 = 0.252)+^{100}Ru(β_2 = 0.215), the deformations of the nuclei decrease and the values of the corresponding Coulomb barriers increase. As a result, the transfer suppresses the capture process in these reactions at the sub-barrier energies. The suppression becomes stronger with decreasing energy (Figs. 4 and 5).

Figures 6 and 7 show the excitation functions for the reactions ^{18}O+^{74}Ge,^{112,118,124}Sn and ^{32}S+^{112,116}Sn. For the ^{32}S-induced reactions, $Q_{2n} > 0$. For the projectile ^{18}O there is a large range of positive Q_{2n}-values, for example, varying from 1.4 MeV for ^{18}O+^{124}Sn up to 5.5 MeV for ^{18}O+^{112}Sn. The agreement between the calculated results and the experimental data [19, 21] is rather good. As seen in Fig. 7, the cross sections increase systematically with the target mass number and run nearly similarly down to the lowest energy treated. In the reactions ^{32}S(β_2 = 0.312)+^{112}Sn(β_2 = 0.123)→^{34}S(β_2 = 0.252)+^{110}Sn(β_2 = 0.122), ^{32}S(β_2 = 0.312)+^{116}Sn(β_2 = 0.112)→^{34}S(β_2 = 0.252)+^{114}Sn(β_2 = 0.121), ^{18}O(β_2 = 0.1) + ^{74}Ge(β_2 = 0.283)→^{16}O(β_2 = 0) + ^{76}Ge(β_2 = 0.262), ^{18}O(β_2 = 0.1)+^{112}Sn(β_2 = 0.123)→^{16}O(β_2 = 0)+^{114}Sn(β_2 = 0.121), ^{18}O(β_2 = 0.1)+^{118}Sn(β_2 = 0.111)→^{16}O(β_2 = 0)+^{120}Sn(β_2 = 0.104), and ^{18}O(β_2 = 0.1)+^{124}Sn(β_2 = 0.095)→^{16}O(β_2 = 0)+^{126}Sn(β_2 = 0.09) the $2n$-transfer suppresses the capture process (Figs. 6 and 7). The sub-barrier capture cross sections for the systems ^{18}O+^{A}Sn studied here do not show any strong dependence on the mass number of the target isotope. Our results show that the cross sections for reactions ^{16}O+^{76}Ge (^{16}O+^{114,120,126}Sn) [$Q_{2n} < 0$] and ^{18}O+^{74}Ge (^{18}O+^{112,118,124}Sn) are very similar (Fig. 6). Just the same behavior

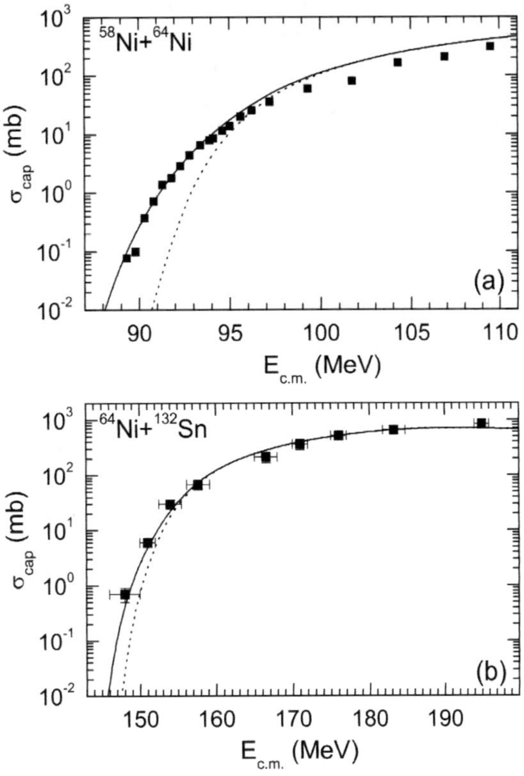

Figure 2. The same as in Fig. 1, but for the reactions ^{58}Ni+^{64}Ni (a) and ^{64}Ni+^{132}Sn (b). The experimental data are from Refs. [15, 16].

was observed in the recent experiments ^{16,18}O+^{76,74}Ge [19].

3.2. Influence of neutron pair transfer on capture

The choice of the projectile-target combination is crucial in the understanding of pair transfer phenomenon in the capture process. In the capture reactions with $Q_{1n} < 0$ and $Q_{2n} > 0$, the two-step sequential transfer is almost closed before capture. So, choosing properly the reaction combination, one can reduce the successive transfer in the process [26].

In the reactions ^{40}Ca + ^{48}Ca,^{116,124}Sn, ^{64}Ni + ^{58}Ni,^{132}Sn, ^{32}S + ^{102,104}Ru,^{104,106}Pd, and ^{18}O + ^{112,118,124}Sn, $1n$-neutron transfer is closed ($Q_{1n} < 0$) and Q_{2n}-values for the $2n$-transfer processes are positive (Figs. 1, 2, 4, 5, and 7). The enhancement or suppression arises not from the coherent successive transfer of two single neutrons, but from the direct transfer of one spatially correlated pair (the simultaneous transfer of two neutrons). Our results show that the capture (fusion) cross section of the reactions under consideration can be described by assuming the preformed dineutron-like clusters in the ground state of the nuclei ^{18}O, ^{48}Ca, ^{64}Ni, ^{116,124,132}Sn, ^{102,104}Ru, and ^{104,106}Pd. Since the dominance of the dineutron-like clusters is found in the surface of double magic, semimagic, and nonmagic nuclei, one can conclude that this effect is general for all stable and radioactive nuclei. Note that the strong spatial two-neutron correlation and the strong surface enhancement of the neutron pairing in the cases of a slab, a semi-infinite nuclear matter, and the finite superfluid nuclei are well known. Previously, the importance of the neutron pair transfer in the capture (fusion) process was stressed in Refs. [8, 18, 23, 24, 25].

One can make unambiguous statements regarding the neutron pair transfer process in the

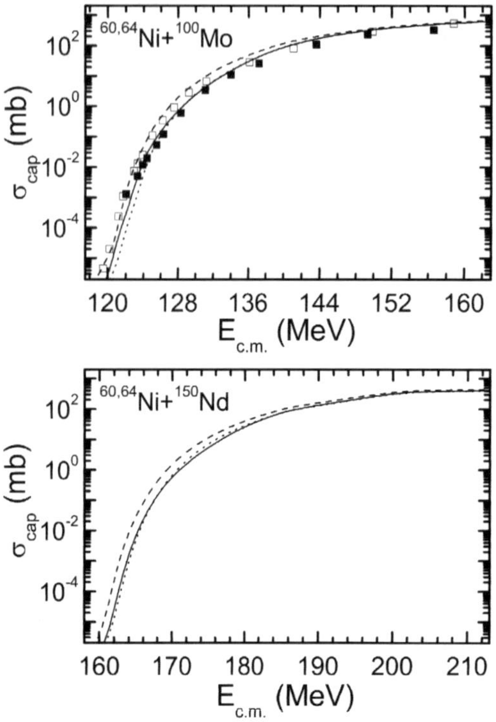

Figure 3. The same as in Fig. 1, for the indicated reactions ^{60}Ni + ^{100}Mo,^{150}Nd (solid lines), and ^{64}Ni + ^{100}Mo,^{150}Nd (dashed lines). For the reactions ^{60}Ni + ^{100}Mo and ^{60}Ni + ^{150}Nd, the calculated capture cross sections without the neutron transfer are shown by dotted lines. The experimental data for the reactions ^{60}Ni + ^{100}Mo (closed squares) and ^{64}Ni + ^{100}Mo (open squares) are from Ref. [17].

reactions ^{40}Ca + ^{62}Ni [$Q_{1n} = -2.23$ MeV, $Q_{2n} = 1.43$ MeV], ^{40}Ca + ^{64}Ni [$Q_{1n} = -1.29$ MeV, $Q_{2n} = 3.45$ MeV], ^{40}Ca + ^{114}Sn [$Q_{1n} = -1.94$ MeV, $Q_{2n} = 1.8$ MeV], ^{40}Ca + ^{118}Sn [$Q_{1n} = -1.55$ MeV, $Q_{2n} = 3.56$ MeV], ^{40}Ca + ^{120}Sn [$Q_{1n} = -0.75$ MeV, $Q_{2n} = 4.25$ MeV], ^{40}Ca + ^{122}Sn [$Q_{1n} = -0.45$ MeV, $Q_{2n} = 4.86$ MeV], ^{58}Ni + ^{62}Ni [$Q_{1n} = -1.6$ MeV, $Q_{2n} = 1.94$ MeV], ^{60}Ni + ^{64}Ni [$Q_{1n} = -1.84$ MeV, $Q_{2n} = 1.95$ MeV], ^{64}Ni + ^{128}Sn [$Q_{1n} = -1.8$ MeV, $Q_{2n} = 1.6$ MeV], and ^{64}Ni + ^{130}Sn [$Q_{1n} = -1.52$ MeV, $Q_{2n} = 2.1$ MeV]. There is a considerable difference between the sub-barrier capture cross sections with and without taking into consideration the neutron pair transfer in these reactions [26]. After two-neutron transfer, the deformation of light nucleus strongly increases and the capture cross section enhances. The neutron pair transfer induces the effect of the quadrupole deformation in the light nucleus. The study of the capture reactions following the neutron transfer will provide a good test for the effects of the neutron pair transfer.

3.3. Neutron pair transfer phenomenon in heavy-ion sub-barrier reactions

The Time-Dependent Hartree-Fock (TDHF) plus BCS approach [27, 28] has been recently used [28, 26] to extract the one- and two-neutron transfer probabilities (P_{1n}, P_{2n}) in heavy-ion scattering reactions. It was shown that, when the energy is well below the Coulomb barrier, the one-nucleon channel largely dominates. This is further illustrated here for the reactions ^{40}Ca + ^{116,124,130}Sn that have been discussed above and where the tin isotopes are superfluid. In Fig. 8, the one- and two-neutron transfer probabilities are displayed as functions of $B_0 - E_{\rm c.m.}$ for the

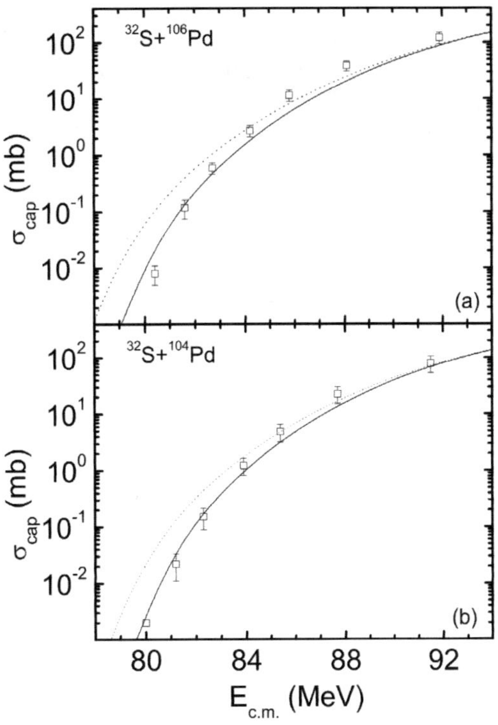

Figure 4. The same as in Fig. 1, but for the reactions ^{32}S+^{106}Pd (a) and ^{32}S+^{104}Pd (b). The experimental data are from Ref. [18].

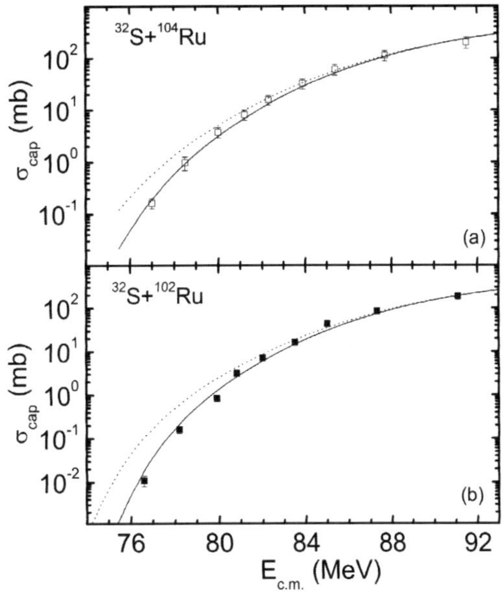

Figure 5. The same as in Fig. 1, but for the reactions ^{32}S+^{104}Ru (a) and ^{32}S+^{102}Ru (b). The experimental data are from Ref. [18].

sub- and near-barrier binary collisions of ^{40}Ca and tin isotopes. The Coulomb barrier (capture threshold energy) B_0 is deduced from the mean-field transport theory. This barrier are equal to

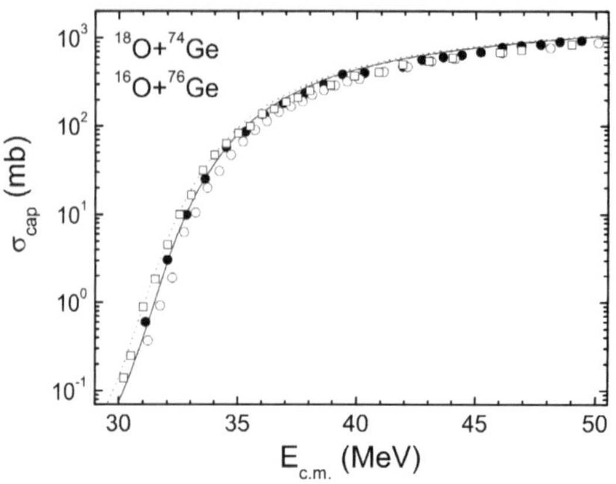

Figure 6. (Color online) The calculated (solid line) capture cross sections vs $E_{\text{c.m.}}$ for the reactions ^{16}O+^{76}Ge and ^{18}O+^{74}Ge (the curves coincide). For the ^{18}O+^{74}Ge reaction, the calculated capture cross sections without neutron transfer are shown by dotted line. The experimental data for the reactions ^{16}O+^{76}Ge (open circles) and ^{18}O+^{74}Ge (open squares) are from Ref. [19]. The experimental data for the ^{16}O+^{76}Ge reaction (solid circles) are from Ref. [20].

116.41 ± 0.07 (^{116}Sn), 114.69 ± 0.04 (^{124}Sn) and 113.92 ± 0.02 (^{130}Sn) MeV. It was found that the calculated B_0 are insensitive to the introduction of pairing and in a good agreement with the barriers extracted from the experimental data [28]. Note that the presented calculation are shown for the mixed pairing interaction only. The use of other interaction (surface or volume) leads to similar conclusions. Figure 8 gives an interesting insight in the one- and two-neutron transfers. As seen, a strong enhancement of P_{1n} and P_{2n} occurs with increasing bombarding energy. Since the enhancement of P_{2n} is stronger than that of P_{1n}, these probabilities become close to each other with decreasing $B_0 - E_{\text{c.m.}}$. This is indeed observed experimentally in Refs. [29, 30] where it was found that P_{2n} grows faster than P_{1n} with decreasing $B_0 - E_{\text{c.m.}}$ at energy relatively far below the Coulomb barrier. In all cases, as the energy approaches the capture barrier energy, there exist an energy range where $P_{2n} > P_{1n}$ dominates (shaded area). We also note that the energy windows where the two-nucleon channel becomes dominant increases as the neutron nucleus become more exotic.

This evidently supports our assumption about important role of the two-neutron transfer (compared to the one-neutron transfer) in the capture process, because in the TDHF calculation the scattering trajectory of two heavy ions at energy near the Coulomb barrier is close to the capture trajectory. Note that in the capture process the system trajectory crosses the barrier position $R = R_b$ at any energies. The results of our calculations predict that there is the crossing point of P_{2n} and P_{1n} at energy very close to the Coulomb barrier. Just before reaching R_b the neutron-pair transfer becomes the dominant channel. Thus, our assumption about two-neutron transfer before the capture is correct. The transfer more than two neutrons mainly occurs at $R < R_b$, i.e., just after the capture.

3.4. Neutron transfer in reactions with weakly bound nuclei

After the neutron transfer in the reactions ^{13}C+^{232}Th($\beta_2 = 0.261$)→^{14}C($\beta_2 = -0.36$)+^{231}Th($\beta_2 = 0.261$) [$Q_{1n} = 1.74$ MeV], ^{15}C+^{232}Th($\beta_2 = 0.261$)→^{14}C($\beta_2 = -0.36$)+^{233}Th($\beta_2 = 0.261$) [$Q_{1n} = 3.57$ MeV] the deformations of the target or projectile nuclei in these reactions and in the ^{14}C+^{232}Th($\beta_2 = 0.261$) ($Q_{1n,2n} < 0$) reaction are the same. In Fig. 9 the calculated cross sections slightly increase with the mass number of C, and are nearly

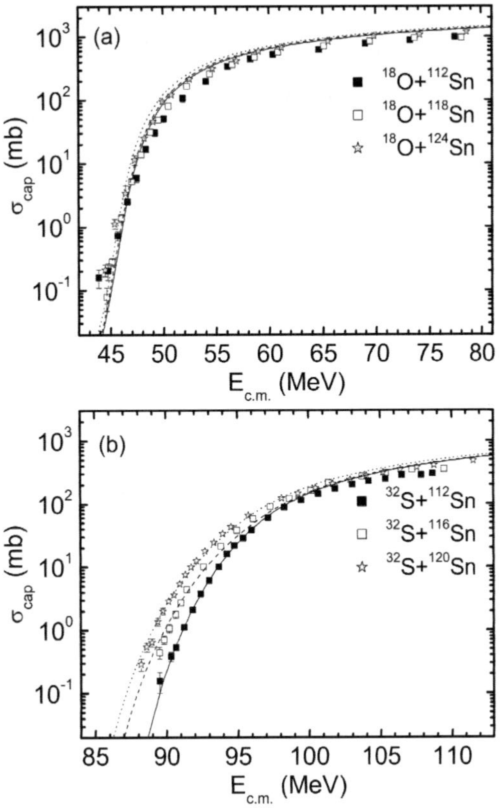

Figure 7. The calculated capture cross sections vs $E_{\rm c.m.}$ for the reactions ^{18}O+^{112,118,124}Sn (solid, dashed and dotted lines, respectively) (a) and ^{32}S+^{112,116,120}Sn (solid, dashed and dotted lines, respectively) (b). The experimental data (symbols) are from Ref. [21, 22].

parallel down to the lowest energy treated. There is a relatively good agreement between the calculated results [6] and the experimental data [31, 32] for the reactions ^{12,13,14}C+^{232}Th, but the experimental enhancement of the cross section in the ^{15}C+^{232}Th reaction at sub-barrier energies cannot be explained with our and other [31] models. Because we take into account the neutron transfer (^{15}C→^{14}C), one can suppose that this discrepancy is attributed to the influence of the breakup channel [1] which is not considered in our model. However, it is unclear why the breakup process influences only two experimental points at lowest energies. Different deviations of these points in energy from the calculated curve in Fig. 9 create doubt in an influence of the breakup on the kinetic energy. So, additional experimental and theoretical investigations are desirable.

The question is whether the fusion of nuclei involving weakly bound neutrons is enhanced or suppressed at low energies. This question can been addressed to the systems $^{12-15}$C+^{208}Pb [33]. After the neutron transfer in the reactions ^{13}C+^{208}Pb($\beta_2=0$)→^{14}C($\beta_2=-0.36$)+^{207}Pb($\beta_2=0$) [$Q_{1n}=1.74$ MeV], ^{15}C+^{208}Pb($\beta_2=0$)→^{14}C($\beta_2=-0.36$)+^{209}Pb($\beta_2=0.055$) [$Q_{1n}=3.57$ MeV] the deformations of the light nuclei are the same as in the ^{14}C+^{208}Pb($\beta_2=0$) [$Q_{1n,2n}<0$] reaction. The heavy nuclei are almost spherical. This means that the slopes of the excitation functions are almost the same (Fig. 10). As in the case of the ^{15}C+^{232}Th reaction, we do not expect enhancement of the capture cross section in the ^{15}C+^{208}Pb reaction owing to the neutron transfer. The same effect was observed in Ref. [33]. The study of the reactions ^{15}C+^{208}Pb,^{232}Th at sub-barrier energies provides a good test for the verification of the effect of weakly bound

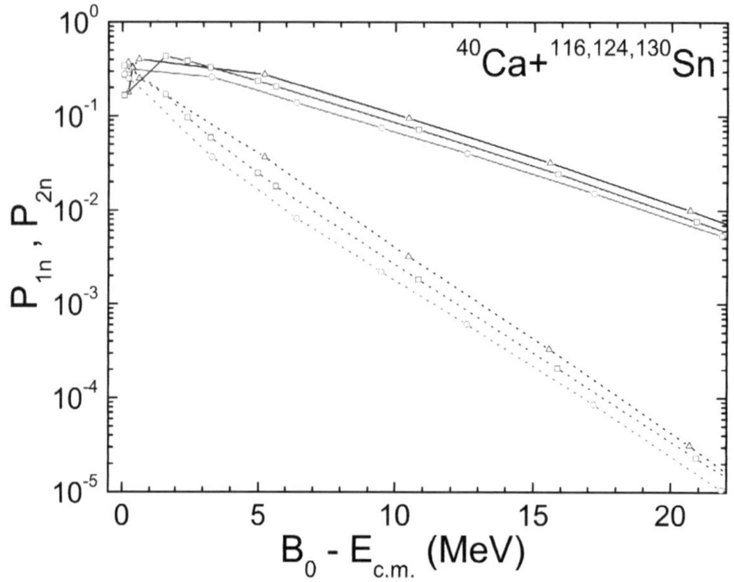

Figure 8. (Color online) The calculated one- (symbols connecting by solid lines) and two-neutron (symbols connecting by dotted lines) transfer probabilities vs $B_0 - E_{\rm c.m.}$ for the reactions ^{40}Ca+^{116}Sn (circles), ^{40}Ca+^{124}Sn (triangles), and ^{40}Ca+^{130}Sn (squares).

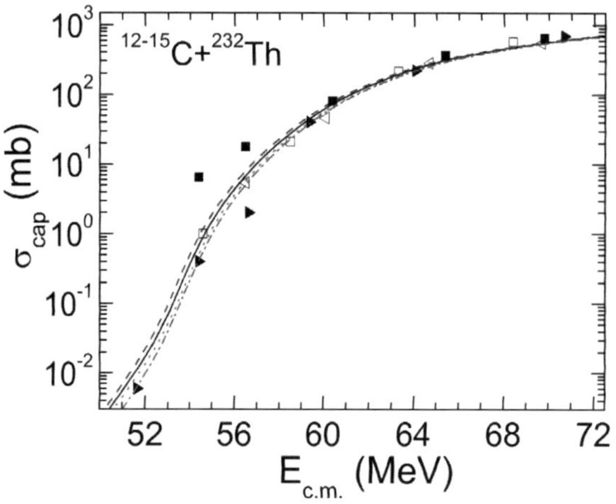

Figure 9. (Color online) The calculated (lines) and experimental (symbols) capture cross sections vs $E_{\rm c.m.}$ for the reactions ^{12}C+^{232}Th (dash-dotted line, solid triangles), ^{13}C+^{232}Th (dotted line, open triangles), ^{14}C+^{232}Th (solid line, open squares), and ^{15}C+^{232}Th (dashed line, solid squares). The experimental data are from Refs. [31, 32].

nuclei on fusion and capture because it reveals the role of other effects besides neutron transfer.

After the $2n$-transfer in the reactions ^{6}He+^{206}Pb→^{4}He($\beta_2 = 0$)+^{208}Pb($\beta_2 = 0.055$) [$Q_{2n} =$ 13.13 MeV], ^{9}Li+^{86}Zn→^{7}Li($\beta_2 \approx 0.4$)+^{70}Zn($\beta_2 = 0.248$) [$Q_{2n} = 9.60$ MeV] they become equivalent to the reactions ^{4}He+^{208}Pb and ^{7}Li+^{70}Zn. Therefore, the slopes of the excitation functions in the reactions with ^{6}He (^{9}Li) and ^{4}He (^{7}Li) should be similar. This conclusion supports the experimental data of Ref. [35], where the authors concluded that the fusion enhancement in the ^{6}He+^{206}Pb reaction (with respect to the ^{4}He+^{208}Pb reaction) is rather

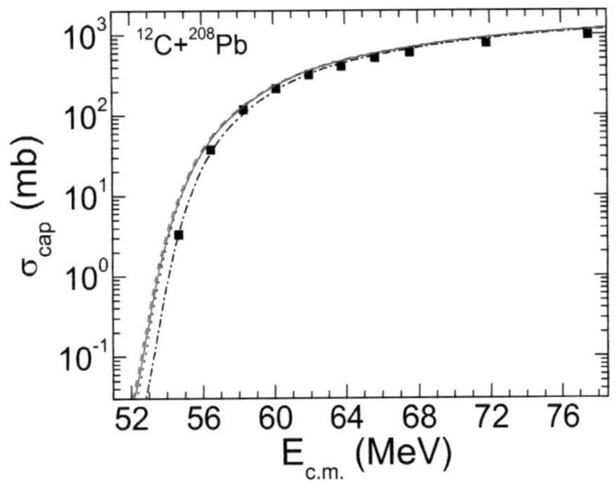

Figure 10. The calculated (lines) and experimental (symbols) capture cross sections vs $E_{\rm c.m.}$ for the reactions ^{12}C+^{208}Pb (dash-dotted line), ^{13}C+^{208}Pb (dotted line), ^{14}C+^{208}Pb (solid line), and ^{15}C+^{208}Pb (dashed line). The experimental data (solid squares) for the ^{12}C+^{208}Pb reaction are from Ref. [34].

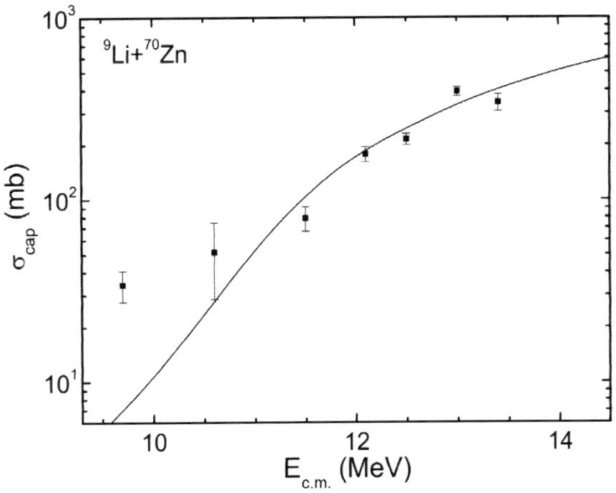

Figure 11. The calculated (solid line) and experimental (symbols) capture cross sections vs $E_{\rm c.m.}$ for the reaction ^{9}Li+^{70}Zn. The experimental data are from Ref. [36].

small or absent.

By assuming that the $2n$-transfer process occurs, we calculated the capture cross sections for the ^{9}Li+^{70}Zn reaction (Fig. 11). At lowest energies, the calculated cross section is by factor of $\sim$ 5 less than the experimental value. The experimental data are reproduced by the model [37] where two-neutron transfer from the ^{70}Zn leads to ^{11}Li halo structure and molecular bond between the nuclei in contact enhances the fusion cross section. However, the two-neutron transfer ^{9}Li+^{70}Zn$\to^{7}$Li+^{72}Zn with Q_{2n} = 8.6 MeV is much energetically favorable than the two-neutron transfer ^{9}Li+^{70}Zn$\to^{11}$Li+^{68}Zn with $Q_{2n} = -15.4$ MeV. These observations deserve further experimental and theoretical investigations including the breakup channel.

4. Summary

The quantum diffusion approach was applied to study the role of the neutron transfer with positive Q-value in the capture reactions at sub-, near- and above-barrier energies.

We demonstrated a good agreement of the theoretical calculations with the experimental data. We found, that the change of the magnitude of the capture cross section after the neutron transfer occurs due to the change of the deformations of nuclei. The effect of the neutron transfer is an indirect effect of the quadrupole deformation. When after the neutron transfer the deformations of nuclei do not change or slightly decrease, the neutron transfer weakly influences or suppresses the capture cross section. Good examples for this effect are the capture reactions ^{60}Ni + ^{100}Mo,^{150}Nd, ^{18}O + ^{64}Ni,^{112,114,116,118,120,122,124}Sn,^{204,206}Pb, and ^{32}S+^{96}Zr,^{94,96,98,100}Mo,^{100,102,104}Ru,^{104,106,108,110}Pd,^{112,114,116,118,120,122,124}Sn. at sub-barrier energies. Thus, the general point of view that the sub-barrier capture (fusion) cross section strongly increases because of the neutron transfer with a positive Q-values has to be revised. The neutron transfer effect can lead to a weak influence of halo-nuclei on the capture. We demonstrated the important role of two-neutron transfer channel in the heavy-ion scattering at sub-barrier energies close to the Coulomb barrier. One can suggest the experiments ^{40}Ca + ^{116,124}Sn,^{48}Ca to check our predictions.

This work was supported by DFG, NSFC, RFBR, and JINR grants. The IN2P3(France)-JINR(Dubna) and Polish - JINR(Dubna) Cooperation Programmes are gratefully acknowledged.

[1] Canto L F, Gomes P R S, Donangelo R and Hussein M S 2006 *Phys. Rep.* **424** 1
[2] Sargsyan V V, Adamian G G, Antonenko N V and Scheid W 2010 *Eur. Phys. J.* A **45** 125
[3] Sargsyan V V, Adamian G G, Antonenko N V, Scheid W and Zhang H Q 2011 *Phys. Rev.* C **84** 064614
[4] Sargsyan V V, Adamian G G, Antonenko N V, Scheid W and Zhang H Q 2012 *Phys. Rev.* C **85** 024616
[5] Sargsyan V V, Adamian G G, Antonenko N V, Scheid W and Zhang H Q 2013 *Eur. Phys. J.* A **49** 54
[6] Sargsyan V V, Adamian G G, Antonenko N V, Scheid W and Zhang H Q 2012 *Phys. Rev.* C **86** 014602
[7] Szilner S et al 2011 *Phys. Rev.* C **84** 014325
[8] Dasso C H, Landowne S and Winther A 1983 *Nucl. Phys.* A **405** 381
[9] Adamian G G, Nasirov A K, Antonenko N V and Jolos R V 1994 *Phys. Part. Nucl.* **25** 583
[10] Washiyama K, Lacroix D and Ayik S 2009 *Phys. Rev.* C **79** 024609
[11] Raman S, Nestor C W, Jr and Tikkanen P 2001 *At. Data Nucl. Data Tables* **78** 1
[12] Aljuwair H A et al 1984 *Phys. Rev.* C **30** 1223
[13] Jiang C L et al 2010 *Phys. Rev.* C **82** 041601(R)
[14] Scarlassara F et al 2000 *Nucl. Phys.* A **672** 99
[15] Beckerman M, Salomaa M, Sperduto A, Molitoris J D and DiRienzo A 1982 *Phys. Rev.* C **25** 837
[16] Liang J F et al 2008 *Phys. Rev.* C **78** 047601
[17] Scarlassara F et al 2011 *EPJ Web Conf.* **17** 05002
[18] Pengo R et al 1983 *Nucl. Phys.* A **411** 255
[19] Jia H M et al 2012 *Phys. Rev.* C **86** 044621
[20] Aguilera E F, Kolata J J and Tighe R J 1995 *Phys. Rev.* C **52** 3103
[21] Jacobs P, Fraenkel Z, Mamane G, and Tserruya L 1986 *Phys. Lett.* B **175** 271
[22] Tripathi V et al 2001 *Phys. Rev.* C **65** 014614
[23] Stefanini A M et al 1986 *Nucl. Phys.* A **456** 509
[24] Borges A M et al 1992 *Phys. Rev.* C **46** 2360
[25] Kalkal S et al 2011 *Phys. Rev. C* **83** 054607
[26] Sargsyan V V, Scamps G, Adamian G G, Antonenko N V and Lacroix D 2013 *Phys. Rev.* C **88** 064601
[27] Simenel C 2013 *Clusters in Nuclei* Vol.3 Beck C (ed.), *Lecture Notes in Physics* **875** 95
[28] Scamps G and Lacroix D 2013 *Phys. Rev.* C **87** 014605
[29] Corradi L et al 2011 *Phys. Rev.* C **84** 034603
[30] Evers M et al 2011 *Phys. Rev.* C **84** 054614
[31] Alcorta M et al 2011 *Phys. Rev. Lett.* **106** 172701
[32] Mein J M et al 1997 *Phys. Rev.* C **55** R995
[33] Keeley N and Alamanos N 2007 *Phys. Rev.* C **75** 054610
[34] Mukherjee A et al 2007 *Phys. Rev.* C **75** 044608
[35] Wolski R et al 2011 *Eur. Phys. J.* A **47** 111
[36] Vinodkumar A M et al 2009 *Phys. Rev.* C **80** 054609
[37] Balantekin B and Kocak G 2008 *AIP Conf. Proc.* **1072** 289

Impact of nuclear structure on production of superheavy nuclei

G G Adamian[1], N V Antonenko[1], A N Bezbakh[1], T M Shneidman[1,2], W Scheid[3]

[1]Joint Institute for Nuclear Research, 141980 Dubna, Russia
[2]Institute of Theoretical Physics, CAS, Beijing 100190, China
[3]Institut für Theoretische Physik der Justus-Liebig-Universität, D-35392 Giessen, Germany

E-mail: antonenk@theor.jinr.ru

Abstract. The calculations performed with the modified two-center shell model reveal quite strong shell effects at $Z = 120 - 126$ and $N = 184$ as in the self-consistent mean-field treatments. If our prediction of the structure of heaviest nuclei is correct, than one can expect the production of evaporation residues $Z = 120$ in the reactions ^{50}Ti+^{249}Cf and ^{54}Cr+^{248}Cm in near future. The nuclear level densities were calculated for the nuclei of α-decay chains containing ^{296,298,300}120. The minima of the level density parameter clearly indicate the strong shell effect at $Z = 120$.

1. Introduction

An extension of the nuclear chart to the elements with $Z > 100$ and a study of their properties provide us an important insight into the determination of an island of stability. The experimental trend of nuclear properties (Q_α-values and half-lives) as well as the production cross sections of superheavy elements (SHE) reveal an increasing stability of nuclei approaching $N = 184$, and indicate quite a large shell effects behind $Z = 114$ [1, 2, 3, 4, 5, 6, 7, 8]. This means that the predictions of relativistic and nonrelativistic mean-field models [9, 10, 11, 12] seem to be valid. In accordance with these self-consistent models a center of the island of stability is expected at Z=120-126 and N=172 or 184. In the (N, Z)–plane, the line, along which all new SHE were discovered in the actinide-based reactions with ^{48}Ca beam, just approaches this region [1, 3].

The structure of superheavies crucially influences the evaporation residue cross sections in the actinide-based complete fusion reactions [13]. The difference in the nuclear properties predicted for heaviest nuclei with various microscopic models is caused by the scanty experimental data in this region. Nuclear models contain a number of parameters which are fixed for the best description of known nuclei. Existing microscopic–macroscopic approaches [14, 15, 16, 17, 18, 19, 20] supply the basis for the intensive calculations of the properties of heavy nuclei. These approaches differ by the parametrization of nuclear shape and by the single-particle potential used. In Refs. [19, 20] we proposed the microscopic–macroscopic approach based on the two-center shell model (TCSM) [21]. The parameters were set so to describe the spins and parities of the ground state of known heavy nuclei. This approach has been used in Ref. [22] to reveal the trends in the shell effects and Q_α values with Z in heaviest nuclei.

The experimental evaporation residue cross sections σ_{xn} in the ^{48}Ca-induced complete fusion reactions do not depend strongly on the atomic number Z of SHE and are on the picobarn level. As known, the cross section of compound nucleus formation strongly decreases with increasing $Z_1 \times Z_2$. Since the absolute value of evaporation residue cross section is ruled by the product of complete fusion cross section and survival probability, the loss in the formation probability of compound nucleus in actinide-based reactions can be compensated by the gain in the survival probability of SHE.

2. Stability of SHE produced in actinide-based complete fusion reactions

The cross section of the production of SHE as the evaporation residues in the xn-evaporation channel is written as a sum over all partial waves J

$$\sigma_{xn}(E_{\rm c.m.}) = \sum_J \sigma_{fus}(E_{\rm c.m.}, J) W_{xn}(E_{\rm c.m.}, J), \quad (1)$$

$$\sigma_{fus}(E_{\rm c.m.}, J) = \int_0^{\pi/2}\int_0^{\pi/2} d\cos\Theta_1 d\cos\Theta_2 \; \sigma_c(E_{\rm c.m.}, J, \Theta_i) P_{CN}(E_{\rm c.m.}, J, \Theta_i).$$

Here, the averaging over the orientations of statically deformed interacting nuclei (Θ_i (i=1,2) are the orientation angles with respect to the collision axis) is taken into consideration. For the correct description of the experimental data, the partial capture cross section σ_c, fusion P_{CN} and survival W_{sur} probabilities should be properly calculated [23, 24, 25, 26, 27, 28, 29, 30, 31, 32]. The value of $\sigma_c(E_{\rm c.m.}, J, \Theta_i) = \frac{\pi\hbar^2}{2\mu E_{\rm c.m.}}(2J+1)T(E_{\rm c.m.}, J, \Theta_i)$ defines the transition of the colliding nuclei over the Coulomb barrier with the probability T and the formation of dinuclear system (DNS) when the kinetic energy $E_{\rm c.m.}$ and angular momentum J of the relative motion are transformed into the excitation energy and angular momentum of the DNS. The capture (transition) probability $T(E_{\rm c.m.}, J, (E_{\rm c.m.}, J)) = (1+\exp[2\pi(V_J(R_b,\Theta_i) - E_{\rm c.m.})/\hbar\omega_J(\Theta_i)])^{-1}$ is calculated with the Hill-Wheeler formula. The effective nucleus-nucleus potential

$$V_J(R,\Theta_i) = V_N(R,\Theta_i) + V_C(R,\Theta_i) + \hbar^2 J(J+1)/(2\Im)$$

is calculated as a sum of nuclear V_N, Coulomb V_C and centrifugal interactions and approximated near the Coulomb barrier at $R = R_b$ by the inverted harmonic-oscillator potential with the barrier height $V_J(R_b,\Theta_i)$ and frequency $\omega_J(\Theta_i)$. In the entrance channel the moment of inertia is $\Im = \mu R^2$. The nuclear potential V_N is calculated with the double-folding model using a nuclear radius parameter r_0=1.15 fm and a diffuseness a=0.54 fm for ^{48}Ca and a=0.56 fm for the actinide targets [33]. The quadrupole deformation parameters of actinides are taken from Ref. [34].

The DNS model [23, 24, 25, 26, 27, 28, 29, 30, 31, 32] is successful in describing the complete fusion reactions especially related to the production of heavy and superheavy nuclei. In the DNS model the compound nucleus is reached by a series of transfers of nucleons from the light nucleus to the heavy one. The dynamics of the DNS is considered as a combined diffusion in the degrees of freedom of the mass asymmetry $\eta = (A_1 - A_2)/(A_1 + A_2)$ (A_1 and A_2 are the mass numbers of the DNS nuclei) and of the internuclear distance R. The diffusion in R occurs towards the values larger than the sum of the radii of the DNS nuclei and finally leads to the quasifission (decay of the DNS). After the capture stage, the probability of complete fusion

$$P_{CN} = \lambda_\eta^{Kr}/(\lambda_\eta^{Kr} + \lambda_{\eta_{sym}}^{Kr} + \lambda_R^{Kr})$$

depends on the competition between the complete fusion in η, diffusion in η to more symmetric DNS and quasifission. This competition can strongly reduce the value of $\sigma_{fus}(E_{\rm c.m.}, J)$ and, correspondingly, the value of $\sigma_{xn}(E_{\rm c.m.})$. Since the initial DNS is in the conditional minimum of potential energy surface, we use a two-dimensional (in η and R coordinates) Kramers-type expression for the quasistationary rates λ_{η}^{Kr} of fusion, $\lambda_{\eta_{sym}}^{Kr}$ of symmetrization of the DNS, and λ_{R}^{Kr} of quasifission through the fusion barrier in η, the barrier in η towards symmetric DNS, and quasifission barrier in R, respectively. These barriers are given by the potential energy U of the DNS which is calculated as the sum of binding energies B_i of the DNS nuclei and of the nucleus-nucleus potential V_J: $U = B_1 + B_2 + V_J$. The binding energies for known and unknown nuclei are taken from Ref. [35] and Ref. [14] (the finite range droplet model), respectively. The uncertainty of calculated P_{CN} is within the factor of 2.

The survival probability $W_{xn}(E_{\rm c.m.}, J)$ estimates the competition between fission and neutron evaporation in the excited compound nucleus. In expression (1) the contributing angular momentum range is limited by W_{xn}. In the case of highly fissile SHE, W_{xn} is a narrow function of J different from zero in the vicinity of J=0 for all bombarding energies. The angular momentum dependence can be separated as

$$W_{xn}(E_{\rm c.m.}, J) = W_{xn}(E_{\rm c.m.}, J=0)\exp[-\sum_{i=1}^{x}\frac{\Delta B_i^{rot}}{T_i}] \approx W_{xn}(E_{\rm c.m.}, J=0)\exp[-\frac{J^2}{J_m^2(x)}], \quad (2)$$

where $\Delta B_i^{rot} = \hbar^2 J(J+1)[\frac{1}{2\theta_{\rm g}^i} - \frac{1}{2\theta_{\rm s}^i}]$ ($\theta_{\rm g}^i$ ($\theta_{\rm s}^i$) is the moment of inertia in the ground state (at the saddle point) in i-th evaporation step, $\theta_{\rm g,s}^i \approx \theta_{\rm g,s}^j \approx \theta_{\rm g,s}$, $i \neq j$), $J_m(x) = \phi(x)J_m(x=1)$, $J_m(x=1) = (T_1/[\hbar^2/(2\theta_{\rm g}) - \hbar^2/(2\theta_{\rm s})])^{1/2}$ (T_1 is the thermodynamical temperature in $1n$-channel) and $\phi(x) = (1 + \frac{1}{\sqrt{2}} + ... + \frac{1}{\sqrt{x}})^{-1}$. Since in heaviest nuclei the nuclear structure is drastically changed when the nucleus moves from the ground state to the saddle point, the values of $\theta_{\rm g}^i$ and $\theta_{\rm s}^i$ are expected to be rather different in spite of small difference in the deformation parameters between the ground state and saddle point. Thus, $W_{xn}(E_{\rm c.m.}, J)$ is cut at higher angular momenta by a gaussian-like factor. The width of this cutoff $J_m(x)$ weakly decreases with increasing x. For the reactions leading to SHE, $J_m(x=1) = 10$ is used at $E_{\rm c.m.}$ near the Coulomb barrier [31, 32]. $J_m(x=1) = 15$ was used in the actinide region [23]. These values of J_m correspond to the impact parameters less than 1 fm. In all heavy-ion complete fusion reactions above the Coulomb barrier, we have $J_m < J_{crit}$, where J_{crit} is the critical angular momentum which restricts the capture cross section. Therefore, only a limited part of the angular momentum distribution of compound nucleus appreciably contributes to the evaporation residue cross section.

Using Eqs. (1) and (2), and replacing the sum over J by the integral, we obtain the following approximate factorization:

$$\sigma_{xn}(E_{\rm c.m.}) = \sigma_{fus}^{eff}(E_{\rm c.m.})W_{xn}(E_{\rm c.m.}, J=0), \quad (3)$$

where the effective fusion cross section

$$\sigma_{fus}^{eff}(E_{\rm c.m.}) = \frac{\pi\hbar^2}{\mu E_{\rm c.m.}}\int_0^{\infty}\int_0^{\pi/2}\int_0^{\pi/2}\frac{dJd\cos\Theta_1 d\cos\Theta_2\; Je^{-\frac{J^2}{J_m^2(x)}}P_{CN}(E_{\rm c.m.}, J, \Theta_i)}{1+\exp[2\pi(V_J(R_b,\Theta_i) - E_{\rm c.m.})/\hbar\omega_J(\Theta_i)]} \quad (4)$$

contains the angular momentum dependence of survival probability. Using Eq. (3), one can extract the value of survival probability at the zero angular momentum from experimental cross section $\sigma_{xn}^{exp}(E_{\rm c.m.})$ as

$$W_{xn}(E_{\rm c.m.}, J=0) = \sigma_{xn}^{exp}(E_{\rm c.m.})/\sigma_{fus}^{eff}(E_{\rm c.m.}). \quad (5)$$

With the reduction to the zero angular momentum the survival probability becomes independent of the projectile-target combination.

The fusion probability and, correspondingly, the effective fusion cross section $\sigma_{fus}^{eff}(E_{\rm c.m.})$ decreases by about 2 orders of magnitude with increasing the charge number of compound nucleus from Z=112 to Z=118. The strong fusion hindrance is due to a competition between complete fusion and quasifission in the DNS.

In Fig. 1 the extracted values of W_{3n} and W_{4n} with Eq. (5) deviates from the expected magic proton number Z=114. This indicates an increase of the stability of SHE beyond Z=114. The experimental error-bars result the error-bars in the deduced W_{xn}. Since the fission barrier is determined by the shell correction energy, the absolute value of the shell correction energy is expected to be increased with Z. The shell correction energy strongly depends on that how the neutron and proton numbers of the compound nucleus are closed to the magic proton and neutron numbers. The found experimental trend of the Q_α-values in α-decay chains also indicates the monotonic increase of the amplitude of the ground state shell correction energy with charge number in the region Z=112-118 [4]. One can expect an increasing stability of nuclei approaching the closed neutron N=184 shell. However, in Fig. 1 $W_{3n}(^{296}_{180}\mathrm{Lv}) < W_{3n}(^{297}_{179}118)$. This probably indicates that Z=114 is not a proper proton magic number and the next doubly magic nucleus beyond ^{208}Pb is the nucleus with $Z \geq$120. The shell closure at $Z \geq$120 may influence stronger on the stability of the SHE than the sub-shell closure at Z=114. Note that the experimental uncertainties seem to be too small to overcome the trends presented in Fig. 1.

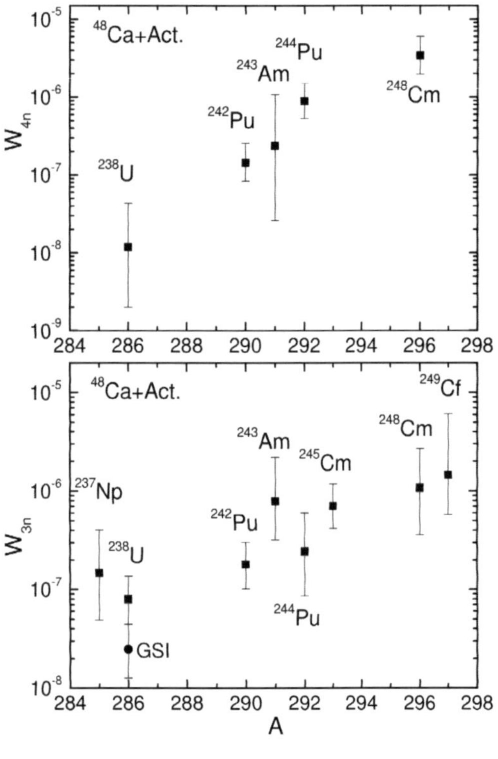

Figure 1. The survival probabilities of SHE in $3n$-and $4n$-channels, extracted with Eq. (5) and experimental σ_{xn}^{exp} from Refs. [1], as functions of mass number of the compound nucleus. For the reaction ^{48}Ca+^{238}U, the experimental σ_{3n}^{exp} from GSI [4] is used as well.

3. Microscopic-macroscopic approaches based on the TCSM

The microscopic-macroscopic approaches provide a powerful tool for systematic calculations and predictions which are important for the experiments planned. We use the shape parametrization adopted for the TCSM [21] for finding the single-particle levels at the ground state of nucleus. The mirror symmetric shape parametrization used in this model effectively includes all even multipolarities. Calculating the quadrupole and hexadecapole moments, one can find the relationship between the deformation parameters used in the TCSM and the parameters of quadrupole β_2 and hexadecapole β_4 deformation used in the models of Refs. [14, 15, 16, 17, 18]. The ground state of the nucleus is resulted from the calculation of the potential energy surface as a function of deformation parameters [20]. Low-lying states in odd-mass nuclei are essentially determined by the unpaired nucleon. The contribution of an odd nucleon, occupying a single-particle state $|\mu\rangle$ with energy e_μ, to energy of a nucleus is described by the one-quasiparticle energy $E_\mu = \sqrt{(e_\mu - e_F)^2 + \Delta^2}$. Here, the Fermi energy e_F and the pairing-energy gap parameter Δ are calculated with the BCS approximation. The values of Δ obtained in our calculations differ from those in Refs. [14, 15, 16, 17, 18] within 0.1 MeV.

In the Nilsson-Strutinsky approach used, the dependencies of the parameters of **ls** and l^2 terms on A and $N - Z$ are modified [20] for the correct description of the ground state spins and parities of known odd actinides. As found, this modification weakly influences the potential energy surface. The order of calculated single-particle levels seems to be close to the one in Ref. [36] for the same values of quadrupole and hexadecapole moments. Although we did not fit the parameters of the model to describe precisely the nuclear binding energies $B(Z, A)$ consisting of macroscopical and microscopical parts, the calculated $Q_\alpha(Z, A) = B(Z, A) + 28.296 - B(Z - 2, A - 4)$ values for the ground state to ground state α-decay differ within 0.3 MeV from the experimental data that is comparable with other approaches.

To demonstrate the quality of our calculations with the TCSM, the calculated energies of one-quasineutron states for ^{245}Cm are compared in Fig. 2 with the experimental data [37]. The discrepancy in energy with well established states does not exceed 300 keV that is quite satisfactory. The Nilsson (asymptotic) quantum numbers $[Nn_z\Lambda]$ are assigned to each state. In addition, we calculated the one-quasiparticle states with the Quasiparticle-phonon model (QPM) [36] and Skyrme-Hartree-Fock-Bogolyubov (SHFB) approach with indicated Skyrme forces [38] (Fig. 2). One can see that all approaches provide the similar quality of the description of the experimental data. Therefore, they can be used to analyze the structure of heavy nuclei which are not well studied yet. The performed modification of the TCSM for nuclear structure calculations seem to be well confident.

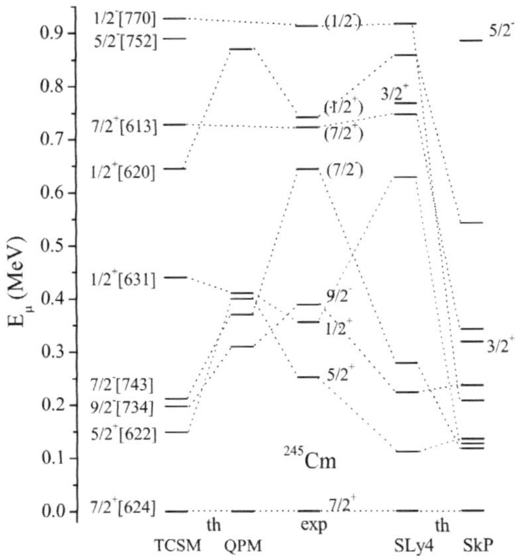

Figure 2. Comparison between the experimental (exp.) [37] and calculated (th.) one-quasineutron spectra for the nucleus ^{245}Cm. The calculations within the TCSM, QPM, SHFB with SLy4 and SkP parameterizations are presented.

The calculated mass excesses $M_{\rm th}$, neutron separation energies B_n, shell corrections $\delta E_{sh} = -B_f$, and Q_α–values are listed in Ref. [22] for nuclei with $105 \leq Z \leq 126$. We treat only the isotopes of superheavy nuclei which can be reached in complete fusion reactions with available

Journal of Physics: Conference Series **515** (2014) 012002 doi:10.1088/1742-6596/515/1/012002

projectiles and targets. As seen in Fig. 3, the calculated Q_α are in a good agreement with the available experimental data. The shell at $N = 162$ is less pronounced in our calculations than in Refs. [14, 15, 16, 17, 18] The shell effects at $Z = 114$ and $N = 172 - 176$ provide rather weak dependence of Q_α on N. The strong role of the shell at $N = 184$ is reflected in the well pronounced minimum of Q_α. For comparison, the Q_α–values predicted in Ref. [14] are shown in Fig. 4. As in our case, the dependence of Q_α on N becomes weaker at $N = 172 - 176$ with the data of Refs. [14, 15, 16, 17, 18]. The small role of $N = 184$ and $Z = 120 - 126$ is seen in Fig. 4. The phenomenological model [39] results no shell effects at $N = 162$ and at $N = 172 - 176$. However, as in our calculations, it provides a strong evidence of the shell closure at $N = 184$.

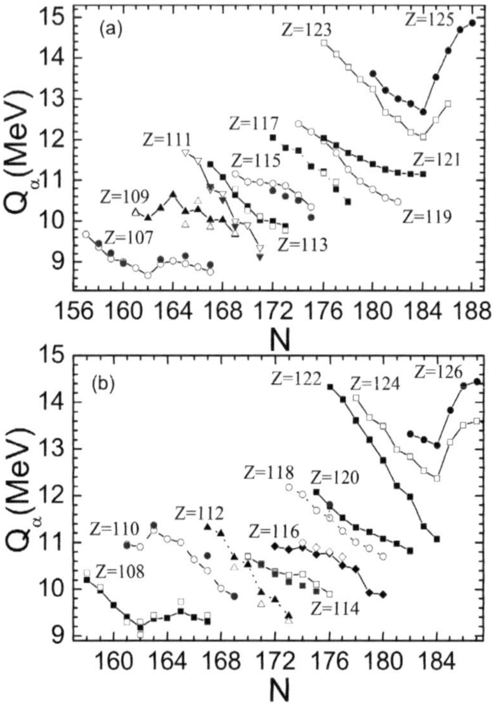

Figure 3. Calculated α–decay energies (symbols connected by lines) are compared with available experimental data (symbols) [1, 2, 4, 40] for nuclei with $Z \geq 108$.

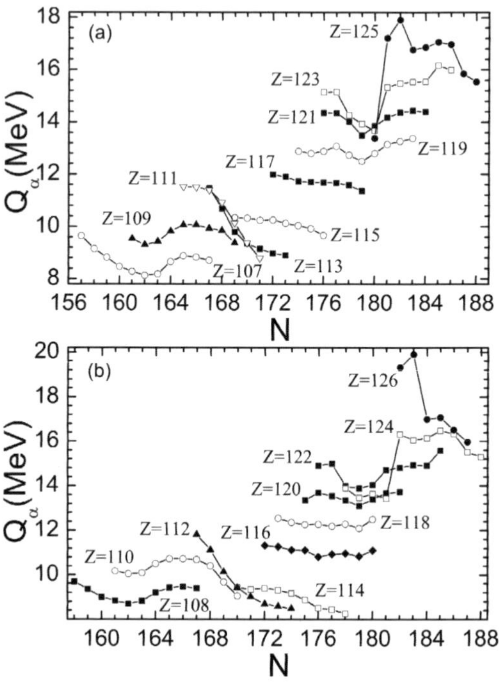

Figure 4. α–decay energies (symbols connected by lines) calculated with the macroscopic-microscopic model [14] for nuclei with $Z \geq 108$.

The value of survival probability strongly depends on $B_f - B_n$, the difference between the height B_f of the fission barrier and the neutron separation energy B_n. At fixed charge number the predicted values of B_n steadily decrease in the region of $N \geq 170$ with increasing N. The values of B_n predicted with different models vary within 0.5 MeV and the shell effects or B_f cause the difference in the dependencies of $B_f - B_n$ on N in Figs. 5 and 6.

The shell corrections prevent the fission of superheavies. The value of B_f is mainly determined by the amplitude of the shell correction in the ground state for nuclei with $Z \geq 105$. As a result, B_f strongly depends on the neutron and proton numbers of the compound nucleus, especially, on how close they are to the magic numbers. As seen in Fig. 6, the model [14] provides the closed proton shell at $Z = 114$ and the fission barrier grows with N up to $N = 178 - 180$. At fixed neutron number and $Z > 114$, the height of the fission barrier decreases with increasing deviation of Z from 114.

As seen in Fig. 5, our macroscopic-microscopic approach provides the shell at $Z = 114$. However, the shell effects at $Z = 120 - 126$ are rather strong. The fission barrier increases when N approaches $N = 184$. Since for nuclei with $Z = 120 - 126$ the values of Q_α are minimal at $Z = 120$ (Fig. 3) where the fission barriers are rather high (Fig. 5), the nuclei with $Z = 120$ and $N = 180 - 184$ are expected to be the most stable nuclei beyond those with $Z = 114$ and $N = 176 - 178$. The shell closure at $Z = 120$ is expected in accordance with the relativistic mean-field model.

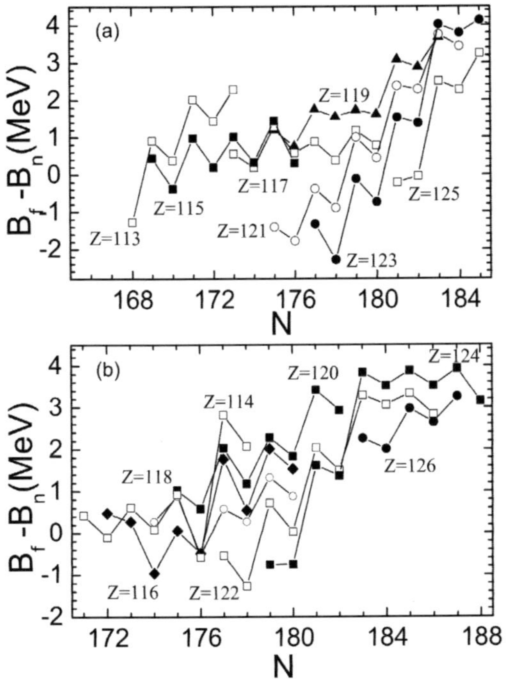

Figure 5. The isotopic dependence of the value of $B_f - B_n$ with our data. The fission barrier B_f is assumed to be an absolute value of the shell correction in the ground state of the nucleus. The results for the isotopes related to the indicated charge number Z are shown by symbols connected by lines.

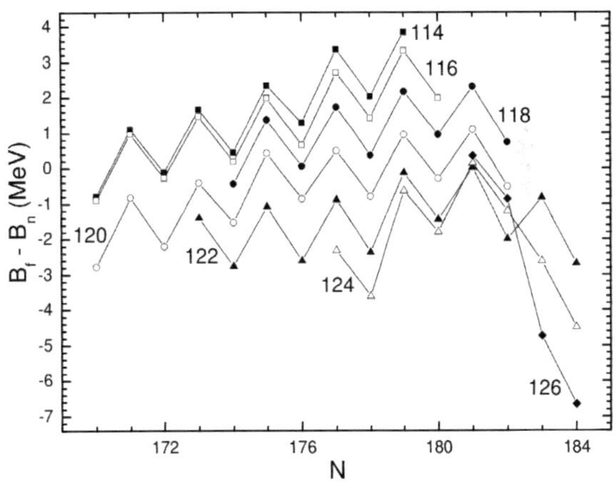

Figure 6. The isotopic dependence of the value of $B_f - B_n$ with the data of mass table [14]. The fission barrier B_f is assumed to be an absolute value of the shell correction in the ground state of the nucleus. The results for the isotopes related to the indicated even charge number Z are shown by symbols connected by lines.

In Fig. 7, the energies of lowest two-quasiproton states for the nuclei of α-decay chains of the SHE $^{308,310,312}126$ are calculated within our model. While for nuclei with $110 \leq Z \leq 118$ the first two-quasiproton states have energies smaller than 1.21 MeV, in nuclei with $Z = 120$ and $Z = 126$ the lowest two-quasiproton state are at about (1.68–1.91) MeV and (1.54–1.59) MeV, respectively. This indicates the larger gap in the proton single-particle spectra of nuclei with $Z = 120$ and 126. So, the proton shell effects become stronger beyond $Z = 114$. The α-decay chain starting from $^{298}120$ will likely terminate at ^{282}Cn by spontaneous fission [1]. The α-decay chain starting from $^{296}120$ probably will terminate at ^{284}Fl by spontaneous fission.

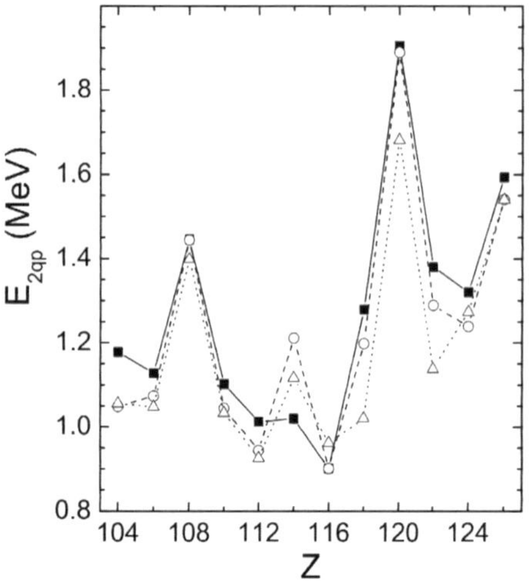

Figure 7. Calculated energies of the lowest two-quasiproton states in the indicated nuclei of α-decay chains of $^{308}126$ (solid squares), $^{310}126$ (open circles), and $^{312}126$ (open triangles) nuclei.

4. Intrinsic level densities in heaviest nuclei

Employing the saddle-point method, the intrinsic level density $\rho(U)$ of nucleus with Z protons, N neutrons, and excitation energy U is expressed as

$$\rho = \frac{\exp[S(\beta, \lambda_Z, \lambda_N)]}{(2\pi)^{3/2}\sqrt{D}}. \tag{6}$$

Here, S is the entropy, $\beta = T^{-1}$ is the inverse temperature, λ_Z and λ_N are the chemical potentials for protons and neutrons, respectively, and D is the determinant of the matrix comprised of the second derivatives of the entropy [41, 42]

$$D = \begin{vmatrix} \frac{\partial^2 S}{\partial \beta^2} & \frac{\partial^2 S}{\partial \beta \partial \mu_Z} & \frac{\partial^2 S}{\partial \beta \partial \mu_N} \\ \frac{\partial^2 S}{\partial \beta \partial \mu_Z} & \frac{\partial^2 S}{\partial \mu_Z^2} & 0 \\ \frac{\partial^2 S}{\partial \beta \partial \mu_N} & 0 & \frac{\partial^2 S}{\partial \mu_N^2} \end{vmatrix} \tag{7}$$

with $\mu_k = \beta\lambda_k$ $(k = N, Z)$.

In the superfluid model adopted here, the entropy S is expressed as

$$S = 2 \sum_{k=Z,N} \sum_{\nu} \left\{ \ln[1 + \exp(-\beta E_{k\nu})] + \frac{\beta E_{k\nu}}{1 + \exp(\beta E_{k\nu})} \right\}, \tag{8}$$

where proton $(k = Z)$ and neutron $(k = N)$ quasi-particle energies $E_{k\nu} = \sqrt{(\varepsilon_{k\nu} - \lambda_k)^2 + \Delta_k^2}$ are calculated using the single-particle levels $\varepsilon_{k\nu}$ obtained with the TCSM. The Fermi energies λ_k and correlation functions Δ_k $(k = N, Z)$, are calculated at the thermodynamic equilibrium of nucleus. At given temperature T, they are determined by solving the system of equations

$$\begin{aligned} Z &= \sum_{\nu} \left(1 - \frac{\varepsilon_{Z\nu} - \lambda_Z}{E_{Z\nu}} \tanh[\frac{1}{2}\beta E_{Z\nu}]\right), \\ N &= \sum_{\nu} \left(1 - \frac{\varepsilon_{N\nu} - \lambda_N}{E_{N\nu}} \tanh[\frac{1}{2}\beta E_{N\nu}]\right), \\ \frac{2}{G_Z} &= \sum_{\nu} \frac{\tanh[\beta E_{Z\nu}/2]}{E_{Z\nu}}, \\ \frac{2}{G_N} &= \sum_{\nu} \frac{\tanh[\beta E_{N\nu}/2]}{E_{N\nu}}, \end{aligned} \tag{9}$$

where G_Z and G_N are the constants of pairing interaction [43]. Here, the sums run over all the single-particle levels considered.

The total $E_{Z,N}$ and excitation U energies of the nucleus at temperature T are calculated as

$$E_{Z,N}(T) = \sum_{k=Z,N} \left\{ \sum_{\nu} \varepsilon_{k\nu} \left(1 - \frac{\varepsilon_{k\nu} - \lambda_k}{E_{k\nu}} \tanh \frac{1}{2}\beta E_{k\nu}\right) - \frac{\Delta_k^2}{G_k} \right\},$$
$$U = E_{Z,N}(T) - E_{Z,N}(0). \tag{10}$$

Intrinsic level density is strongly influenced by nuclear shell structure. For magic or nearly magic nuclei, the level density is smaller than for mid-shell nuclei at the same excitation energy. This effect is related to the large single-particle spacings in the nuclei with the closed shell in the ground state. So, irregularities of the single-particle spectra are responsible for the shell corrections and peculiarities of the intrinsic level density. Note that the expectation of strong shell effects at Z=120 is consistent with the predictions of relativistic shell models [9, 10, 11, 12] as well as with the recent calculations with modified Nilsson potential [44].

Using the relation between the excitation energy U and entropy S,

$$a = S^2/(4U), \tag{11}$$

one can extract the level density parameter. The value of $a(U)$ smoothly approaches the asymptotic value with increasing U. The growth of a with U is sharper for the isotopes with smaller shell corrections.

We found that the best fit of the calculated intrinsic level density with the Fermi-gas expression

$$\rho_{FG}(U) = \frac{\sqrt{\pi}}{12a^{1/4}U^{5/4}} \exp 2\sqrt{aU}, \tag{12}$$

is achieved if one uses the level density parameter (11) and excitation energy U, where S and U are calculated as shown above.

The level density parameters a for the nuclei with $Z \geq 100$ are of especial interest to look for the position of the next proton shell closure beyond $Z = 82$. Employing the method of the extraction of the level density parameter, we analyze the dependencies of a on A, Z, and N for three α-decay chains containing the SHE $^{296,298,300}120$ which could be synthesized in near future with available projectiles and targets. The results are presented in Fig. 8.

The lower and the middle parts of Fig. 8 present, respectively, the dependence of the shell corrections δE_{sh} and level density parameters a on the atomic number A for superheavy nuclei considered. The data of Ref. [22] are used for δE_{sh}. One can see the strong correlation of these dependencies. The larger negative shell corrections result in the decrease of the value of a with respect to the neighborhood nuclei. At Z=108 and 120 there are minima of a in all chains. This reflects quite strong proton shell effects at Z=108 and 120. At Z=120, the minima of a are the deepest and well pronounced. The similar behavior of a occurs near Z=82. As in Ref. [22], we conclude here that the modified TCSM provides the proton shell closure at Z=120. The sub-shell at Z=114 exists but provides weaker shell effect than at Z=120. For nuclei with Z=124-128, the minima of a are due to the neutron shell at N=184.

At small excitation energy (U=10 MeV), one can use the parameterizations $a \approx A/(12-14)$ MeV for the nuclei with $Z < 116$ and $a \approx A/(14-17)$ MeV for the nuclei with $Z > 116$. At higher excitation energy (U=60 MeV), the correlation between δE_{sh} and a is destroyed. Based on the study of the dependencies of a on δE_{sh} and U, one can use the following parametrization [45] of the level density parameter:

$$a(A,U) = \tilde{a}(A) \left[1 + \frac{1 - \exp\{-U/E'_D\}}{U} \delta E_{sh}\right], \tag{13}$$

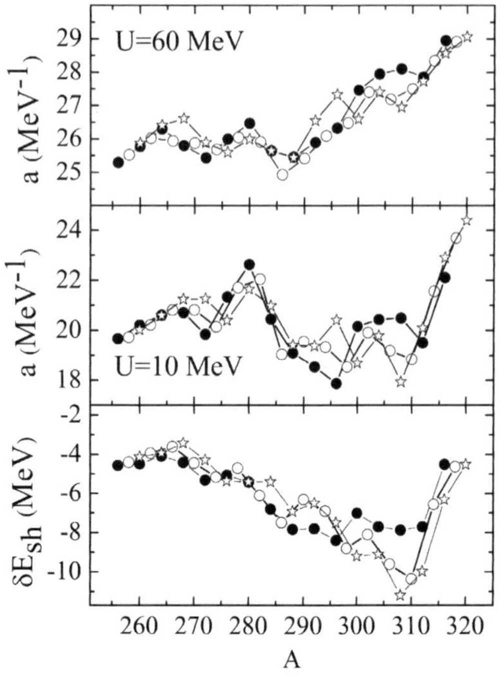

Figure 8. Calculated ground-state shell corrections δE_{sh} (lower part), the level density parameter a at U=10 (middle part) and 60 MeV (upper part), obtained with Eq. (11), as functions of mass number A. The nuclei from alpha-decay chains containing ^{296}120, ^{298}120, ^{300}120 are marked by closed circles, open circles, and stars, respectively.

where $\tilde{a}(A)$ is the parameter smoothly depending on A. It defines a at large excitations when the shell effects are washed out. By analyzing the level density parameters with the Eq. (11), the value of the damping parameter E'_D=17.3 MeV is found. The corresponding asymptotic level density parameter $\tilde{a}(A)$ can be fitted with the following functions [45]:

$$\tilde{a}(A) = \alpha A + \beta A^2, \tag{14}$$

where the constants α=0.118 MeV^{-1} and β=-0.53$\times$10^{-4} MeV^{-1} are found with the least square method. These values are close to those proposed in Ref. [45].

5. Impact of SHE properties on evaporation residue cross sections

Using our predictions of nuclear properties, we calculated the evaporation residue cross sections in the reactions ^{48}Ca,^{50}Ti,^{54}Cr,^{58}Fe,^{64}Ni+^{238}U,^{244}Pu,^{248}Cm,^{249}Cf (Fig. 9). In comparison to our previous calculations with the mass table of Ref. [14], in Fig. 9 the values of σ_{ER} decreases slower with increasing Z. The stronger shell effect revealed here for nuclei with $Z > 118$ result larger survival probabilities and larger values of σ_{ER}. In the reactions ^{48}Ca+^{238}U,^{248}Cm,^{249}Cf the experimental values of σ_{ER}^{3n} are 0.5–2.5 pb, about 1 pb, and 0.5 pb [1], respectively. Thus, the difference between the calculated and experimental σ_{ER} are within the experimental and theoretical uncertainties. A good description of existing data allows us to be confident in the predictions for the reactions with heavier projectiles.

With ^{50}Ti beam the values of σ_{ER} for the nuclei with $Z = 114 - 118$ are expected to be 5–10 times smaller than those resulted by ^{48}Ca beam. The main reason for this is the decrease of P_{CN} with mass asymmetry in the entrance channel of reaction. With ^{50}Ti the nucleus ^{295}120 is predicted to be produced with the cross section of 23 fb. In the ^{54}Cr+^{248}Cm reaction the compound nucleus would have 3 neutrons more than in the ^{50}Ti+^{249}Cf reaction. Therefore, the decrease of P_{CN} is partly negated by the increase of W_{sur} and the nucleus ^{298}120 could be produced with the cross section of 10 fb. For the production of nuclei with $Z = 122 - 126$, ^{64}Ni beam would lead to largest cross sections, 1–8 fb.

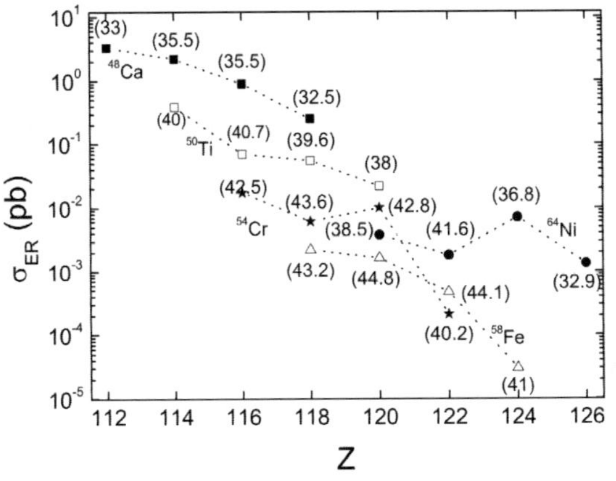

Figure 9. The evaporation residue cross sections in the maxima of excitation functions versus charge number Z for the reactions ^{48}Ca,^{50}Ti,^{54}Cr,^{58}Fe,^{64}Ni+^{238}U,^{244}Pu,^{248}Cm, ^{249}Cf with the predicted properties of SHE. The excitation energies of compound nuclei are given in brackets.

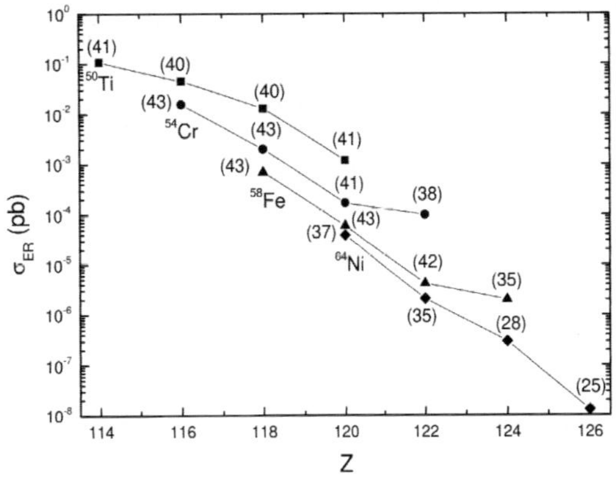

Figure 10. The same as in Fig. 9, but with using the predicted properties of SHE from Ref. [14].

The evaporation residue cross-sections at the maxima of excitation functions and corresponding optimal excitation energies calculated with the mass table of Ref. [14] are presented in Fig. 10 for the reactions ^{50}Ti,^{54}Cr,^{58}Fe,^{64}Ni+^{238}U,^{244}Pu,^{248}Cm,^{249}Cf. The values of $\sigma_{\rm ER}$ decrease by about 2–3 orders of magnitude with increasing the charge number of the target from 92 to 98. The reactions with lighter targets are more favorable for larger $\sigma_{\rm ER}$. The main reason of the fall-off of $\sigma_{\rm ER}$ with Z of the compound nucleus is the strong decrease of the fusion probability P_{CN}. Only the projectiles ^{50}Ti and ^{54}Cr results in the production cross-section of Z =114, 116, and 118 on the level of present experimental possibilities.

6. Summary

The calculations performed with the modified TCSM and the DNS model reveal quite strong shell effects at $Z = 120 - 126$. So, our macroscopic-microscopic treatment qualitatively leads to the results close to those in the mean-field treatments. The strong shell effect is at $N = 184$. The shell structure of heaviest nuclei predicted with macroscopic-microscopic model crucially depends on the parameters of single-particle potential and spin-orbit interaction. If our predictions of the structure of heaviest nuclei are correct, than one can expect the production of evaporation residues 120 in the reactions ^{50}Ti+^{249}Cf and ^{54}Cr+^{248}Cm with the cross sections 23 and 10 fb. The $Z = 120$ nuclei with $N = 178 - 182$ are expected to have Q_α about 12.1–11.2 MeV and lifetimes 1.7 ms–0.16 s in accordance with our predictions. These Q_α are in fair agreement with Ref. [39] and about 2 MeV smaller than in Refs. [14, 15, 16, 17, 18]. The experimental measurement of Q_α for at least one isotope of $Z = 120$ nucleus would help us to set proper shell model for the SHE with $Z > 118$. Note that the definition of maxima of the excitation functions provides a good test for the predictions of the models as well.

Acknowledgments

This work was supported by RFBR and DFG. The IN2P3(France)-JINR(Dubna) and Polish-JINR(Dubna) Cooperation Programmes are gratefully acknowledged.

References

[1] Oganessian Yu Ts 2007 *J. Phys.* G **34** R165.
[2] Oganessian Yu Ts et al. 2010 *Phys. Rev. Lett.* **104** 142502
[3] Oganessian Yu Ts et al. 2013 *Phys. Rev.* C **87** 014302
[4] Hofmann S et al. 2007 *Eur. Phys. J.* A **32** 251
[5] Hofmann S (2009) *Lec. Notes Phys.* **764** 203
[6] Adamian G G, Antonenko N V and Sargsyan V V 2009 *Phys. Rev.* C **79** 054608
[7] Adamian G G, Antonenko N V, Sargsyan V V and Scheid W 2010 *Nucl. Phys.* **A834** 345c
[8] Dong J, Zuo W and Scheid W (2011) *Phys. Rev. Lett.* **107** 012501
[9] Ring P 1996 *Prog. Part. Nucl. Phys.* **37** 193
[10] Bender M, Heenen P H and Reinhard P G 2003 *Rev. Mod. Phys.* **75** 121
[11] Meng J, Toki H, Zhou S G, Zhang S Q, Long W H and Geng L S 2006 *Prog. Part. Nucl. Phys.* **57** 470
[12] Li J J, Long W H, Margueron J and Van Giai N 2013 *arXiv: nucl-th* **1303.2765v1**
[13] Adamian G G, Antonenko N V and Scheid W 2009 *Eur. Phys. J.* A **41** 235
[14] Möller P, Nix J R, Myers W D and Swiatecki W J (1995) *At. Data Nucl. Data Tables* **59** 185
[15] Muntian I, Patyk Z, Sobiczewski A 2001 *Acta. Phys. Pol.* B **32** 691; 2003 ibid **34** 2141
[16] Muntian I, Hofmann S, Patyk Z and Sobiczewski A 2003 *Acta. Phys. Pol.* B **34** 2073
[17] Parkhomenko A, Muntian I, Patyk Z and Sobiczewski A 2003 *Acta. Phys. Pol.* B **34** 2153
[18] Parkhomenko A and Sobiczewski A 2005 *Acta. Phys. Pol.* B **36** 3095
[19] Adamian G G, Antonenko N V and Scheid W 2010 *Phys. Rev.* C **81** 024320
[20] Adamian G G, Antonenko N V, Kuklin S N and Scheid W 2010 *Phys. Rev.* C **82** 054304
[21] Maruhn J and Greiner W 1972 *Z. Phys.* A **251** 431
[22] Kuzmina A N, Adamian G G, Antonenko N V and Scheid W 2012 *Phys. Rev.* C **85** 014319
[23] Antonenko N V, Cherepanov E A, Nasirov A K, Permjakov V P and Volkov V V 1993 *Phys. Lett.* B **319** 425
[24] Volkov V V 2004 *Particles and Nuclei* **35** 797
[25] Adamian G G, Antonenko N V, Scheid W and Volkov V V 1998 *Nucl. Phys.* A **633** 409
[26] Jolos R V, Muminov A I and Nasirov A K 1999 *Eur. Phys. J.* A **4** 245
[27] Giardina G, Hofmann S, Muminov A I and Nasirov A K 2000 *Eur. Phys. J.* A **8** 205
[28] Nasirov A, Kim K, Mandaglio G, Giardina G, Muminov A and Kim Y 2013 *Eur. Phys. J.* A **49** 147
[29] Adamian G G, Antonenko N V and Scheid W 2000 *Nucl. Phys.* A **678** 24
[30] Adamian G G and Antonenko N V 2005 *Phys. Rev.* C **72** 064617
[31] Zubov A S, Adamian G G, Antonenko N V, Ivanova S P and Scheid W 2002 *Phys. Rev.* C **65** 024308
[32] Adamian G G, Antonenko N V and Scheid W 2004 *Phys. Rev.* C **69** 014607; **69** 044601
[33] Adamian G G et al. 1996 *Int. J. Mod. Phys.* E **5** 191
[34] Raman S, Nestor C W and Tikkanen P 2001 *At. Data Nucl. Data Tables* **78** 1
[35] Audi G and Wapstra A H 1993 *Nucl. Phys.* A **565** 1
[36] Soloviev V G 1976 *Theory of Complex Nuclei* (Oxford: Pergamon Press)
[37] *http://www.nndc.bnl.gov/ensdf/*
[38] Adamian G G, Antonenko N V, Kuklin S N, Lu B N, Malov L A and Zhou S G 2011 *Phys. Rev.* C **84** 024324
[39] Liran S, Marinov A and Zeldes N 2000 *Phys. Rev.* C **62** 047301; **63** 017302
[40] Stavsetra L et al. 2009 *Phys. Rev. Lett.* **103** 132502
[41] Decowski P, Grochulski W, Marcinkowski A, Siwek K and Wilhelmi Z 1968 *Nucl. Phys.* A **110** 129
[42] Adeev G D and Cherdantsev P A 1975 *Yadernaya Fizika* **21** 491
[43] Bezbakh A N, Shneidman T M, Adamian G G and Antonenko N V 2014 *to be published*
[44] Zhang Z H, Meng J, Zhao E G and Zhou S G 2013 *Phys. Rev.* C **87** 054308
[45] Ignatyuk A V, Smirenkin G N and Tishin A S 1975 *Yadernaya Fizika* **21** 485

The (^{18}O,^{16}O) reaction: a bridge from direct to dissipative dynamics

M. Cavallaro[1a], M. Colonna[a], F. Cappuzzello[a,b], C. Agodi[a], M. Bondì[a,b], D. Carbone[a], A. Cunsolo[a], M. De Napoli[c], A. Foti[b,c]
[a] INFN – LNS, Catania, Italy
[b] Dipartimento di Fisica e Astronomia, Università degli Studi di Catania, Catania, Italy
[c] INFN – Sezione di Catania, Catania, Italy

E-mail: manuela.cavallaro@lns.infn.it

Abstract. A study of the (^{18}O,^{16}O) two-neutron transfer reaction at 84 MeV incident energy was pursued at the Catania INFN-LNS laboratory on several targets. In the energy spectra obtained from the light targets, several known low lying and resonant states of the product nuclei are observed. Instead, the spectra from the heavier targets are dominated by the appearance of a broad bump, indicating that more dissipative processes are contributing. In this case, calculations based on semi-classical transport theories seem to indicate the possible occurrence of dissipative reaction mechanisms.

1. Introduction

Direct transfer reactions represent essential tools in nuclear physics since they have the basic property to select specific degrees of freedom in the complex many body nuclear system. For this reason, they have always played a crucial role in the investigation of the nuclear structure. For example, a subject of great interest is the connection between transfer probabilities in two-neutron transfer reactions and pairing correlations in nuclei [1], [2], [3].

In the past, spectroscopic investigations of single-particle as well as cluster configurations above the Fermi level were mainly conducted by light ions induced reactions. Moreover the potentiality of heavy-ion two-neutron transfer reactions has not been fully exploited till now [4], [5] also due to the experimental difficulties in producing high-resolution spectra covering a wide energy and angular range [6]. Nevertheless, studies with heavy projectiles demonstrate their reliability for quantitative analyses, as long as the multistep processes remain comparatively weak [7] or are explicitly accounted for.

An interesting case, which has been focused in a recent paper [1], is given by transfer reactions driven by ^{18}O beam. In particular it turned out that spectroscopic factors can be

1 To whom any correspondence should be addressed.

extracted for both one- and two-neutron configurations, opening the door to future application of heavy-ion reactions as spectroscopic tools.

In this context, a systematic exploration of the response of atomic nuclei to the (^{18}O,^{16}O) and (^{18}O,^{17}O) probes has started at the INFN-LNS in Catania at bombarding energy slightly above the Coulomb barrier.

In the present manuscript a specific attention is paid to the ^{64}Ni(^{18}O,^{16}O)^{66}Ni reaction, which shows hints for the evolution of the reaction mechanism from a pure direct to a more dissipative process.

2. The experiment

The ^{64}Ni(^{18}O,^{16}O)^{66}Ni reaction was studied at the INFN-LNS laboratory in Catania. The ^{18}O beam was delivered at 84 MeV total incident energy by the Tandem accelerator. The experiment was performed on a ^{64}Ni self-supporting solid target 113 μg/cm^2 thick. The ^{16}O ejectiles were momentum analysed and detected at forward angles by the MAGNEX magnetic spectrometer [8], [9], [10], [11].

Exploiting the large momentum acceptance of MAGNEX (25%), energy spectra of the reaction products were obtained up to about 30 MeV excitation energy. Four angular setting were explored, with the spectrometer optical axis located at θ_{opt} = 6°, 12°, 18° and 24°, which, considering the MAGNEX solid angle (50 msr), correspond to a measured angular range 4° < θ_{lab} < 30°. The application of the trajectory reconstruction technique [12], [13] based on the knowledge of the 3D magnetic field maps [14], [15] and the use of the differential algebra [16], [17] did allow to get energy spectra with energy resolution of about 150 keV and angular distributions with angular resolution better than 0.3°.

The measured kinetic energy is shown in Fig. 1 as a function of the scattering angle for the identified ^{16}O ions emitted by the ^{64}Ni (^{18}O,^{16}O) ^{66}Ni reaction. The data from four different angular setting of the spectrometer are plotted and the yields for each setting are normalized by an arbitrary scaling factor.

In the upper part of the plot one can notice the general trend of the data characterized by a kinetic energy decreasing as a function of the scattering angle. For example, the maximum kinetic energy locus, corresponding to the ^{66}Ni ground state is clearly visible. This behaviour is not surprising, since it is due to the two-body kinematics.

A different trend is observed in the lower part of the plot, where the kinetic energy decreases with decreasing the angle. For these events, the initial relative motion is partially damped and a part of energy is converted from relative motion into internal degrees of freedom. This is not a kinematic effect but can be interpreted as the beginning of a more dissipative process, where the trajectories are deflected more and more from the grazing trajectory towards smaller angles and the final kinetic energy of the system decreases due to an effective nuclear friction force [18], [19].

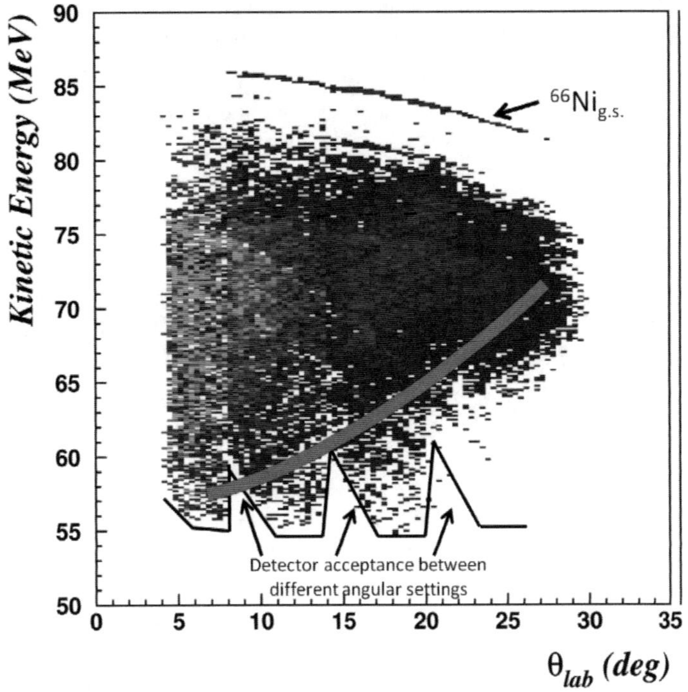

Figure 1. Measured kinetic energy as a function of the scattering angle for the identified ^{16}O ions emitted by the ^{64}Ni (^{18}O,^{16}O) ^{66}Ni reaction. Data from four different angular setting of the spectrometer at θ_{opt} = 6°, 12°, 18° and 24° are plotted and the yields for each setting are normalized by an arbitrary scaling factor. The black solid line in the lower part of the plot represents the limit of the detector acceptance. The red line is drawn just to guide the eye to show the trend of the events where the kinetic energy decreases with decreasing the angle.

3. Preliminary dynamical analysis

The reaction path of the ions involved in the collision was investigated in a semi-classical approach [20] by the Vlasov transport equation, where the semi-classical one-body distribution function $f(\mathbf{r}, \mathbf{p}, t)$ follows a time evolution in the self-consistent mean-field potential U:

$$\frac{\partial f}{\partial t} + \frac{\mathbf{p}}{m}\frac{\partial f}{\partial \mathbf{r}} - \frac{\partial U}{\partial \mathbf{r}}\frac{\partial f}{\partial \mathbf{p}} = 0$$

The potential U is derived from a Skyrme-like effective interaction. Moreover, a fluctuating term is implemented in the dynamics, through stochastic spatial density fluctuations. Stochasticity is essential to get distributions, as well as to allow the growth of dynamical instabilities, see Ref. [21] for more details. In order to map the particle occupation number at each time step, Gaussian phase-space wave packets (test particles) were considered. In the performed simulations 100 test particles per nucleon were used for an accurate description of the mean field dynamics. A number of 100 events were run for each set of impact parameters.

Two typical examples of time evolution of the space density distributions projected on the reaction plane for two events of different impact parameter are shown in Fig. 2. In the more peripheral reaction (on the left side), the two fragments remain separate and a typical grazing dynamics is visible. The more central collisions (on the right side), instead, is characterized by a fusion process.

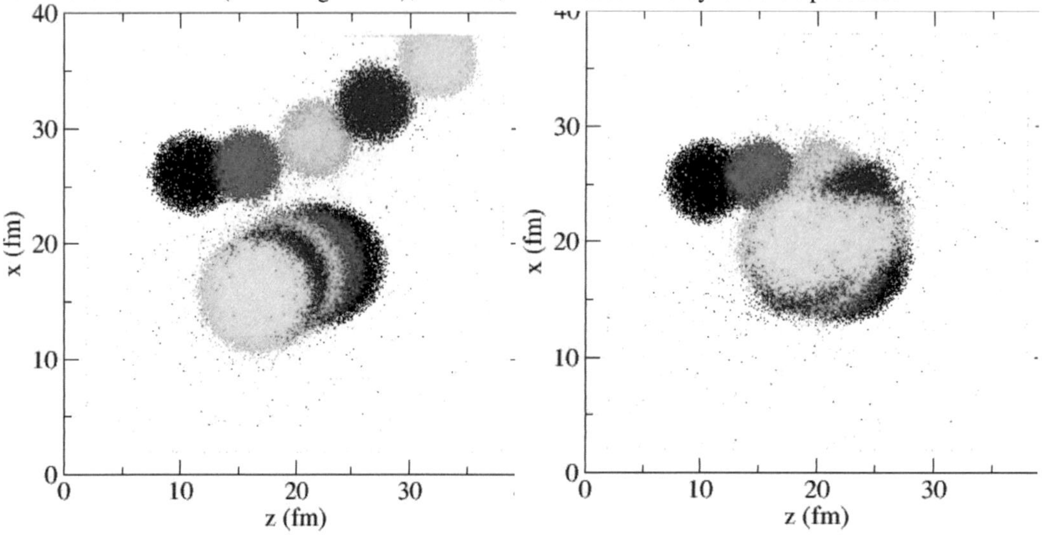

Figure 2. Time evolution of the space density distributions for the reaction $^{18}O + ^{64}Ni$ at 84 MeV incident energy for impact parameter b = 7 fm (left panel) and b = 6 fm (right panel). The different colours represent different time steps.

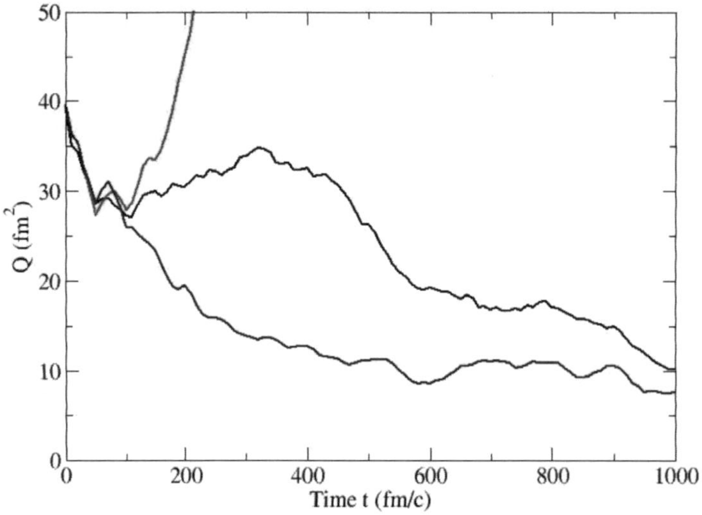

Figure 3. Time evolution of the space quadrupole moment for the reaction $^{18}O + ^{64}Ni$ at 84 MeV incident energy for impact parameter b = 6.6 fm. The different colours represent different events.

In the present semi-classical approach, the time evolution of the quadrupole moment in coordinate space was also extracted for each event:

$$Q(t) = \langle 2z^2(t) - x^2(t) - y^2(t) \rangle$$

where the z axis is along the rotating projectile/target direction and the x axis is on the reaction plane. In Fig. 3 the time evolution of the quadrupole moment calculated for some events of the studied reaction at the same impact parameter b = 6.6 fm is plotted. The competition among different mechanisms appears. Positive values of $Q(t)$ slope correspond to a quadrupole deformation velocity of the dinuclear system that is going to a breakup exit channel. Thus the red curve represents a typical binary process where the final system breaks in two products, while the blue curve is a fusion event. The black curve, where the quadrupole moment increases and after decreases, is an event of transition where the dynamics can evolve towards a binary process or a more dissipative one. Thus we expect to explore the conditions of kinetic energy and deflection angle observed in the data (red line in Fig.1).

4. Conclusions

A preliminary analysis based on the semi-classical approach of the Vlasov transport equation was performed to describe the ^{64}Ni (^{18}O,^{16}O) ^{66}Ni reaction studied at 84 MeV incident energy by the MAGNEX spectrometer.

The time evolution of the space density distribution and of the quadrupole moment for a number of events at different impact parameters was studied. The results show that a transition from direct binary events to more dissipative processes is starting for this system in the present energy regime.

The use of the same approach to different targets excited by the same (^{18}O,^{16}O) reaction is planned in the near future in order to point out the systematic evolution of the reaction mechanism and establish the condition where these reactions are good spectroscopic probes.

5. References

[1] M. Cavallaro, et al., Phys. Rev. C 88, 054601 (2013).
[2] F. Cappuzzello, et al., Phys. Lett. B 711, 347 (2012).
[3] I. Tanihata, et al., Phys. Rev. Lett. 100, 192502 (2008).
[4] P. D. Bond, et al., Phys. Rev. C 16, 177 (1977).
[5] M. C. Mermaz, et al., Phys. Rev. C 20, 2130 (1979).
[6] S. Szilner, et al., Nucl. Phys. A 779, 21 (2006).
[7] S. Kahana and A. J. Baltz, in Advances in Nuclear Physics, edited by M. Baranger and E. Vogt (Plenum Press, New York, 1977), Vol. 9, pp. 1–122.
[8] F. Cappuzzello et al., MAGNEX: an innovative large acceptance spectrometer for nuclear reaction studies in: Magnets: Types, Uses and Safety, Nova Publisher Inc., New York, 2011, pp 1-63.
[9] A. Cunsolo et al., Nucl. Instrum. Methods A 484, 56 (2002).
[10] M. Cavallaro, et al., Eur. Phys. J. A (2012) 48: 59.
[11] F. Cappuzzello, et al., Nucl. Instr. and Meth. A 621, 419 (2010).
[12] M. Cavallaro, et al., Nucl. Instr. and Meth. A 637, 77 (2011).
[13] F. Cappuzzello, et al., Nucl. Inst. and Meth. A 638, 74 (2011).
[14] A. Lazzaro, et al., Nucl. Instr. and Meth. A 585, 136 (2008).
[15] A. Lazzaro, et al., Nucl. Instr. and Meth. A 591, 394 (2008).
[16] K. Makino, M. Berz, Nucl. Instr. and Meth. A 427, 338 (1999).
[17] M. Berz, K. Makino, COSY INFINITY, Version 8.1, Department of Physics and Astronomy and

NSCL, Michigan State University, East Lansing, USA, 2001.
[18] J. Wilczynski, Phys. Lett. B 47, 484 (1973).
[19] V.V. Volkov, Phys. Rep. 44, 93 (1978).
[20] V. Baran, et al., Phys. Rep. 410, 335 (2005).
[21] C. Rizzo, et al., Phys. Rev. C 83, 014604 (2011).

Study of the break-up channel in ^{11}Li+^{208}Pb collisions at energies around the Coulomb barrier.

J P Fernández-García[1,2,3], M Cubero[4,5,6], M Rodríguez-Gallardo[1], L Acosta[3,7], M Alcorta[2], M A G Alvarez[1,2], M J G Borge[4], L Buchmann[8], C A Diget[9], H A Falou[10], B R Fulton[9], H O U Fynbo[11], D Galaviz[12], J Gómez-Camacho[1,2], R Kanungo[10], J A Lay[1,13], M Madurga[4], I Martel[7], A M Moro[1], I Mukha[14], T Nilsson[15], A M Sánchez-Benítez [7,12], A Shotter[16], O Tengblad[4], and P Walden[8]

[1] Departamento de FAMN, Universidad de Sevilla, Apartado 1065, E-41080 Seville, Spain.
[2] Centro Nacional de Aceleradores, Universidad de Sevilla/Junta de Andalucía/CSIC, E-41092 Seville, Spain.
[3] Istituto Nazionali di Fisica Nucleare-Laboratori Nazionali del Sud (INFN-LNS), I-95123 Catania, Italy.
[4] Instituto de Estructura de la Materia-CSIC, E-28006 Madrid, Spain.
[5] Centro de Investigación en Ciencias Atómicas, Nucleares y Moleculares (CICANUM), CR-2060 San José, Costa Rica.
[6] Instituto de Fisica, Universidade Federal Fluminense, Av. Litoranea s/n, Niterói, R.J., Brazil.
[7] Departamento de Física Aplicada, Universidad de Huelva, E-21071 Huelva, Spain.
[8] TRIUMF, Vancouver, British Columbia V-6T2A3, Canada.
[9] Departament of Physics, University of York, YO1O-5DD Heslington, York, United Kingdom.
[10] Department of Astronomy and Physics, Saint Marys University, Halifax B3H3C3, Canada.
[11] Department of Physics and Astronomy, Aarhus University, DK-8000 Aarhus, Denmark.
[12] Centro de Fisica Nuclear da Universidade de Lisboa (CFNUL), 1649-003 Lisbon, Portugal.
[13] Dipartimento di Fisica e Astronomia "Galileo Galilei", Università di Padova e INFN, Sezione di Padova, via Marzolo, 8, I-35131 Padova, Italy.
[14] GSI Helmholtzzentrum fr Schwerionenforschung GmbH, Planckstr, 1, D-64291 Darmstadt, Germany.
[15] Fundamental Physics, Chalmers University of Technology, S-41296 Gteborg, Sweden.
[16] School of Physics and Astronomy, University of Edinburgh, EH9 3JZ Edinburgh, United Kingdom.

E-mail: `fernandez@lns.infn.it`

Abstract. We present a study of ^{11}Li+^{208}Pb collisions at energies around the Coulomb barrier (E_{lab} = 24.3 and 29.8 MeV), measured at the post-accelerated beam facility, ISAC II, at TRIUMF (Vancouver, Canada). A remarkably large yield of ^{9}Li has been observed, a result that is attributed to the weak binding of the ^{11}Li nucleus. The angular distribution of this ^{9}Li yield, relative to the elastic one, has been analysed in terms of first-order semiclassical calculations as well as four-body and three-body Continuum-Discretized Coupled-Channels (CDCC) calculations, based on a three-body and di-neutron model of the ^{11}Li nucleus, respectively. The calculations reproduce well the trend of the data and support the existence of a large concentration of $B(E1)$ strength at very low excitation energies. The connection of this large $B(E1)$ with a possible low-lying dipole resonance is discussed.

[3] Present address: Istituto Nazionali di Fisica Nucleare-Laboratori Nazionali del Sud (INFN-LNS), I-95123 Catania, Italy.

1. Introduction

Since the pioneering experiments with Radioactive Ion Beams (RIB) performed in the 80s at Berkeley by Taninata and collaborators [1], it has been possible to extend the study of stable nuclear reactions to neutron- and proton-rich nuclei. Such developments allow us to investigate the properties of the nuclear interaction, the reaction mechanisms and the nuclear structure of these particular systems, and assess the validity of the structure models developed for stable light nuclei (N/Z≈1) when the neutron/proton asymmetric increases.

One of the most studied exotic nuclei is ^{11}Li (N/Z=2.7). This is an unstable system ($\tau_{1/2}$= 8.75(13) ms [2]), with a Borromean structure consisting of a ^{9}Li core surrounded by two-weakly bound neutrons (S_{2n}=369.15(65) KeV [3]), in which none of the binary subsystems (n-n or ^{9}Li-n) is bound. The remarkably large reaction cross section observed by Tanihata in 1985 for this nucleus, was interpreted by Hansen and Jonson as due to a high probability of the outermost neutrons to be at large distances from the ^{9}Li core, which they referred to as a halo structure [4]. A Glauber analysis of ^{11}Li elastic data on protons at intermediate energies yielded a matter radius of 3.62(19) fm [5], significant larger than that obtained in the same experiment for the ^{9}Li nucleus (2.43(7) fm). We can refer also to the halo size, which is estimated in the same reference as 6.54 fm, and is comparable to the size of the much heavier ^{208}Pb nucleus.

As a consequence of its weak binding, neutron-halo nuclei exhibit a large $B(E1)$ strength at low excitation energies. In the case ^{11}Li, this has been experimentally confirmed by several Coulomb dissociation experiments in complete kinematics [6–9]. Several theoretical calculations have also confirmed this low-lying large $B(E1)$ strength [10–16].

Due to this large $B(E1)$ at small excitation energies, the ^{11}Li nucleus can be easily polarized (and eventually broken-up) by a strong external electric field, such as that produced by a heavy target nucleus. This can be understood because the ^{9}Li core, positively charged, will move in the field direction, while the weakly-bound neutrons of the halo will not be affected by the electric field. This effect shows up as a strong reduction of the elastic cross section with respect to the Rutherford cross section accompanied by a larger yield of break-up fragments. These effects have been observed and studied, for example, in ^{6}He+^{208}Pb reactions [17–21]. The study of collisions of weakly-bound nuclei provide a very valuable information on the structure of these nuclei and, in particular, on the $B(E1)$ strength.

With the motivation of studying these phenomena in the case of ^{11}Li, in a recent experiment the elastic and inclusive breakup of ^{11}Li+^{208}Pb was measured for the first time at energies below (24.3 MeV) and around (29.8 MeV) the Coulomb barrier ($V_B \approx 28$ MeV) at the RIB facility of TRIUMF (TRI-University Meson Facility), in Vancouver (Canada) [15, 22]. The data revealed a remarkably large yield of ^{9}Li, presumably coming from the breakup of the projectile. The angular distribution of the ^{9}Li cross section, relative to the ^{11}Li cross sections, was compared with first-order semiclassical calculations, based on the theory of Alder and Winther for Coulomb excitation, and it was concluded that, in order to reproduce the magnitude of this observable, a very large $B(E1)$ strength close to the breakup threshold was required, in qualitative agreement with the most recent Coulomb dissociation experiment [9].

These conclusions were corroborated with more sophisticated four-body continuum-discretized coupled-channels (CDCC) calculations, based on a three-body model of the ^{11}Li nucleus. Within this model, the large $B(E1)$ close to the threshold is enhanced due to the presence of a dipole resonance, in agreement with some theoretical works [11, 12, 14].

In this contribution, we describe the referred experiment and present additional calculations based on a simple three-body CDCC calculation, in which the ^{11}Li is described within a simple di-neutron model, with an effective two-neutron separation energy. These calculations are compared with the semiclassical and four-body CDCC calculations presented in Refs. [15, 22].

2. Experimental set-up

The experiment, coded as E1104, was carried out at one of the beam lines of the post-accelerated beam facility, ISAC-II, at TRIUMF with a high beam intensity, around 4300 ions of ^{11}Li per second. As displayed in Fig. 1, the experimental set-up was composed by four silicon telescopes covering the angular range between 10° and 140° (see table 1).

Each telescope system consists on two independent detectors: a thin detector, to measure the energy loss (ΔE) by the nuclear reaction fragments, coupled to a thicker detector, to measure the remaining energy (E) deposited by the fragments. Such set-up allows us to separate different fragments in terms of mass and charge in a bi-dimensional diagram ΔE versus ΔE+E (see Figs. 2 and 3).

Table 1. Angular range of each telescope system.

Telescope (ΔE+E)	Angular range
T1: DSSSD+PAD	10°- 40°
T2: DSSSD+PAD	30°- 60°
T3: SSSSD+DSSSD	50°- 100°
T4: SSSSD+DSSSD	90°- 140°

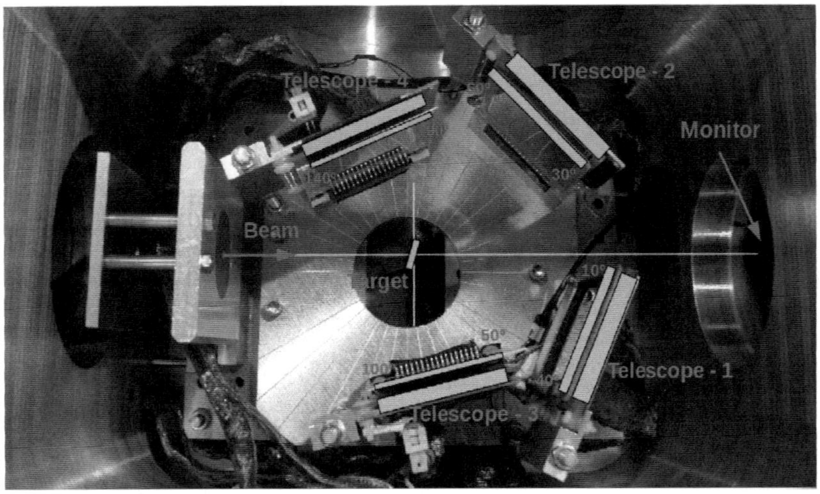

Figure 1. Picture of the experimental setup, with indications of different elements and angles.

The telescopes 1 and 2, covering the forward angles, were composed by a Double Sided Silicon Strip Detector (DSSSD), 40 μm thin detector, and a 500 μm thicker PAD detector. For the backward angles, a 20 μm thin Single Sided Silicon Strip Detector (SSSSD) and a 60 μm thicker DSSSD, were used to form the telescope 3 and 4. Each telescope system was divided into 16 vertical and 16 horizontal strips, that results on 256 pixels. In total, considering four telescopes, our detection system was composed by 1024 (3x3 mm^2 active area) pixels, which must be considered as independent detectors during the analysis procedure.

To reduce the straggling energy effects, a thin 1.45 mg/cm^2 ^{208}Pb target was used, which was positioned at 75° with respect to the beam direction (see Fig. 1). With these experimental conditions, data were accumulated during 82 hours for the incident energy below the Coulomb barrier, 24.3 MeV, and 119 h for 29.8 MeV incident energy. More details can be found in Refs. [23, 24].

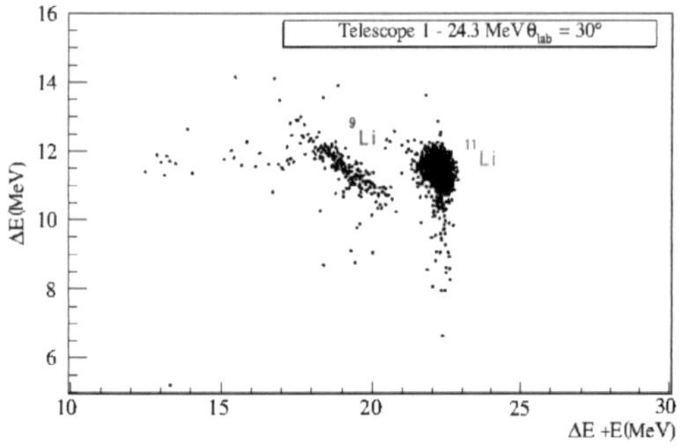

Figure 2. Experimental bi-dimensional diagram for the telescope 1 at the incident energy 24.3 MeV, integrated for the pixels corresponding to the angle 30°±1.5°.

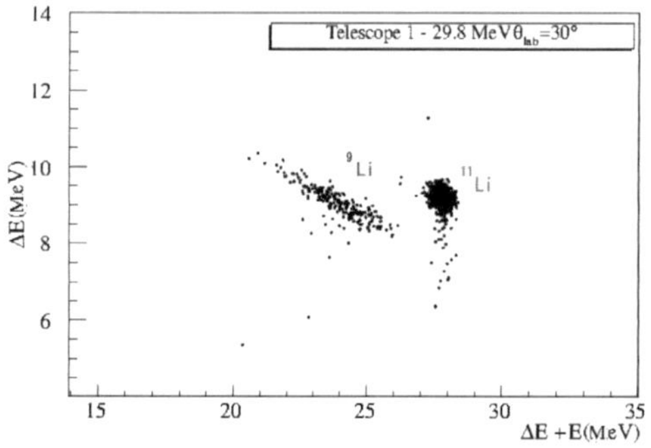

Figure 3. Experimental bi-dimensional diagram for the telescope 1 at the incident energy 29.8 MeV, integrated for the pixels corresponding to the angle 30°±1.5°.

In Figs. 2 and 3, the data obtained in the reaction ^{11}Li+^{208}Pb below and above the Columb barrier are shown in the bi-dimensional diagrams ΔE versus ΔE+E for the telescope 1 integrating all pixels corresponding to the angle 30±1.5°. Two different processes, elastic scattering and break-up, are well distinguished.

3. Results and discussion.

The angular distributions of the differential elastic scattering and the break-up cross sections at the two measured energies ($E_{\rm lab}$=24.3 and 29.8 MeV) are presented in Figs. 4 and 5, respectively. The latter are represented in terms of the so-called breakup probability, defined as the ratio

$$P_{BU}(\theta) = \frac{N_{BU}(\theta)}{N_{BU}(\theta) + N_{Elast}(\theta)}, \quad (1)$$

where $N_{BU}(\theta)$ and $N_{Elast}(\theta)$ are the number of break-up and elastic events, respectively.

The measured elastic cross section (Fig. 4) exhibits a very strong reduction with respect to the Rutherford prediction, even at energies well below the Coulomb barrier (V_B ≈28 MeV). This behavior is mainly a consequence of the large breakup probability of the ^{11}Li nucleus, as confirmed by the large production of ^{9}Li observed in the experiment (see Fig. 5).

In order to get further insight into the mechanisms responsible for the suppression of the elastic cross section and the sizable break-up probability, the experimental data are compared with two different theoretical formalisms.

First, semiclassical calculations, performed in Ref. [15] and based on the Alder and Winter approximation of Coulomb excitation [25], were considered. In this model, a classical treatment of the trajectory of the incident particle is considered, while its excitations are treated quantum-mechanically. Due to the large $B(E1)$ strength of ^{11}Li at low excitation energies reported by several Coulomb dissociation experiments [6–9], we expect a dominance of the electric dipole

Coulomb couplings at small angles and/or low incident energies. For a pure $E1$ process, the first-order semi-classical theory gives the break-up probability

$$P_{BU}(E1,\theta) = \left(\frac{Z_A e}{\hbar v a_0}\right)^2 \frac{2\pi}{9}\int_{\varepsilon_b}^{\infty} d\varepsilon \frac{dB(E1)}{d\varepsilon}[I^2_{1,\,1}(\theta,\,\varepsilon) + I^2_{1,\,-1}(\theta,\,\varepsilon)], \quad (2)$$

where $I_{1,\,\pm1}(\theta,\,\varepsilon)$ are the Coulomb integrals [25] and Z_A the target charge.

The main physical ingredient entering this expression is the $B(E1)$ distribution connecting the ground state of the ^{11}Li nucleus with the continuum states, $dB(E1)/d\varepsilon$. If, for example, the experimental $dB(E1)/d\varepsilon$ obtained by Nakamura *et al* [9] is used, one obtains the result given by the dotted line in Fig. 5. This calculation reproduces well the trend of the data for $\theta_{\rm lab} \lesssim 40°$ and confirms the dominance of the dipole Coulomb mechanism. However, the magnitude of the breakup probability is somewhat underestimated, suggesting that either *i)* other mechanisms are contributing (e.g. higher-order Coulomb couplings, nuclear breakup, etc) or *ii)* the $dB/d\varepsilon$ strength in ^{11}Li is even larger than that reported by Nakamura *et al* . It is worth noting that these calculations consider only Coulomb couplings and hence they are not expected to be valid at large scattering angles, where nuclear effects are possibly present.

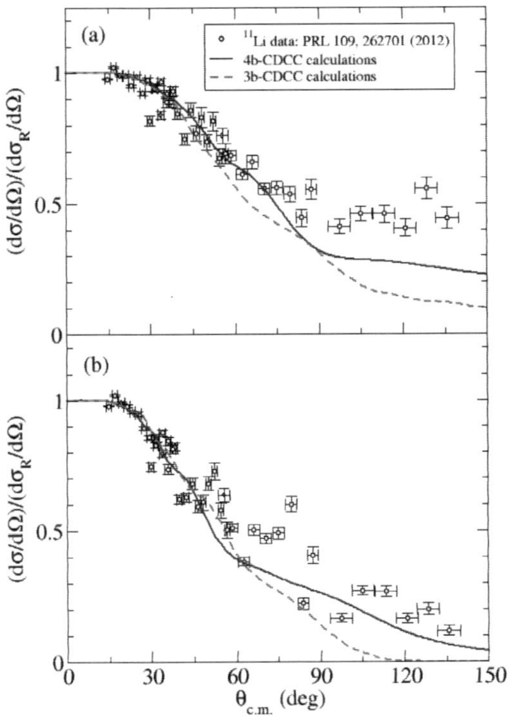

Figure 4. Elastic scattering angular distribution in the center of mass frame at 24.3 (a) and 29.8 (b) MeV.

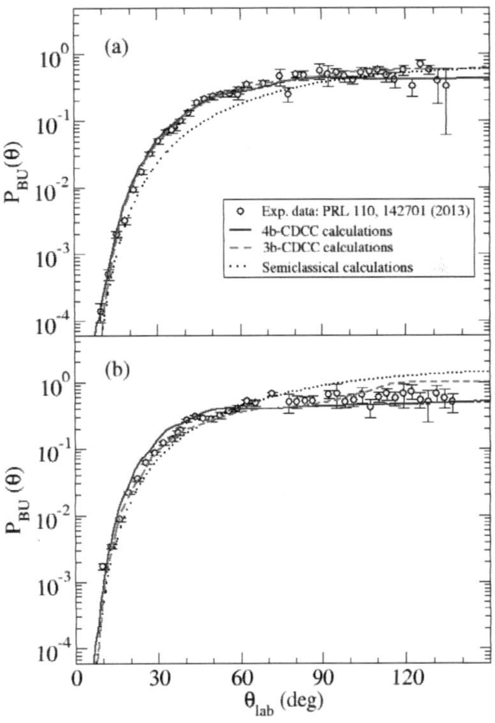

Figure 5. Break-up probability as a function of the laboratory angle at 24.3 (a) and 29.8 (b) MeV.

Since these semiclassical calculations consider only Coulomb couplings, and only to first order, we have performed also more complete three-body (3b) and four-body (4b) CDCC calculations. The CDCC formalism takes into account both nuclear and Coulomb couplings to all orders and hence are not restricted to small angles. Furthermore, the CDCC calculations provide also the elastic cross sections, in addition to the breakup observables. The four-body CDCC calculations (denoted 4b-CDCC for shortness) make use of a three-body model of ^{11}Li with an inert and spinless ^{9}Li core. This model gives rise to a low-lying dipole resonance, but its position cannot be accurately predicted by the model, since it depends critically on the effective three-body force required in this model to account for effects beyond the three-body assumption. In particular, for the neutron configuration $J^{\pi}_{nn} = 0^{+}$, the strength of the three-body force is adjusted to reproduce the experimental two-neutron separation energy. The three-body CDCC calculations (3b-CDCC hereafter) use a two-body (di-neutron) model for the ^{11}Li nucleus (^{9}Li+2n). In both cases, the set of coupled equations are solved with the coupled-channels code FRESCO [26].

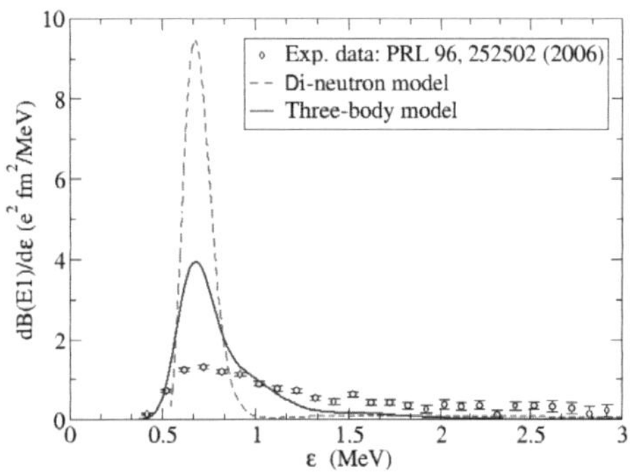

Figure 6. $B(E1)$ distribution of ^{11}Li as a function of the excitation energy.

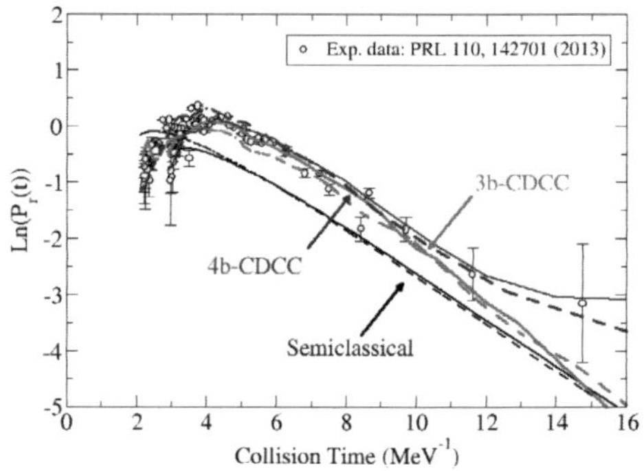

Figure 7. Natural logarithm of the reduced break-up probability as a function of the collision time.

The 4b-CDCC results for the elastic and breakup cross sections are compared with the experimental data in Figs. 4 and 5 (solid lines). As mentioned above, the $B(E1)$ distribution obtained in the three-body model of ^{11}Li (used in the projectile states in the 4b-CDCC calculations) depends critically on the position of the dipole resonance predicted by this model, which can be varied in the model tuning the effective three-body interaction. The calculations shown in Figs. 4 and 5 correspond to the resonance at an excitation energy of $E_x = 0.69$ MeV above the ground state (0.32 MeV above the breakup threshold). With this value, both the elastic cross section and breakup probability are very well described. Thus, the high break-up probability of ^{11}Li can be explained considering a direct break up mechanism. For the 3b-CDCC calculations, the 2n-^{9}Li interaction was adjusted in order to give the correct rms separation between the center of mass of the two halo neutrons and the core (according to the prediction of the three-body model of ^{11}Li) and a dipole resonance at $E_x = 0.69$ MeV. The results are shown in Figs. 4 and 5 by dashed lines. The breakup probability is also well reproduced, although some underestimation is observed, at backward angles, in the elastic cross sections.

In Fig. 6, the two $B(E1)$ distributions obtained with the di-neutron (red dashed line) and three-body model (blue solid line) of ^{11}Li are compared with the experimental distribution deduced from the latest Coulomb dissociation experiment performed at RIKEN [9]. A large narrow strength at low excitation energies of the $B(E1)$ distributions, considering a di-neutron model of ^{11}Li, is observed. This behavior affects the elastic cross section and the break-up probability, producing the underestimation of the experimental data mentioned before. However, the two models of ^{11}Li, which reproduce the experimental data of Refs. [15] and [22] (Figs. 5 and 4, respectively), suggest more strength of the $B(E1)$ distribution than the one reported from the RIKEN experiment [9].

In general, the breakup probability depends on the structure of the projectile nucleus (^{11}Li in our case) as well as on the characteristics of the reaction (target, energy of the collision, scattering angle, nuclear and Coulomb forces). Thus, disentangling structure information, such as the $B(E1)$ distribution, from the measured cross section is a difficult task. In [15], we proposed a novel procedure that permits constraining the behaviour of the $B(E1)$ distribution in a model-independent way. For that purpose, it is convenient to introduce the so-called *reduced breakup probability*, defined as

$$P_r(t) \equiv P_{BU}(E1,\theta)\frac{9t^2(\hbar v)^3 a_0}{16\pi^2(Z_A e)^2\sin(\theta/2)}. \qquad (3)$$

Within the semiclassical framework, the breakup probability is given by Eq. (2) and hence the reduced breakup probability yields

$$P_r(t) = \int d\varepsilon \frac{dB(E1)}{d\varepsilon}\varepsilon e^{-t\varepsilon}t^2 , \qquad (4)$$

where t is the so-called *collision time* and is given by the equation

$$t = \frac{a_0}{\hbar v}(\pi + \frac{2}{\sin(\theta/2)}). \qquad (5)$$

So, when the first-order semi-classical approximation is valid, the reduced breakup probability is completely determined by the $B(E1)$ distribution, and is independent on the scattering energy, or the target properties. Expression (4) can be simplified further for small scattering angles (i.e. large collision times) since in that case the exponential appearing in this expression cancels the contribution of large excitation energies and hence the result of the integral is mostly determined by the values of the $B(E1)$ close to the threshold. Approximating the $B(E1)$ by a linear function [15], one can perform the integral explicitly to obtain

$$P_r(t) \approx b e^{-\varepsilon_b t}, \qquad (6)$$

where ε_b can be interpreted as an *effective break-up threshold*, while the parameter b is associated with the slope of the $B(E1)$ distribution at low excitation energies. Further details can be found in Ref. [15].

In Fig. 7, the experimental reduced breakup probabilities, obtained from the present breakup data, are compared with the theoretical results, obtained with the 3b-CDCC (red lines) and 4b-CDCC (blue lines) calculations. The solid and dashed lines correspond to the calculations at 24.3 and 29.8 MeV, respectively. The corresponding calculations are in good agreement between themselves and with the experimental data. Calculations and experimental data are very well fitted by the scaling property predicted by Eq. (6) [15]. The reduced breakup probabilities obtained with the semiclassical calculations, making use of the experimental $B(E1)$ distribution of [9], are also shown (black lines) in Fig. 7. These calculations are found to underestimate the experimental reduced breakup probability, thus suggesting that the actual $B(E1)$ strength could be even larger than that reported in Ref. [9].

4. Conclusions

A simultaneous analysis of the elastic and break-up channels of the ^{11}Li+^{208}Pb reaction at incident energies below (24.3 MeV) and around (29.8 MeV) the Coulomb barrier ($V_B \approx$28 MeV) has been presented. The weakly bound structure of ^{11}Li gives rises to interesting phenomena in the collision with a heavy target like ^{208}Pb, such as a large suppression of the elastic cross section (with respect to the Rutherford formula) and a large yield of ^{9}Li fragments. Semiclassical calculations, based on the first-order theory of Alder and Winther for Coulomb excitation for a pure $E1$ excitation, and whose main ingredient is the $B(E1)$ distribution of ^{11}Li, confirm that the large ^{9}Li cross section arises from a Coulomb breakup mechanism, mainly due to the strong dipole Coulomb couplings stemming from the strong Coulomb interaction with the heavy target.

The data have been also compared with three-body (3b) and four-body (4b) CDCC calculations. These calculations include explicitly the coupling to the breakup channels due to the Coulomb and nuclear interactions. The 4b-CDCC calculations rely on a pure three-body model of ^{11}Li (^{9}Li+n+n), while the 3b-CDCC calculations use a simple two-body model (^{9}Li+2n). These calculations reproduce simultaneously the elastic and breakup data, although the results are very sensitive to the exact location of the dipole resonance in ^{11}Li predicted by this three-body model. Although the total $B(E1)$ strength predicted by the di-neutron model is similar to that obtained in the more realistic three-body model, most of the $B(E1)$ strength is concentrated in the region of the dipole resonance. The 3b-CDCC calculations based on this model reproduce reasonably well the breakup probability and the elastic cross section. Both sets of calculations suggests a large $B(E1)$ strength at low excitation energies, even larger than that obtained from previous Coulomb dissociation experiments [9].

The analysis of the experimental energy and angular distributions of the ^{9}Li particles coming from the reaction ^{11}Li+^{208}Pb , currently in progress, will allow us to study, in detail, the break-up channels, which will help us to understand better the mechanisms that produce the break-up of the projectile. Due to the difficulties to calculate these observables within a full four-body reaction model, these calculations will be firstly done with the much simpler di-neutron model. These results will be presented elsewhere.

References

[1] Tanihata I *et al.* 1985 *Phys. Lett. B* **160** 380
[2] Audi G, Wapstra A H and Thibault C 2003 *Nucl. Phys. A* **729** 337–676
[3] Smith M *et al.* 2008 *Phys. Rev. Lett.* **101** 202501
[4] Hansen P G and Jonson B 1987 *Euro. Phys. Lett.* **4** 409
[5] Egelhof P *et al.* 2002 *Eur. Phys. J. A* **15** 27–33
[6] Ieki K *et al.* 1993 *Phys. Rev. Lett.* **70** 730
[7] Shimoura S, Nakamura T, Ishihara M, Inabe N, Kobayashi T, Kubo T, Siemssen R, Tanihata I and Watanabe Y 1995 *Phys. Lett. B* **348** 29–34
[8] Zinser M *et al.* 1997 *Nucl. Phys. A* **619** 151
[9] Nakamura T *et al.* 2006 *Phys. Rev. Lett.* **96** 252502
[10] Korsheninnikov A A *et al.* 1997 *Phys. Rev. Lett.* **78** 2317–2320
[11] Ikeda K 1992 *Nucl. Phys. A* **538** 355–365
[12] Garrido E, Fedorov D V and Jensen A S 2002 *Nucl. Phys. A* **700** 117
[13] Myo T, Kato K, Toki H and Ikeda K 2007 *Phys. Rev. C* **76** 024305
[14] Pinilla E C, Descouvemont P and Baye D 2012 *Phys. Rev. C* **85** 054610
[15] Fernández-García J P *et al.* 2013 *Phys. Rev. Lett.* **110** 142701
[16] Kikuchi Y, Myo T, Kato K and Ikeda K 2013 *Phys. Rev. C* **87** 034606
[17] Kakuee O R *et al.* 2006 *Nucl. Phys. A* **765** 294–306
[18] Sánchez-Benítez A *et al.* 2008 *Nucl. Phys. A* **803** 30
[19] Escrig D *et al.* 2007 *Nucl. Phys. A* **792** 2–17
[20] Fernández-García J P, Rodríguez-Gallardo M, Alvarez M A G and Moro A M 2010 *Nucl. Phys. A* **840** 19–38
[21] Fernández-García J P, Alvarez M A G, Moro A M and Rodríguez-Gallardo M 2010 *Phys. Lett. B* **693** 310–315
[22] Cubero M *et al.* 2012 *Phys. Rev. Lett.* **109** 262701
[23] Cubero M *et al.* 2011 *Eur. Phys. J, web of conf.* **17** 16002
[24] Borge M J G *et al.* 2012 *Journal of Physics: Conf. Series* **381**
[25] Alder K and Winther A 1975 *Electromagnetic excitation: Theory of Coulomb excitation with heavy ions* (North-Holland)
[26] Thompson I J 1988 *Computer Physics reports* **7** 167–212

ECHIC2013 IOP Publishing
Journal of Physics: Conference Series **515** (2014) 012005 doi:10.1088/1742-6596/515/1/012005

Collisions induced by halo and weakly bound nuclei around the Coulomb barrier: experimental results at INFN-LNS

P. Figuera
INFN Laboratori Nazionali del Sud, Catania, Italy

E-Mail: figuera@lns.infn.it

Abstract. The study of nuclear collisions involving halo or weakly bound nuclei, at energies around the Coulomb barrier, had a considerable interest in the last decade since the peculiar structure of such nuclei can deeply affect the reaction dynamics. In this paper will summarize some of the experimental results obtained by our group at INFN-LNS over the last years in collisions induced by the halo nuclei ^{6}He and ^{11}Be and the stable weakly bound nuclei ^{6}Li and ^{7}Li. Very strong entrance channel effects have been observed in elastic scattering, fusion and direct processes comparing collision induced by the ^{6}He and ^{11}Be halo nuclei with the ones induced on the same target by their cores ^{4}He and ^{10}Be. Collisions induced by the stable weakly bound nuclei ^{6}Li, ^{7}Li shows some peculiarities in comparison to the ones induced by well bound nuclei, such as absence of usual threshold anomaly in the optical potential and strong competition of complete fusion with incomplete fusion and transfer in the heavy residue production cross sections. Our experimental results are compared with the ones of other authors, in order to give an overview of our present understanding of the discussed topic.

1. Introduction

The study of nuclear collisions around the Coulomb barrier induced by halo and, more in general, weakly bound nuclei such as the stable ^{6}Li, ^{7}Li, ^{9}Be, has been the object of many publications in the last decade. In fact, it has been shown that the very low breakup threshold of such nuclei combined with the halo or cluster structure of the ground state, can strongly affect the reaction dynamics around the Coulomb barrier. [e.g. 1-3].

One expects that, direct reaction processes such as breakup or transfer can be favoured by the low breakup threshold coupled with the cluster or halo structure of the ground state. In addition, since the continuum of such nuclei is very close to the ground state, coupling to continuum effects become important both for elastic and the different reaction channels, and a complete theoretical description of such collisions requires complex Continuum Discretized Coupled Channels (CDCC) calculations.

Fusion excitation functions may be affected by static and dynamic effects. Static effects are due to the fact that the diffuse surface of these nuclei affects the shape of the projectile-target potential reducing the average Coulomb barrier, thus leading to a possible enhancement of the fusion cross section. Dynamic effects on fusion, due to the coupling of the different reaction channels, present here a new aspect due to the effects of coupling to continuum. Moreover, fusion reactions in collisions induced by weakly bound nuclei are complicated by the fact that, owing to the large breakup probability, in addition to complete fusion (CF) one can have a non negligible contribution of incomplete fusion (ICF) following breakup of the projectile.

In the last decade our group at INFN-LNS Catania performed different experiments to investigate on the above discussed topics [e.g. 4-12]. In the following, our experimental results will be summarized and compared to the ones of other authors in order to give and overview on what has been observed in such studies, underlining the need of new better quality data for a deeper understanding of the discussed topic.

2. Elastic scattering and direct reactions with halo nuclei: the ^{6}He and ^{11}Be on ^{64}Zn cases

Elastic scattering with halo nuclei around the Coulomb barrier has been measured by different authors on a wide range of target masses. Most of the data are relative to ^{6}He scattering, since this beam has been available for long time at different facilities, with reasonable currents. ^{6}He+^{208}Pb scattering was measured at different energies around the barrier [13,14], and showed a suppression in the region of the Coulomb nuclear interference peak when compared with alpha scattering at the same energy, indicating the presence of long range absorption for ^{6}He. Optical Model (OM) analysis allowed to reproduce the data only using very large imaginary diffuseness. Similar results were observed for ^{6}He scattering on other heavy targets such as for instance ^{197}Au [15]. The ^{6}He scattering data on ^{208}Pb and ^{197}Au were also reproduced within the CDCC approach in [13,16]. ^{11}Li+^{208}Pb elastic scattering around the Coulomb barrier was reported in [17] and compared with the scattering of the ^{11}Li core ^{9}Li on the same target. A huge suppression of the ^{11}Li elastic angular distribution with respect to the ^{9}Li one, much larger than the one observed for ^{6}He, was found. CDCC calculations confirm that, also for this system, coupling to continuum via the Coulomb dipole interaction gives an important contribution to the disappearance of the Coulomb nuclear interference peak. Very few experiments have been performed concerning elastic scattering of halo nuclei on medium mass targets.

At the radioactive beam facility of Louvain la Neuve we measured elastic scattering and direct reactions for collisions induced by the 2n halo nucleus ^{6}He ($S_{2n} \approx 0.9$ MeV), and its core ^{4}He at different energies around the Coulomb barrier [e.g. 4,5]. Some corresponding experimental results are summarized in figure 1. Light particle energy spectra for the ^{6}He induced collision show a large yield of α particles coming from transfer an breakup, which are not observed in the ^{4}He induced collision. The elastic scattering angular distribution for ^{6}He is suppressed when compared to the one for ^{4}He at the same E_{cm}, although the Coulomb nuclear interference pattern is not as much reduced as it was observed for heavy targets. As a consequence, the total reaction cross section for ^{6}He is much larger (about a factor two) than the one for ^{4}He at the same E_{cm} energy. The angular distribution for the large α particle yield observed with ^{6}He is strongly peaked and very different than the one expected for evaporated α particles, confirming that these α particles are produced in transfer and breakup processes. The corresponding transfer and breakup integrated cross section is saturating about 80% of the total reaction cross section, a result which is very different to what observed in collisions induced by well bound nuclei. Similar results have been observed at all measured energies. Our ^{6}He scattering data were reproduced within the CDCC approach [e.g. 16] confirming the importance of coupling to continuum effects.

In the following years we measured, at the REX-ISOLDE facility of CERN, scattering and direct reactions for collisions induced by the 1n halo nucleus ^{11}Be ($S_n \approx 0.5$ MeV) and its core ^{10}Be at the same center of mass energy E_{cm} = 24.5 MeV [8,9]. As shown in figure 2(a), a suppression of the elastic angular distribution in the region of the Coulomb nuclear interference peak, much stronger than the one observed for ^{6}He on the same target, was found for ^{11}Be. The ^{11}Be elastic scattering angular distribution has been reproduced within the OM, using as bare potential the one that reproduces ^{10}Be scattering, and adding a phenomenological Dynamic Polarization Potential (DPP) having the shape of a Woods-Saxon derivative. A good fit was obtained with a DPP having a large diffuseness of the order of 3.5 fm in agreement with [18]. As shown in figure 2(b), the ^{11}Be data were also reproduced within the CDCC frame, showing that the suppression of the Coulomb nuclear interference peak is due to the combined effect of Coulomb and nuclear couplings to continuum.

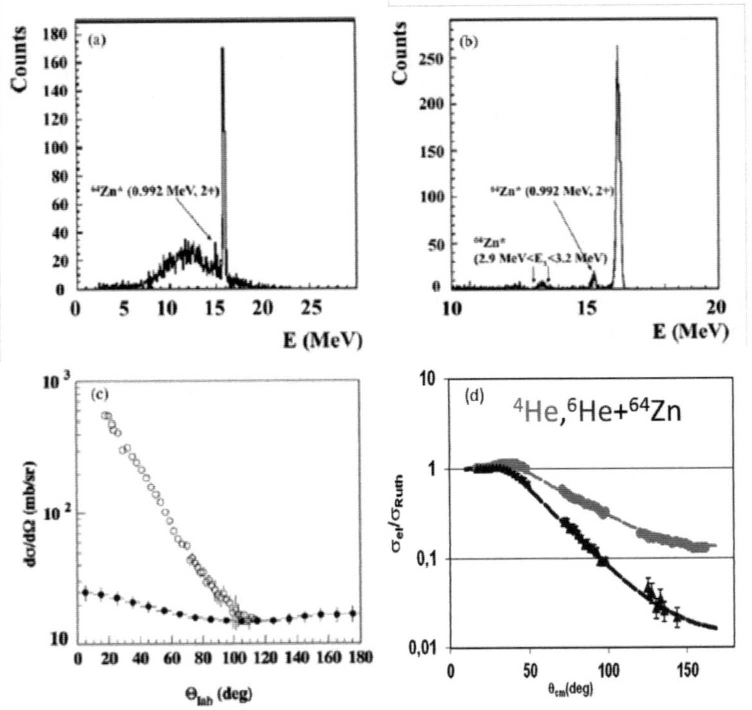

Figure 1. (Color online) (a) Light charged particle energy spectrum for ^{6}He+^{64}Zn at θ_{lab}=63^{0} and E_{lab}= 18 MeV. (b) Same as (a) but for ^{4}He+^{64}Zn. (c) Experimental α particle angular distribution at E_{lab}= 18 MeV (open symbols) compared with the one for evaporated α, calculated within the statistical model. (d) Elastic scattering angular distributions for the collisions ^{6}He+^{64}Zn (blue triangles) and ^{4}He+^{64}Zn (red circles) at the same E_{cm} = 12.4 MeV.

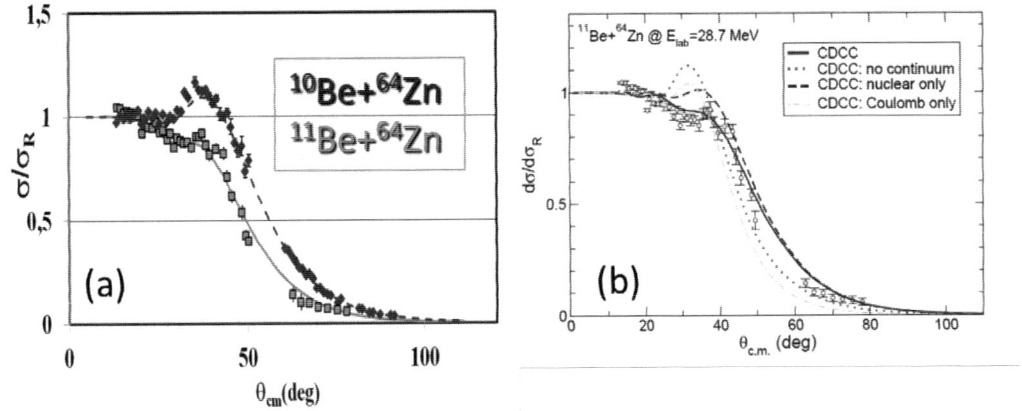

Figure 2. (Color online) (a) Elastic scattering angular distributions for the collisions ^{11}Be+^{64}Zn (green squares) and ^{10}Be+^{64}Zn (blue diamonds) at the same E_{cm}. The continuous lines are the optical model fits. (b) Experimental elastic scattering angular distributions for ^{11}Be+^{64}Zn (symbols) compared with different CDCC calculations.

The total reaction cross section for the collision induced by ^{11}Be is more than a factor two larger than the one for ^{10}Be at the same $E_{cm.}$ In the ^{11}Be+^{64}Zn collision we measured a large yield of ^{10}Be nuclei produced in transfer and breakup, saturating a large fraction of the total reaction cross section. Although presently available results concerning scattering of halo nuclei around the Coulomb barrier start to show some common features, most of available data are with ^{6}He beams and, in some cases, the results are limited by poor beam quality. Therefore, additional data with different good quality beams, including p-halo ones, are needed to build a wider systematics and allow a deeper understanding of the discussed topic.

3. Fusion reactions with halo nuclei: the ^{4,6}He+^{64}Zn case

Different papers have been published in the literature concerning fusion studies around the Coulomb barrier in collision induced by halo nuclei [e.g. 19-22]. In the last years we measured, in two independent experiments performed at the radioactive beam facility of Louvain La Neuve, the fusion excitation functions around the Coulomb barrier for the systems ^{4,6}He+^{64}Zn [5,6]. The excitation functions have been measured by using an activation technique and all experimental details are reported in [5,6]. Different ^{64}Zn targets followed by Nb catchers were irradiated with ^{4}He beams having energies in the range 8÷10 MeV. The Evaporation Residues (ER) produced in the reaction decay by electron capture (EC) and are implanted in the irradiated target and following catcher. After the end of the activation, we were looking off line at the atomic X rays following the electron capture decay of the residues. This technique allows to charge identify the ER via their characteristic X ray energy. Moreover, following the decay of the targets for a long time, the contribution of the different isotopes can be separated due to their different half lives, allowing a complete mass and charge identification.

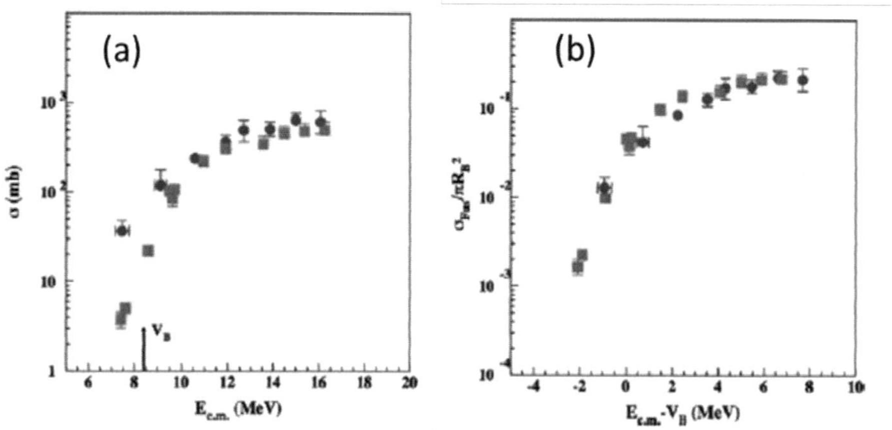

Figure 3. (Color online) (a) Fusion excitation functions for ^{6}He+^{64}Zn (blue circles) and ^{4}He+^{64}Zn (purple squares). The arrow indicates the position of the Coulomb barrier. (b) Reduced fusion excitation functions for ^{6}He+^{64}Zn (blue circles) and ^{4}He+^{64}Zn (purple squares) to eliminate static effects. See text for details.

In Figure 3(a) the ^{6}He+^{64}Zn fusion excitation function is compared with the one for ^{4}He+^{64}Zn. As one can see, an enhancement of fusion for the collision induced by the halo nucleus ^{6}He is observed in the region around the barrier ($V_B \approx$ 9MeV). As discussed in the introduction, fusion enhancement effects in collisions induced by halo nuclei may be due to a combination of static and dynamic effects. Static effects, due to the extended matter distribution in ^{6}He, can be removed by comparing reduced fusion excitation functions i.e. ($\sigma/\pi R_B^2$) versus (E_{cm}-V_B) were V_B and R_B are the barrier and barrier radius extracted from potentials calculated with realistic densities [23]. As shown in figure 3(b), by performing such comparison, the observed enhancement disappears and we therefore conclude that it can be explained as mainly due to static effects. Enhancement of fusion cross section around the barrier in collisions induced by halo nuclei, with respect to collisions induced by the halo nucleus core, has been found for different systems in the literature [e.g. 6, 21-23]. Recently, a consistent systematic comparison of many published fusion data, in collisions induced by halo and weakly bound nuclei on targets of different mass, was published [e.g. 23-25]. Such systematic study suggests that, for collisions induced by halo nuclei, once static effects are removed, the residual effect of coupling to continuum or transfer is a suppression of fusion above barrier and, most probably, a residual enhancement of fusion below the barrier. Unfortunately most of presently available data do not explore the sub-barrier region with reasonable errors and new data would be necessary for a complete understanding of this topic. This, however, is very difficult due to the low intensities of halo nuclei beams presently available.

4. Elastic scattering of ^{6,7}Li+^{64}Zn and threshold anomaly in the Optical Potential

The study of collisions induced by stable weakly bound nuclei such as ^{6}Li(S_α=1.47 MeV), ^{7}Li(S_α=2.47 MeV) and ^{9}Be(S_n=1.66 MeV), which do not have a halo structure, has also a considerable interest to investigate on the problems discussed in the introduction. When compared with halo nuclei, they have larger breakup threshold and do not show the same extended mass distribution. However, coupling to continuum effects with such nuclei are still expected to be important and, since beams of such stable nuclei are available with high currents, high quality data can be collected.

One of the problems which has been investigated in the literature concerns the energy dependence of the optical potential in collisions induced by such nuclei. It is well known that the optical potential extracted by the analysis of the elastic scattering of heavy ions, involving tightly bound nuclei, shows a variation with energy in the region of the Coulomb barrier, a phenomenon known as the threshold anomaly (TA) (see, e.g. [26]). The strength of the imaginary potential decreases sharply as the incident energy decreases toward the Coulomb barrier energy, while the real part of the potential increases and shows a bell-shaped peak in the same energy region. The TA has been ascribed to the energy dependence of the dynamic polarization potential that originates from the coupling of the elastic scattering to the non-elastic channels. The coupling to non-elastic channels results in an attractive real polarization potential. The decrease of the imaginary potential can be understood as due to the closure of non-elastic channels at energies near and below the Coulomb barrier.

Recently, much work has been devoted to the study of the TA in elastic scattering of weakly bound nuclei, such as ^{6}Li and ^{7}Li. For these projectiles the breakup channel is expected to be important even at energies below the Coulomb barrier. The coupling to the breakup produces a repulsive polarization potential [e.g. 27] and, therefore the usual TA in the OP may disappear.

In order to investigate on this topic, we measured elastic scattering angular distributions at several energies around the barrier for the systems ^{6,7}Li+^{64}Zn. The measurements have been performed at the Catania tandem by using an array of five ΔE(10 μm)-E(200 μm) telescopes mounted on a rotating arm, in order to measure the angular distributions in a wide angular range. As an example, the elastic angular distributions for the system ^{6}Li+^{64}Zn are shown in figure 4. As one can see, at energies below the Coulomb barrier (V_B ~13 MeV in the present system) the difference between the elastic and the Rutherford cross sections is rather small, therefore good quality data are needed to perform such investigations. The experimental angular distributions have been reproduced via optical model fits using a renormalized optical potential both for the real and imaginary part, leaving the renormalization coefficients as free parameters. As an example, the obtained trend of the renormalization coefficients for the ^{6}Li+^{64}Zn case is shown in figure 5 . As one can see, the data show a rather 'flat' behaviour around the Coulomb barrier corresponding to a clear absence of the usual threshold anomaly. Similar conclusions were obtained using a Woods-Saxon imaginary potential. We found Similar results also for the ^{7}Li+^{64}Zn system. Details of the used experimental procedure are reported in [11]. Absence of the usual threshold anomaly in the OP, in collisions induced by weakly bound nuclei both stable and radioactive has been observed by many other authors in the literature [e.g. 28-32]

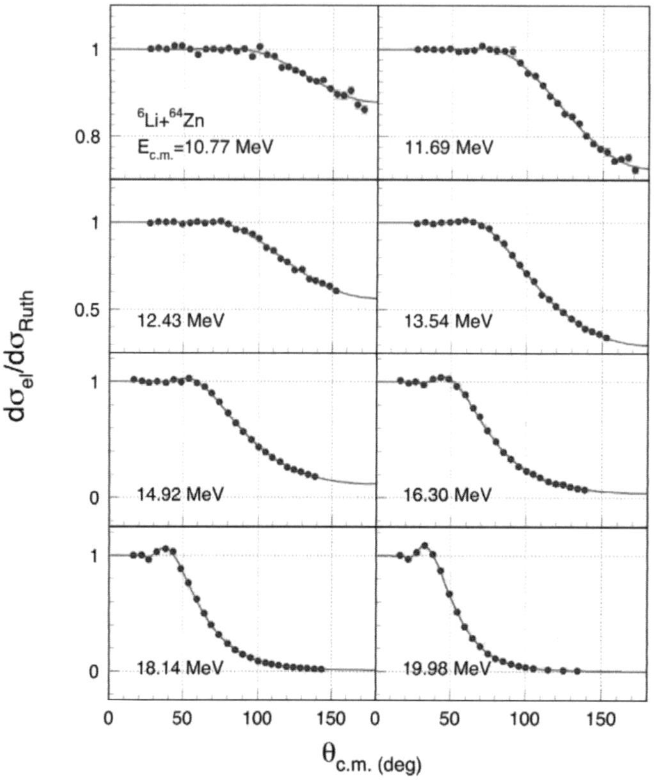

Figure 4. (Color online) Elastic scattering angular distributions measured for the system ^{6}Li+^{64}Zn. The continuous lines are the results of the optical model fits.

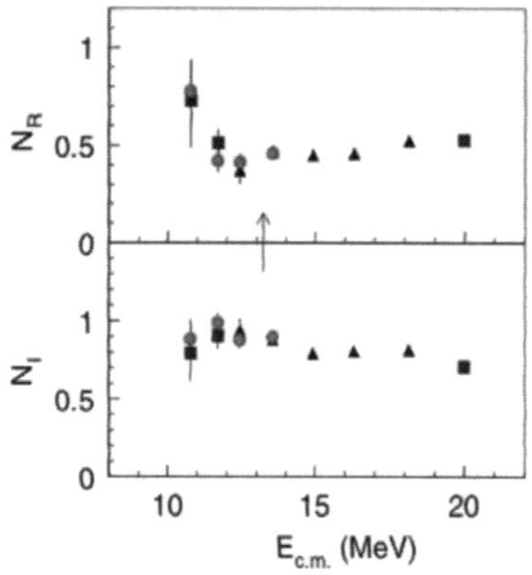

Figure 5. (Color online) Renormalization coefficients for the real and imaginary part of the potential in the collision ^{6}Li+^{64}Zn. The position of the Coulomb barrier is indicated by the arrow. Different symbols refer to independent experiments.

5. Fusion excitation functions with stable weakly bound nuclei: the ^{6,7}Li+^{64}Zn case

Fusion excitation functions in collisions induced by stable weakly bound nuclei have been measured for several systems in a wide target mass range [e.g. 33-40]. For fusion-evaporation reactions induced by weakly bound nuclei on heavy targets [e.g. 33-36], where the emission of charged particles is hindered by the presence of a strong Coulomb barrier, the CF and ICF contributions can be usually clearly separated since they produce evaporation residues (ER) having different charge. The main finding for such systems is that CF is suppressed above the barrier by about 30% with respect to the prediction of Single Barrier Penetration (SBP) calculations or Coupled Channel (CC) calculations not taking into account coupling to continuum. Since the sum of CF and ICF, normally referred as total fusion (TF), is not suppressed, it was suggested that the observed suppression is mainly due to breakup of the projectiles followed by ICF.

For fusion of weakly bound nuclei on medium mass or light targets [e.g. 37-40], the compound nucleus can evaporate charged particles. Therefore, the same residues can be populated by the CF and ICF mechanisms, making their separation much more difficult. For this reason, most of the fusion data for light and medium mass systems are relative to TF cross section which, as for collisions on heavy targets, do not show any suppression or enhancement effect above barrier, whereas below the barrier not much data are available.

Recently we measured excitation functions for heavy residue (HR) production in the collisions ^{6,7}Li+^{64}Zn around the Coulomb barrier, with the aim to extract some information on the competition between CF, ICF and other reaction mechanisms in the HR production. The experiment has been performed at the Catania tandem with the same activation technique mentioned in section 3 and all the experimental details are reported in [10]. Excitation functions for HR production are shown in figure 6, together with the results of SBP calculations and CC calculations including the excitation of different states of projectile and target. As one can see, both calculations reproduce the data above barrier but fail below it. In order to understand which are the possible reaction mechanisms responsible for heavy residue production, the experimental relative yield of the HR was compared with different statistical model calculations performed using the code Cascade.

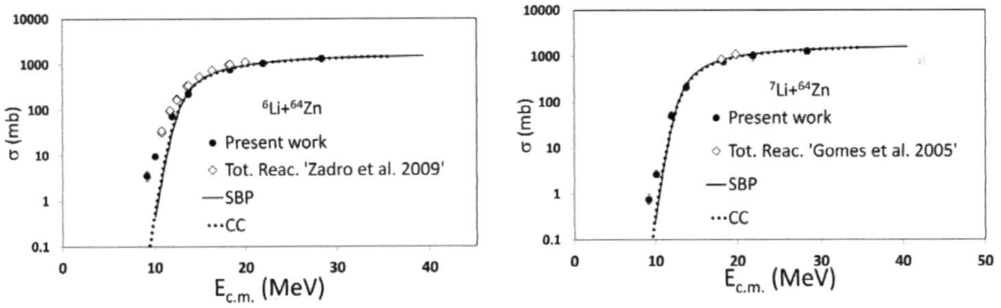

Figure 6. Excitation functions for heavy residue production (full circles) and total reaction (open diamonds) for the systems ^{6}Li+^{64}Zn and ^{7}Li+^{64}Zn. The total reaction data are from refs. [11,37].

Since ^{6}Li and ^{7}Li have weakly bound cluster structures of the ground state with separation energies S_α=1.47 and S_α=2.47 MeV respectively, we expect reactions where only an alpha particle, a deuteron or a tritium are captured by the target. This may happen in two different ways, breakup followed by incomplete fusion (ICF) or a direct cluster transfer (DCT). Since ICF and DCT populate the same nuclei with similar excitation energies, we cannot distinguish between the two modes in the present experiments. Expected relative yields of the HR populated in ICF or DCT have been calculated with Cascade. The excitation energies following the d, t or α capture were estimated assuming an ICF mechanism where the available fragment-target centre of mass energy before ICF occurs is: $(E_{c.m.}-S_\alpha)\times(m_{clu}/m_{proj})$. Here $E_{c.m.}$ is the initial centre of mass energy of the Li+^{64}Zn system, S_α is the α separation energy of the projectile and m_{clu} and m_{proj} are the masses of the captured cluster and of the projectile respectively. Excitation energies estimated for DCT were only slightly smaller than the ones for ICF, giving basically the same relative yield as for ICF. A comparison between the expe at two different energies. As one can see, above the Coulomb barrier ($V_B \sim 13$ MeV), the experimental relative yields are similar to the ones predicted by CF Cascade calculations, showing that CF is the dominating mechanism. However, at energies below the barrier, the HR experimental relative yield is rather different than the one expected for CF, showing that CF is no longer the dominating reaction mechanisms while other processes such as ICF, DCT or single nucleon transfer become very important. As one can see in figure 7, for ^{6}Li+^{64}Zn the relative yield below the barrier is dominated by ^{65}Zn and ^{65}Ga which are not expected in CF but can be produced in ICF/DCT of a deuteron as well as in 1n and 1p transfer reactions respectively. Similar results were also observed for ^{7}Li+^{64}Zn. Present results suggest therefore that, in order to study fusion below the barrier in collisions induced by weakly bound nuclei on medium mass targets, the simple integration of the HR yield is not giving us the TF cross sections since different processes such as DCT and single neutron transfer can give an important contribution. Therefore, one needs to find experimental techniques allowing to properly separate the different reaction mechanisms contributing to the HR yield.

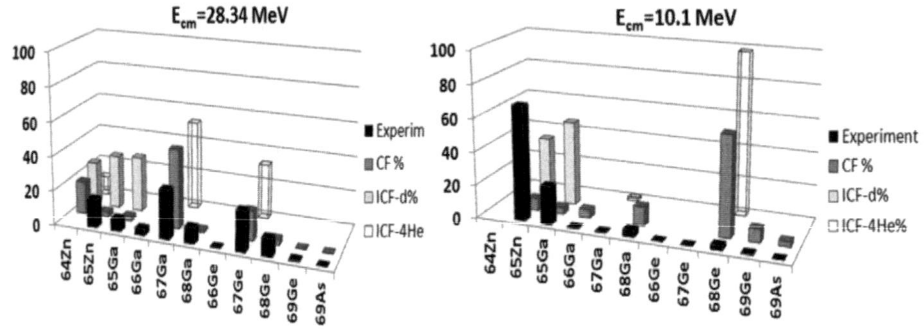

Figure 7. (Color online) Experimental heavy residue relative yield for ^{6}Li+^{64}Zn compared with the prediction of statistical model calculations for complete fusion and deuteron or α capture.

6. Summary and conclusions

In this paper our experimental results, concerning the study of elastic scattering and reactions for halo and stable weakly bound nuclei around the Coulomb barrier on a ^{64}Zn target, have been summarized and compared with the ones of other authors.

We have shown that, in collisions induced by halo nuclei, a strong suppression of the elastic scattering cross section and, correspondingly, a large enhancement of the total reaction cross section has been observed by various authors. It has been found that the observed enhancement of the total reaction cross section is due to a very large yield of transfer and breakup which, contrary to what observed with well bound projectiles, saturates most of the total reaction cross section. In particular, in collisions induced by the 2n halo nucleus ^{6}He, it has been found that the 2n transfer process is the most important reaction channel around the Coulomb barrier. In order to reproduce the measured elastic scattering angular distributions within the OM, one has to use rather large diffuseness of the imaginary part showing the presence of long range absorption. Elastic scattering data have also been reproduced by different authors within the CDCC frame showing that the measured elastic scattering angular distributions cannot be reproduced without taking into account coupling to continuum effects.

Fusion excitation functions in collisions induced by halo nuclei have also been measured by different authors. Presently, available results suggest the presence of a fusion cross section enhancement around the barrier for halo nuclei when compared both to SBP calculations based on potentials which do not take into account the diffuse structure the projectile, or to the excitation functions induced by the halo

nucleus core on the same targets. Although most of the observed effect on fusion can be described as due to static effects, these are not the only ones, and it has been suggested that coupling to continuum or transfer channels is also playing an important role.
Elastic scattering and fusion have also been studied in collisions induced by the stable weakly bound nuclei ^{6}Li, ^{7}Li and ^{9}Be, having low breakup thresholds and a marked cluster structure of the ground state. Elastic scattering with such nuclei has been studied by different authors and reproduced with OM fits. For many of the studied systems, including our ^{6,7}Li+^{64}Zn data, the obtained energy dependence of the real and imaginary part of the OP, extracted from the OM fits, show absence of the usual threshold anomaly. This has been interpreted ad due to the presence of a repulsive polarization potential due to coupling to continuum effects. Fusion excitation functions with stable weakly bound nuclei have also been studied. In collision on heavy targets, were CF and ICF can be easily separated, data show a well established suppression of CF due to ICF following projectile breakup. In collisions with medium mass or light targets CF and ICF cannot be easily separated and most of available data refers to TF. Our ^{6,7}Li+^{64}Zn fusion data suggest that above barrier HR production is dominated by CF. Below the barrier, on the contrary, other processes such as ICF but also direct cluster transfer or single neutron transfer are the dominating processes for HR production.
Although different aspects of the discussed topics have been clarified by the papers published so far, new better quality data would be necessary for a more complete understanding to our opinion.
Almost all the results obtained so far for collisions induced by halo nuclei have been obtained with ^{6}He beams and additional data with different halo nuclei would be necessary to build up a compete systematics. At the same time, new fusion cross sections measurements with halo nuclei, better exploring the sub-barrier region are still needed, since most of the data available do not explore this energy range with reasonable errors.
In the study of collisions induced by stable weakly bound nuclei, new fusion cross sections measurements would be needed, especially on medium mass or light targets, at energies extending well below the barrier. These measurements should possibly be performed with experimental techniques allowing to clearly disentangle among the transfer, CF and ICF mechanisms leading to the production of the detected heavy residues, in order to bring new clear information on fusion and competing reaction mechanisms.

Acknowledgments
Our experimental results discussed in the present review paper, have been obtained over the last decade in different experiments involving many colleagues, belonging to INFN-LNS and other institutions, whose contribution is gratefully acknowledged.

References
[1] Canto L F et al. 2006 *Phys. Rep.* **424** 1
[2] Keeley N et al. 2007 *Prog. Part. Nucl. Phys.* **59** 579
[3] Keeley N et al. 2009 *Prog. Part. Nucl. Phys.* **63** 396
[4] Di Pietro A et al. 2003 Eur*ophys. Lett.* **64** 309
[5] Di Pietro A et al. 2004 *Phys. Rev. C* **69** 044613
[6] Scuderi V et al. 2011 *Phys. Rev. C* **84** 064604
[7] Di Pietro A et al. 2007 *Eur. Phys. J. Spec. Top.* **150** 15
[8] Di Pietro A et al. 2010 *Phys. Rev. Lett.* **105** 022701
[9] Di Pietro A et al. 2012 *Phys. Rev. C* **85** 054607
[10] Di Pietro A et al. 2013 *Phys. Rev. C* **87** 064614
[11] Zadro M et al. 2009 *Phys. Rev. C* **80** 064610
[12] Zadro M et al. 2013 *Phys. Rev. C* **87** 054606
[13] Acosta L et al. 2011 *Phys. Rev. C* **84** 044604
[14] Sanchez Benitez A M et al. 2008 *Nucl. Phys. A* **803** 30
[15] Kakuee O R et al. 2006 *Nucl. Phys. A* **765** 294

Journal of Physics: Conference Series **515** (2014) 012005 doi:10.1088/1742-6596/515/1/012005

[16] Moro A M et al. 2007 *Phys. Rev. C* **75** 064607
[17] Cubrero M et al. 2012 *Phys. Rev. Lett.* **109** 262701
[18] Bonaccorso A et al. 2002 *Nucl. Phys. A* **706** 322
[19] Raabe R et al. 2004 *Nature* **431** 823
[20] Kolata J J et al. 1998 *Phys. Rev. Lett.* **81** 4580
[21] Lemasson A et al. 2009 *Phys. Rev. Lett.* **103** 232701
[22] Alcorta M et al. 2011 *Phys. Rev. Lett.* **106** 172701
[23] Canto L F et al. 2009 *Nucl. Phys. A* **821** 51
[24] Rangel J et al. 2013 *Eur. Phys. Jour. A* **49** 57
[25] Gomes P R S et al. 2009 *Phys. Rev. C* **79** 027606
[26] Satchler G R et al. 1991 *Phys. Rep.* **199** 147
[27] Sakuragi Y et al. 1987 *Phys. Rev. C* **35** 2161
[28] Maciel A M M et al. 1999 *Phys. Rev. C* **59** 2103
[29] Pakou A et al. 2004 *Phys. Rev. C* **69** 054602
[30] Kumawat H et al. 2008 *Phys. Rev. C* **78** 044617
[31] Figueira J M et al. 2010 *Phys. Rev. C* **81** 024613
[32] Deshmukh N N et al. 2011 *Phys. Rev. C* **83** 024607
[33] Dasgupta M et al. 2004 *Phys. Rev. C* **70** 024606
[34] Gomes P R S et al. 2006 *Phys. Rev. C* **73** 064606
[35] Wu Y W et al. 2003 *Phys. Rev. C* **68** 044605
[36] Pradhan M K et al. 2011 *Phys. Rev. C* **83** 064606
[37] Gomes P R S et al. 2005 *Phys. Rev. C* **71** 034608
[38] Anjos R M et al. 2002 *Phys. Lett. B* **534** 45
[39] Sinha M et al. 2010 *Eur. Phys. J. A* **44** 403
[40] Beck C et al. 2003 *Phys. Rev. C* **67** 054602

Fusion of the ^{6}Li+^{120}Sn and ^{7}Li+^{119}Sn and the role of neutron transfer and breakup processes

M. Fisichella[1], A.C. Shotter[2], A. Di Pietro [1], P. Figuera [1], M. Lattuada [1,3], A. Musumarra [1,3], M.G. Pellegriti [1,3], C. Ruiz [4], V. Scuderi [1,3], E. Strano [1,3], D. Torresi [1,3], M. Zadro [5]
[1]INFN - Laboratori Nazionali del Sud, via S.Sofia 62, I-95125 Catania, Italy
[2]School of Physics and Astronomy, JCMB, Mayfield Road, Edinburgh EH9 3JZ, UK
[3]Dipartimento di Fisica ed Astronomia, via S.Sofia 64, I-95125 Catania, Italy
[4]TRIUMF 4004 Wesbrook Mall Vancouver BC V6T 2A3 Canada
[5]Ruđer Bošković Institute, Bijenička cesta 54, HR-10000 Zagreb, Croatia

E-mail: `fisichella@lns.infn.it`

Abstract. Excitation functions have been measured for the fusion of the weakly bound nuclei ^{6}Li and ^{7}Li respectively with ^{120}Sn and ^{119}Sn at energies around the Coulomb barrier by using an activation technique. With such a study we aim to investigate the role played by direct processes as neutron transfer and breakup on fusion. At energies above the barrier, the complete fusion cross section are lower than those predicted by fusion models. This suppression is independent of beam energy and is in agreement with the values reported in literature for heavier systems. At energies below the barrier, no significant enhancement due to the positive Q-value for one and two neutron transfer for the ^{7}Li+^{119}Sn system is observed as compared with the ^{6}Li+^{120}Sn.

1. Introduction

The fusion process at energies close to the Coulomb barrier has been object of extensive experimental and theoretical studies in the last decades. The sensitivity of the fusion process to internal degree of freedom of the colliding nuclei, such as rotation and vibration, has been well recognized [1]. At energies below the Coulomb barrier, the coupling of the relative motion to these degrees of freedom leads to an enhancement of the fusion cross-section, relative to the one expected from the one-dimension barrier penetration model (1D-BPM). It has been shown [2, 3] that the effect of channel coupling is to replace the single fusion barrier by a distribution of barriers of different heights, some of which are lower than the height of the single barrier. Moreover, the importance of neutron transfer in increasing the sub-barrier fusion probabilities was also established, but quantitative understanding of the coupling strengths remained elusive due to the need to rely on complex theoretical calculations. The importance of neutron transfer on nuclear fusion originates from the fact that neutrons are insensitive to the Coulomb field and their transfer might start at larger separations than other particles. Therefore, it is possible that a flow of neutrons between target and projectile has already started [4], before fusion takes place, influencing the subsequent fusion reaction which occurs inside the barrier. Thus, it is expected that, if a system exhibits neutron-transfer reactions with positive Q-values, the sub-barrier fusion cross-section will increase [4]. The idea that neutron transfer process would lead to an enhancement of the fusion cross-section is supported by a semi-classical model [5], according to which an intermediate neutron transfer with positive Q-value may lead to a gain in relative kinetic energy of the colliding nuclei and, thus, to an enhancement of the barrier penetrability and therefore of the fusion cross-section. Most of the experiments performed to study the effects of neutron transfer on the fusion cross-section compared systems with positive Q-value for neutron transfer with their isotopic counterparts with no positive Q-value. By comparing similar systems one could expect that it is easier to isolate the effects of transfer couplings with respect to inelastic channels (which are always present). Experimentally, the study of ^{28}Si+^{94}Zr [6] ^{32}S+^{96}Zr,^{100}Mo,^{110}Pd [7, 8, 9] and ^{40}Ca+^{48}Ca,^{96}Zr,^{124,132}Sn [10, 11, 12, 13, 14] fusion reactions has shown an enhancement in the sub-barrier fusion cross section of these systems with respect to their reference systems. This enhancement has been related to the presence of positive Q-value neutron transfer channels.

In recent works concerning the study of fusion reaction of Sn+Ni and Te+Ni [15], ^{16,17}O+^{76,74}Ge [16] and ^{60}Ni+^{100}Mo [17], similar sub barrier fusion cross sections have been measured even if the studied systems present very different Q-values for multi-neutron transfer. A possible solution of this controversy has been proposed by studying these fusion reactions using the quantum diffusion approach [18]. According to this theory, in systems like the ^{40}Ca+^{96}Zr, the change of the magnitude of the fusion cross section, occurs since the neutron transfer produce a change in the deformation of the interacting nuclei, before the fusion. In particular, the variation of the nuclei's shapes generates a lowering of the Coulomb barrier and consequently an enhancement of the fusion cross section. Instead, in systems like as example the ^{60}Ni+^{100}Mo, the neutron transfer do not modify significantly the deformation of the colliding nuclei and so it weakly influences the fusion cross section. With these calculation the authors were able to well reproduce existing the data. New measurements, particularly on the simple cases of one-neutron ($1n$) and two-neutron ($2n$) transfer with positive Q-values are needed for improving the understanding of this issue.

For further investigate the possible relation between the presence of positive neutron transfer Q-value and the sub-barrier enhancement of fusion cross-section, we measured the fusion cross section of ^{6}Li+^{120}Sn and ^{7}Li+^{119}Sn. These two system are characterized by very similar entrance channels, their fusion leads to same compound nucleus and present different Q-values for $1n$ and $2n$ transfers. They seems well suitable for investigate on the possible role of the neutron transfer on the fusion process, at energies below the Coulomb barrier.

Moreover, our projectiles are weakly bound nuclei, characterised by low-breakup threshold, cluster structure and diffuse matter distribution. Thus, by studying these reactions, at energies above the Coulomb barrier, it is also possible to investigate the role played by breakup on the fusion process. In reactions on heavy targets ($Z > 62$) a reduction of the complete fusion (CF) cross-section has been observed at energies above the barrier, with respect to 1D-BPM calculations or to reactions induced by well bound nuclei [19, 20, 21]. Such a reduction was attributed to the projectile break-up in the strong Coulomb field of the target nucleus. It has been observed that for heavy targets the CF cross-section reduction depends on the break-up threshold of the projectile; it does not seem to depend on the atomic number of the target. On the contrary, fusion reaction on medium and light target does not seem affected by the breakup process, no reduction of the fusion cross section was observed (see as example [22]). In ^{6}Li+^{120}Sn and ^{7}Li+^{119}Sn reactions the discrimination of complete fusion from incomplete fusion is possible since the two processes produce ERs of different charge. Thus, with the study of these systems we aimed also to investigate the relation between the suppression factor and the atomic number of the target in a new mass range.

Journal of Physics: Conference Series **515** (2014) 012006 doi:10.1088/1742-6596/515/1/012006

2. Experimental technique

2.1. Target activation

A fusion reaction induced by a beam of light nuclei on a medium mass target, at energies around the Coulomb barrier, leads to the production of evaporation residues (ER) whose largest fraction does not have enough kinetic energy to leave the target and they cannot be detected directly. In our case, the ER of interest are unstable against electron capture (EC) decay and thus, as in our previous experiments [23, 24, 25], we measured the fusion excitation functions by using an activation technique, based on the measurement of the atomic X-ray emission following the EC decay. The measurement was performed at Laboratori Nazionali del Sud - INFN. Sn targets ($\approx$ 500 μg/cm^2), followed by a ^{93}Nb ($\approx$ 2 mg/cm^2) catchers, were irradiated with Li beams, whose energy was chosen such as to measure the fusion excitation function in a centre of mass energy range from 16 MeV to 24 MeV. The catcher is used to stop the fraction of ER emerging from the target. Possible reactions induced by the beam on the catcher do not represent a problem since the associated X-ray energies are different from the ones corresponding to reactions on the target. To extract the production cross-section it is necessary to measure the number of incident particles on the target. If the ER produced in the reaction have a much longer life time with respect to the irradiation time it is possible to obtain this value simply by integrating the charge accumulated in a Faraday cup, placed downstream the target. In our case the ER of interest have a half-life shorter than, or comparable with, the duration of the activation period. In this case it is necessary to know the beam profile of irradiation. The beam intensity measurement as a function of the time has been performed by detecting the elastically scattered particles from a thin gold foil ($\approx$ 100 $\mu g/cm^2$) placed before the stack on the beam line, using two $1000\mu m$ Surface Barrier silicon detectors. These two monitor detectors were symmetrically placed at $\pm$20 with respect to the beam line and at a distance of about 80 cm from the Au foil. Since the elastic scattering at this angle follows the Rutherford law, the beam intensity can be extracted by the well-known cross-section formula. By using two symmetrical monitors it is possible to reduce systematic errors due to mechanical misalignments and small beam position shifts.

2.2. Characteristic X-rays measurement

After the irradiation, the characteristic X-rays of the ERs, emitted from different targets (together with own catcher), were measured off-line using Pb shielded large area Si(Li) detectors. Each measurement was repeated in order to measure the activity as a function of time. To determine the fusion cross section it is crucial to know the intrinsic efficiency of the detector. We measured the efficiency of our Si(Li) detectors using some calibrated sources, because in the energy range of interest the efficiency for this kind of detectors is strongly dependent on the energy. In figure 1 some typical X-ray spectra for ^{7}Li+^{119}Sn reactions are shown, in the energy region of interest and for different times after irradiation. It is possible to distinguish the k_α and k_β X-rays produced in the decay of Iodine (the only element which can be produced after complete fusion) to Tellurium. In the present experiment, the analysis was performed only on the k_α lines, taking into account that they represent about the 80% of the total X-ray emission. In the spectra it is also possible to distinguish the k_α and k_β peaks of Sb, produced by incomplete fusion of tritium with the target, which then decays by E.C. into Sn. Symilar spectra have been observed forv the ^{6}Li+^{120}Sn reaction. Since, the resolution of the X-ray spectra is such that it is not possible to resolve the isotopic shift, from the X-ray lines only different elements can be identified. Each isotopes contributes to the overall activity with its characteristic half-life. This means that contributions of different isotopes can be unfolded by following the activity of the X-ray lines as a function of the time. In figures **??** typical activity curves for the ^{6}Li+^{120}Sn and ^{7}Li+^{119}Sn reactions are plotted. It is possible to observe three different slopes which characterize these curves, which allow to identify in both the curves the ^{124}I and the ^{123}I produced in the 1 neutron and 2 neutron evaporation channel, respectively. The

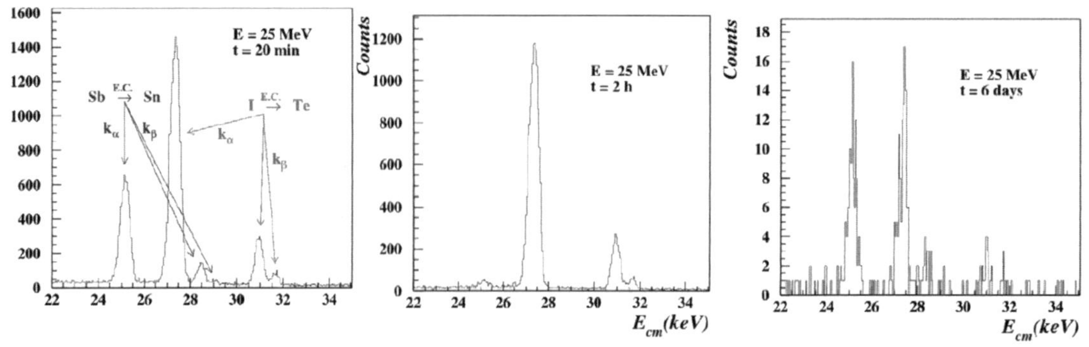

Figure 1. X-ray spectra measured off-line, 20 minutes, 2 hours and 6 days after the end of the irradiation, for the reaction ^{7}Li+^{119}Sn at E_{LAB}= 25 MeV.

third component has been identified as ^{123}Te in the case of ^{6}Li+^{120}Sn reaction and ^{121}Te in the case of ^{7}Li+^{119}Sn, produced in the incomplete fusion of α with the target nucleus. In particular these two nuclei are produced from the 1 neutron and 2 neutron evaporation channels; they are metastable and decay by internal conversion thus emitting X-rays. By fitting the activation

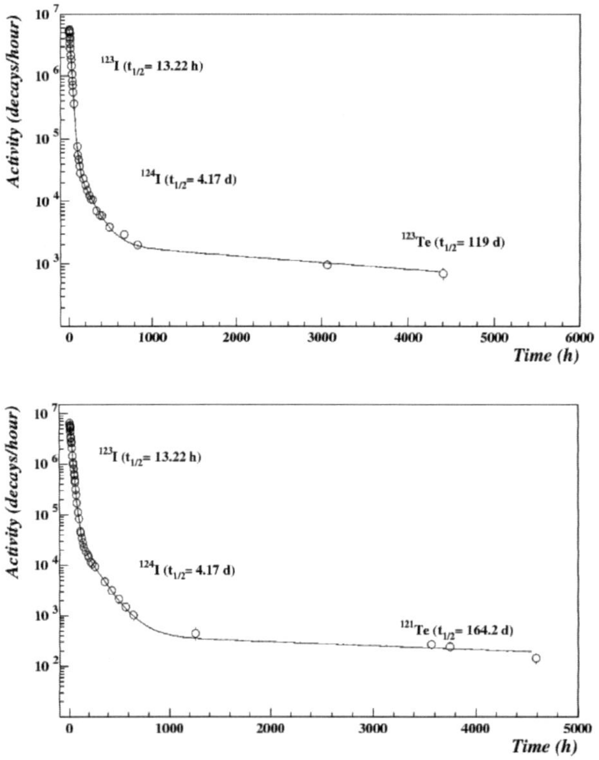

Figure 2. Activity curves for the Iodine isotopes for the ^{6}Li+^{120}Sn (top) and ^{7}Li+^{119}Sn (bottom) at E_{LAB}=25MeV. See details in the text.

curves for each ER one obtains A_{0exp}, which is its activity at the end of the irradiation time. Of course, the experimental activity is only a fraction of the real activity of the residue. The values that we have obtained have to be corrected for the detector efficiency and also for the fluorescence probability, k_{α}, since the electron capture decay is in competition with other decay modes.

In figure 3 the experimental relative yields for the produced residues is shown and compared with the prediction of the statistical model code CASCADE. A good agreement between the experimental data and the calculation predictions has been obtained.

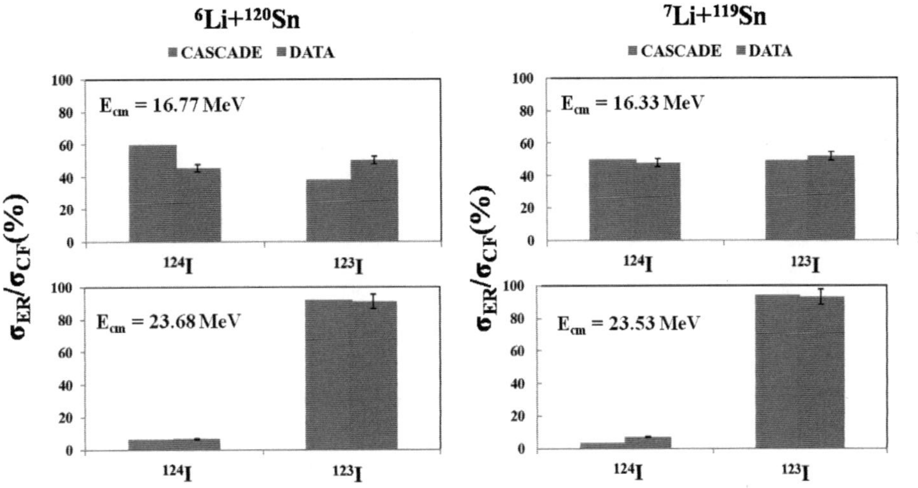

Figure 3. ER production relative yields for the ^{6}Li+^{120}Sn (left) and ^{7}Li+^{119}Sn (right) reactions compared with the prediction of the statistical model CASCADE, at two different energies.

3. Results and conclusions

The complete fusion cross-section has been extracted by summing the contributions of ^{123}I and ^{124}I. In figure 4 the reduced fusion excitation functions for the two studied reactions are reported and compared. The data have been reduced by dividing the fusion cross-section by the square of the barrier radius (R_B^2) and subtracting from the center-of-mass energy the height of the Coulomb barrier V_b. The two fusion excitation functions are different at energies above the barrier. According to previous studies [19, 20, 21] this difference could depend on the different breakup probability of the two projectiles. For both the systems, the suppression factor, obtained by comparing the fusion excitation functions with respect to the the 1D-BPM prediction, is in a agreement with the values reported in literature for heavier systems. It seems that this difference disappears at energies below the barrier, thus sub-barrier fusion enhancement in the ^{7}Li induced collisions imputable to larger n-transfer Q-value cannot be deduced from these data.

In conclusion, by measuring the fusion cross section induced by the weakly bound ^{6}Li and ^{7}Li projectiles on different Sn isotopes, we looked for effects on fusion cross section which can be attributed to the breakup and neutron transfer channels. For both the reaction, the above barrier suppression factors are in agreement with the ones previously measured for heavier targets, confirming further that there is no relation between the CF fusion suppression and the target atomic number. Concerning the possible influence of the positive neutron transfer Q-value on the fusion excitation function at energies below the barrier, no difference imputable to

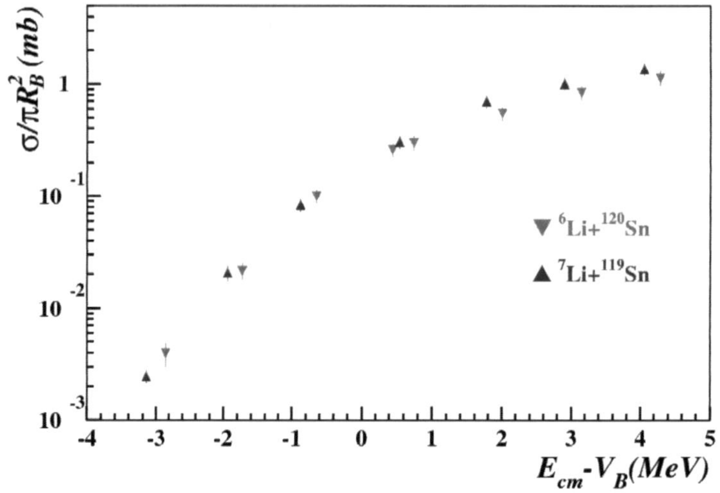

Figure 4. ^{6}Li+^{120}Sn (red reversed triangles) and ^{7}Li+^{119}Sn (blue triangles) fusion excitation function reduced as $\sigma/\pi R_B^2$ vs $E_{cm} - V_B$

the different n-transfer Q-values can be deduced from the present data. To further investigate the possible relation between the n-transfer Q-values with the fusion excitation function, we are going to extend the measure at lower energies. In addition a CDCC calculation not including the contribution of the transfer could helpful to extract this effect.

References

[1] M. Beckerman. Sub-barrier fusion of two nuclei. *Reports on Progress in Physics*, 51(8):1047, 1988.

[2] C.H. Dasso, S. Landowne, and A. Winther. Channel-coupling effects in heavy-ion fusion reactions. *Nuclear Physics A*, 405(2):381–396, 1983.

[3] C.H. Dasso, S. Landowne, and A. Winther. A study of q-value effects on barrier penetration. *Nuclear Physics A*, 407(1–2):221–232, 1983.

[4] P. H. Stelson, H. J. Kim, M. Beckerman, D. Shapira, and R. L. Robinson. Fusion cross sections for ^{46,50}Ti+^{90}zr,^{93}nb and some systematics of heavy-ion fusion at barrier and subbarrier energies. *Phys. Rev. C*, 41:1584–1599, Apr 1990.

[5] V. I. Zagrebaev. Sub-barrier fusion enhancement due to neutron transfer. *Phys. Rev. C*, 67:061601, Jun 2003.

[6] Sunil Kalkal, S. Mandal, N. Madhavan, E. Prasad, Shashi Verma, A. Jhingan, Rohit Sandal, S. Nath, J. Gehlot, B. R. Behera, Mansi Saxena, Savi Goyal, Davinder Siwal, Ritika Garg, U. D. Pramanik, Suresh Kumar, T. Varughese, K. S. Golda, S. Muralithar, A. K. Sinha, and R. Singh. Channel coupling effects on the fusion excitation functions for ^{28}Si+^{90,94}Zr in sub- and near-barrier regions. *Phys. Rev. C*, 81:044610, Apr 2010.

[7] H. Q. Zhang, C. J. Lin, F. Yang, H. M. Jia, X. X. Xu, Z. D. Wu, F. Jia, S. T. Zhang, Z. H. Liu, A. Richard, and C. Beck. Near-barrier fusion of ^{32}S +^{90,96} Zr: The effect of multi-neutron transfers in sub-barrier fusion reactions. *Phys. Rev. C*, 82:054609, Nov 2010.

[8] H.-J. Hennrich, G. Breitbach, W. Kühn, V. Metag, R. Novotny, D. Habs, and D. Schwalm. Differences in angular momentum distributions for near-barrier fusion of 32S+100Mo and 36S+96Mo. *Physics Letters B*, 258(3–4):275–278, 1991.

[9] A. M. Stefanini, D. Ackermann, L. Corradi, J. H. He, G. Montagnoli, S. Beghini, F. Scarlassara, and G. F. Segato. Signatures of multiphonon and neutron transfer couplings in the fusion of ^{36,32}S+^{110}Pd. *Phys. Rev. C*, 52:R1727–R1731, Oct 1995.

[10] M. Trotta, A. M. Stefanini, L. Corradi, A. Gadea, F. Scarlassara, S. Beghini, and G. Montagnoli. Sub-barrier fusion of the magic nuclei ^{40,48}Ca+^{48}Ca. *Phys. Rev. C*, 65:011601, Dec 2001.

[11] H. A. Aljuwair, R. J. Ledoux, M. Beckerman, S. B. Gazes, J. Wiggins, E. R. Cosman, R. R. Betts, S. Saini,

and Ole Hansen. Isotopic effects in the fusion of ^{40}Ca with ^{40,44,48}Ca. *Phys. Rev. C*, 30:1223–1227, Oct 1984.

[12] H. Timmers, D. Ackermann, S. Beghini, L. Corradi, J.H. He, G. Montagnoli, F. Scarlassara, A.M. Stefanini, and N. Rowley. A case study of collectivity, transfer and fusion enhancement. *Nuclear Physics A*, 633(3):421–445, 1998.

[13] F. Scarlassara, S. Beghini, G. Montagnoli, G.F. Segato, D. Ackermann, L. Corradi, C.J. Lin, A.M. Stefanini, and L.F. Zheng. Fusion of 40Ca+124Sn around the Coulomb barrier. *Nuclear Physics A*, 672(1–4):99–110, 2000.

[14] J. J. Kolata, A. Roberts, A. M. Howard, D. Shapira, J. F. Liang, C. J. Gross, R. L. Varner, Z. Kohley, A. N. Villano, H. Amro, W. Loveland, and E. Chavez. Fusion of ^{124,132}sn with ^{40,48}ca. *Phys. Rev. C*, 85:054603, May 2012.

[15] Z. Kohley, J. F. Liang, D. Shapira, R. L. Varner, C. J. Gross, J. M. Allmond, A. L. Caraley, E. A. Coello, F. Favela, K. Lagergren, and P. E. Mueller. Near-barrier fusion of Sn+Ni and Te+Ni systems: Examining the correlation between nucleon transfer and fusion enhancement. *Phys. Rev. Lett.*, 107:202701, Nov 2011.

[16] H. M. Jia, C. J. Lin, F. Yang, X. X. Xu, H. Q. Zhang, Z. H. Liu, L. Yang, S. T. Zhang, P. F. Bao, and L. J. Sun. Fusion of the ^{16}o+^{76}ge and ^{18}o+^{74}ge systems and the role of positive q-value neutron transfers. *Phys. Rev. C*, 86:044621, Oct 2012.

[17] A.M. Stefanini, G. Montagnoli, F. Scarlassara, C.L. Jiang, H. Esbensen, E. Fioretto, L. Corradi, B.B. Back, C.M. Deibel, B. Giovine, J.P. Greene, H.D. Henderson, S.T. Marley, M. Notani, N. Patel, K.E. Rehm, D. Sewerinyak, X.D. Tang, C. Ugalde, and S. Zhu. Fusion of 60Ni + 100Mo near and below the Coulomb barrier. *The European Physical Journal A*, 49(5):1–6, 2013.

[18] V. V. Sargsyan, G. G. Adamian, N. V. Antonenko, W. Scheid, and H. Q. Zhang. Role of neutron transfer in capture processes at sub-barrier energies. *Phys. Rev. C*, 85:024616, Feb 2012.

[19] M. Dasgupta, P. R. S. Gomes, D. J. Hinde, S. B. Moraes, R. M. Anjos, A. C. Berriman, R. D. Butt, N. Carlin, J. Lubian, C. R. Morton, J. O. Newton, and A. Szanto de Toledo. Effect of breakup on the fusion of ^{6}Li, ^{7}Li, and ^{9}Be with heavy nuclei. *Phys. Rev. C*, 70:024606, Aug 2004.

[20] L. R. Gasques, D. J. Hinde, M. Dasgupta, A. Mukherjee, and R. G. Thomas. Suppression of complete fusion due to breakup in the reactions ^{10,11}B $+^{209}$ Bi. *Phys. Rev. C*, 79:034605, Mar 2009.

[21] H. Kumawat, V. Jha, V. V. Parkar, B. J. Roy, S. K. Pandit, R. Palit, P. K. Rath, C. S. Palshetkar, Sushil K. Sharma, Shital Thakur, A. K. Mohanty, A. Chatterjee, and S. Kailas. Fusion reaction studies for the ^{6}li + ^{90}zr system at near-barrier energies. *Phys. Rev. C*, 86:024607, Aug 2012.

[22] Di Pietro, A. Reaction mechanisms in collisions induced by halo and/or weakly bound radioactive nuclei around the barrier. *Eur. Phys. J. Special Topics*, 150:15–20, 2007.

[23] A. Di Pietro, P. Figuera, F. Amorini, C. Angulo, G. Cardella, S. Cherubini, T. Davinson, D. Leanza, J. Lu, H. Mahmud, M. Milin, A. Musumarra, A. Ninane, M. Papa, M. G. Pellegriti, R. Raabe, F. Rizzo, C. Ruiz, A. C. Shotter, N. Soić, S. Tudisco, and L. Weissman. Reactions induced by the halo nucleus ^{6}He at energies around the coulomb barrier. *Phys. Rev. C*, 69:044613, Apr 2004.

[24] V. Scuderi, A. Di Pietro, P. Figuera, M. Fisichella, F. Amorini, C. Angulo, G. Cardella, E. Casarejos, M. Lattuada, M. Milin, A. Musumarra, M. Papa, M. G. Pellegriti, R. Raabe, F. Rizzo, N. Skukan, D. Torresi, and M. Zadro. Fusion and direct reactions for the system ^{6}he + ^{64}zn at and below the coulomb barrier. *Phys. Rev. C*, 84:064604, Dec 2011.

[25] A. Di Pietro, P. Figuera, E. Strano, M. Fisichella, O. Goryunov, M. Lattuada, C. Maiolino, C. Marchetta, M. Milin, A. Musumarra, V. Ostashko, M. G. Pellegriti, V. Privitera, G. Randisi, L. Romano, D. Santonocito, V. Scuderi, D. Torresi, and M. Zadro. Heavy residue excitation functions for the collisions ^{6,7}li+^{64}zn near the coulomb barrier. *Phys. Rev. C*, 87:064614, Jun 2013.

Deep Inelastic Transfer Reactions - A New Way to Exotic Nuclei?

Sophie Heinz[1,a] and Olga Beliuskina[1]

[1] *GSI Helmholtzzentrum für Schwerionenforschung GmbH, 64291 Darmstadt, Germany*

Abstract. We studied deep inelastic multinucleon transfer reactions in collisions of ^{64}Ni+^{207}Pb and ^{48}Ca+^{238}U at energies around the Coulomb barrier. The experiments were performed at the velocity filter SHIP at GSI Darmstadt. One of the goals was to investigate if deep inelastic transfer is superior to fragmentation reactions for producing neutron-rich isotopes in the astrophysically interesting region of nuclei along the magic neutron number N = 126. With both collision systems, rather neutron-rich transfer products were populated, some of them reaching out to the limits of the present chart of nuclides. New isotopes could not be identified. A comparison of the measured transfer cross-sections and yields with those from fragmentation reactions allowed for interesting conclusions.

1 Introduction

In recent years, intense discussions restarted about the possibility to produce new neutron-rich heavy and superheavy nuclei in deep inelastic transfer (DIT) reactions at Coulomb barrier energies. These considerations were triggered by new theoretical model calculations which result in experimentally feasible cross-sections for DIT products in these regions (see e.g. [1, 2]). One region of interest concerns nuclei along the closed neutron shell N = 126 which are of relevance for the astrophysical r-process. These isotopes are presently produced in fragmentation reactions at relativistic energies with heavy projectiles and light target nuclei. However, calculations in macroscopic-microscopic models predict larger cross-sections for these isotopes if they are produced in DIT reactions. Another region of interest is neutron-rich transuranium nuclei which are not reachable in fragmentation reactions or in fusion reactions with stable beams.

Only few experimental data exist on the production of heavy transfer products in the above named regions. In the last decade, collisions of ^{58}Ni+^{208}Pb and ^{64}Ni+^{208}Pb at energies around the Coulomb barrier were studied [3, 4]. In these "thick-target" experiments, the transfer products were stopped in the target. The isotopic identification was performed via in-beam and offline gamma decay spectroscopy. Limit cross-sections of about 20 μb were reached. Concerning target-like transfer products, a vast region of nuclei around Z = 82 was populated. New, previously unknown isotopes were not identified. Also attempts to produce new transactinide nuclei in transfer reactions were made already several decades ago. The most sensitive of those experiments applied chemical separation techniques with subsequent isotope identification via alpha decays. With this method, transfer products up to

[a]e-mail: s.heinz@gsi.de

Z = 101 were observed in reactions of ^{48}Ca or ^{238}U beams with ^{238}U and ^{248}Cm targets [5–7]. Limit cross-sections of 50 nb were reached but also in these experiments no new isotopes were observed.

The expected cross-sections for new neutron-rich transfer products are small. Model predictions result in values of $\sigma < 1\,\mu$b for isotopes along the N = 126 shell and in $\sigma < 1$ nb for new neutron-rich nuclei with Z ≥ 102. As a consequence, similarly efficient separation and detection techniques have to be applied like have been developed for the identification of single atoms in superheavy element experiments. At GSI Darmstadt, we performed studies of DIT reactions in heavy collision systems using the velocity filter SHIP for separation of the target-like transfer products from background events. The isotopic identification was performed via alpha and/or gamma decay spectroscopy in the focal plane of SHIP. In particular when the isotopes were identified via alpha decays, the method turned out to be significantly more sensitive than the techniques applied in previous experiments. In this case we reached limit cross-sections of 0.1 nb [8]. The identification via gamma spectroscopy resulted in cross-section limits of about $10\,\mu$b, similar like in the above named experiments [3, 4]. The sensitivity of our setup allows, in principle, to reach new isotopes in the region of N = 126 as well as in the transuranium region if we assume the cross-sections predicted by theory. Here, we want to concentrate on our studies of DIT reactions leading to nuclei in the region of N = 126. For this, we investigated collisions of ^{48}Ca+^{238}U and ^{64}Ni+^{207}Pb and compared the transfer cross-sections and yields with those from fragmentation reactions.

2 Theoretical Predictions

Extensive calculations on the production of heavy transfer products in the region of N = 126 are available from two different macroscopic-microscopic theories. Both models follow different philosophies concerning the optimum projectile/target combinations for this kind of reactions. The *dinuclear system model* which uses diabatic potentials suggests the application of neutron-rich intermediate heavy projectiles like ^{48}Ca and heavy target nuclei. As a possible target nucleus, ^{238}U is suggested [1] because due to its large neutron number the population of very neutron-rich transfer products is expected. However, with uranium targets a very large flow of nucleons from the target to the projectile nucleus is necessary to populate transfer products around N = 126. The other model, which uses adiabatic potentials and transport (Langevin type) equations of motion, suggests the application of rather heavy neutron-rich projectiles like ^{136}Xe and ^{208}Pb as target material [2]. In this case, the total neutron number of the system is large and one might expect relatively large cross-sections for neutron-rich transfer products. The application of the adiabatic model results in long nuclear interaction times also in very heavy collision systems. This is a necessary condition for a large flow of nucleons between the colliding nuclei. In the dinuclear system model, in contrast, sufficiently long interaction times are only obtained with intermediate heavy projectiles while in very heavy systems the Coulomb repulsion becomes dominant and prevents the sticking of the nuclei. As a consequence, this model predicts only direct transfer reactions in very heavy systems which do not lead to the population of isotopes far from the projectile and target nucleus, respectively.

Calculated cross-sections for the production of nuclei around N = 126 in the suggested collision systems ^{48}Ca+^{238}U [1] and ^{136}Xe+^{208}Pb [2] are shown in fig. 1. In both models, the optimum beam energies for the production of neutron-rich nuclei are just at the Coulomb barrier. Already at such low energies a large flow of nucleons takes place but the excitation energy of the primary transfer products is still moderate (typically a few 10 MeV). The excitation energy has to be kept as low as possible in order to reduce the probability for fission of the excited primary transfer products and to keep the number of evaporated neutrons small (this was also confirmed by our experimental results from the systems ^{58,64}Ni+^{207}Pb [8]). In fig. 1 (a), (b) one can see that in both calculations the cross-sections

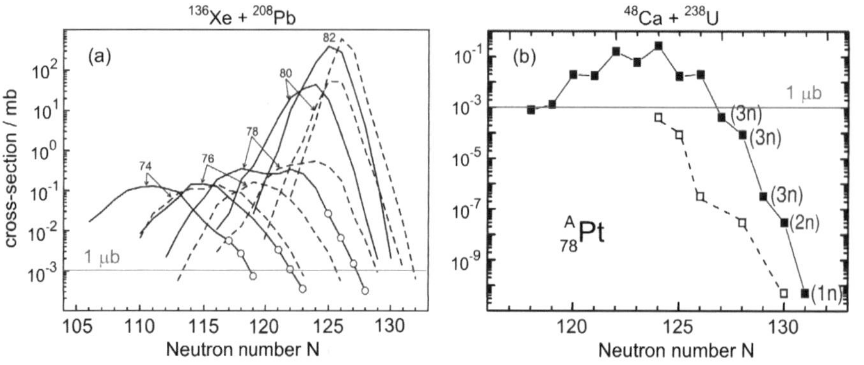

Figure 1. Calculated cross-sections of deep inelastic transfer products in the region of N = 126 from two different theoretical models. The cross-sections are drawn as a function of the neutron number N of the transfer products. (a) Cross-sections of transfer products with Z = 74 − 82 from collisions of ^{136}Xe+^{208}Pb at 5.5 MeV/u, calculated in the macroscopic-microscopic model with adiabatic potentials [2]. Dashed curves represent primary transfer products; full lines represent secondary transfer products. (b) Cross-sections of Pt isotopes (Z = 78) from deep inelastic transfer reactions in ^{48}Ca+^{238}U at 4.7 MeV/u, calculated in the dinuclear system model [1]. Full symbols represent primary transfer products and open symbols represent secondary products. The horizontal lines in both figures mark the cross-section of 1 μb.

of the surviving (i.e. secondary) transfer products with neutron numbers around N = 126 are mostly below microbarns and decrease rather steeply with increasing neutron number. Comparing the two model predictions, the reactions of Xe + Pb result in about (100 − 1000) times larger cross-sections for the same isotopes than reactions of Ca + U and seem therefore more favourable.

3 Experimental Setup and Method

In our experiments we used the velocity filter SHIP for separation of the target-like transfer products from background events. A sketch of the setup is shown in fig. 2 (a detailed description of our method applied for DIT reaction studies can be found in [8]). Reaction products which leave the target at forward angles of (0 ± 2) degree with respect to the beam direction are accepted by the entrance aperture of SHIP. All nuclei which enter SHIP are separated according to their velocities. The ratio of the electric and magnetic field values, E/B, determines at which velocity an ion can pass SHIP. The accepted velocity window at a given setting is $\Delta v/v = 0.1$ (FWHM). For the here described experiments, E and B were chosen such that the target-like transfer products could pass SHIP while the much faster projectiles and projectile-like reaction products were deflected to the beam stop. All reaction products which pass the velocity filter are implanted in a position sensitive 16 strip silicon detector (stop detector) where their time of implantation, position, kinetic energy and radioactive decays (alpha, beta and fission decays) are registered. In addition, a germanium clover detector is mounted behind the stop detector. It consists of four crystals which register gamma rays accompanying the decays of implanted nuclei. With this method, nuclei with half-lives down to about 5 μs can be detected and studied. In order to reduce background in the alpha and gamma spectra, we utilize the structure of the UNILAC beam which consists of 5 ms long beam pulses followed by 15 ms long beam-off periods. With this, background events can strongly be suppressed if the radioactive decays are recorded during beam-off periods.

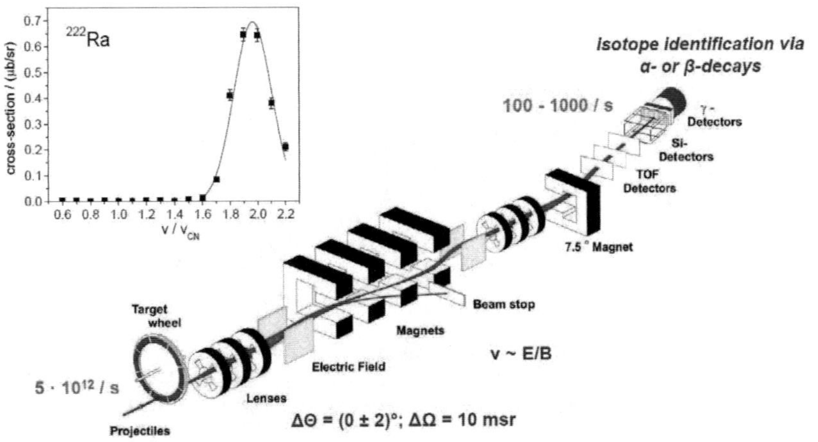

Figure 2. Scheme of the velocity SHIP and its detection system which was used for the study of deep inelastic transfer products. Typical beam intensities at the target and total count rates at the implantation detector during our experiments are also given. Further, an example of a measured velocity spectrum is shown for the case of ^{222}Ra nuclei produced in deep inelastic transfer reactions in ^{48}Ca+^{238}U collisions at 4.90 MeV/u. The velocities in the spectrum are relative to the compound nucleus velocity v_{CN} (i.e. centre-of-mass velocity).

Usually, a broad variety of transfer products is created in the reactions where each of the nuclei has a characteristic velocity, depending on its A and Z. In order to cover this range of different velocities, we varied stepwise the E and B values of the velocity filter. In this way, one obtains the velocity spectrum of each individual isotope. An example of such a spectrum is shown in fig. 2 for the transfer product ^{222}Ra from collisions of ^{48}Ca+^{238}U at 4.90 MeV/u. In this spectrum, the transfer product velocities v are given relative to the compound nucleus velocity v_{CN} (a peak at $v/v_{CN} = 1.0$ would be expected for nuclei resulting from complete fusion reactions). Typically, the spectra of target-like transfer products from reactions with medium heavy projectiles and heavy targets, like discussed here, reveal peak structures like in fig. 2 where the maxima are located at velocities between 1.5 and 2.0 times the compound nucleus velocity v_{CN}. The velocity of the transfer products reflects also the energy dissipation during the reaction. Therefore we used the locations of the peak maxima to calculate the total kinetic energy (TKE) of target-like and projectile-like transfer products in the exit channel. For this we calculated the kinetic energy of the target-like transfer product from the velocity which corresponds to the peak maximum. Then we reconstructed the kinetic energy of the corresponding projectile-like transfer product by considering the reaction as a two-body process and taking into account the energy and momentum conservation laws.

4 Experimental Results and Discussion

4.1 Transfer Products from Collisions of ^{64}Ni+^{207}Pb

We investigated the population of transfer products in reactions of ^{64}Ni+^{207}Pb at six different beam energies between 4.80 and 5.92 MeV/u (the Coulomb barrier is 5.0 MeV/u). Here, we will show the results measured at 5.2 MeV/u because at this energy we observed the largest cross-sections for neutron-rich transfer products. The beam intensity was typically several times 10^{12} particles / s. The targets consisted of 380 μg/cm^2 thick laysers of the chemical compound ^{207}PbS. We measured the

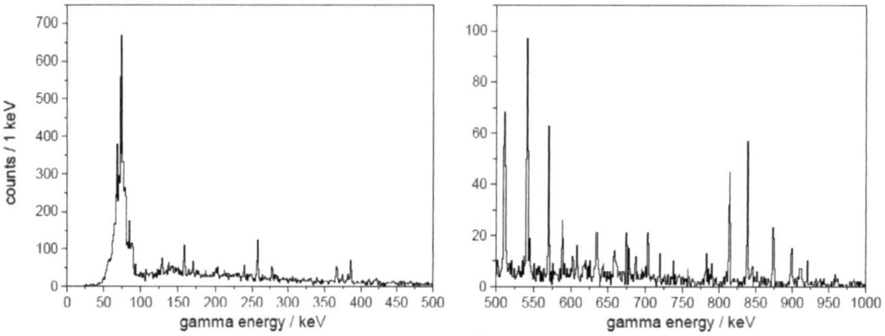

Figure 3. Gamma spectra from collisions of ^{64}Ni+^{207}Pb measured in the focal plane of SHIP during beam-off periods. The energy regions (0 - 500) keV and (500 - 1000) keV are displayed separately. The figures contain the integral data from all investigated beam energies. The isotopes which we identified via these gamma lines are shown in the chart of fig. 4.

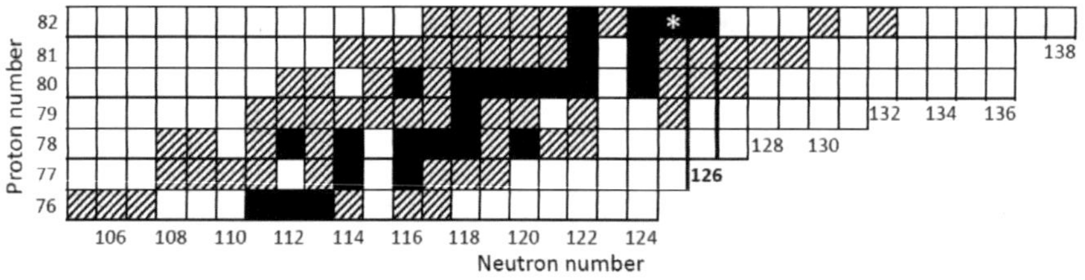

Figure 4. Transfer products identified in collisions of ^{64}Ni+^{207}Pb at 5.2 MeV/u via the gamma lines shown in fig. 3 (hatched squares). The black squares mark stable isotopes. For orientation, the location of the target nucleus ^{207}Pb is marked by the white asterisk. The limits on the neutron-rich side correspond to the limits of the present Karlsruhe chart of nuclides of 2012.

velocity spectra of the transfer proudcts in the range $(1.6 - 2.0)\,v_{CN}$ in steps of $0.1\,v_{CN}$. The total measurement time at 5.2 MeV/u was about four hours. In this experiment, we identified transfer products in the region of Z = 76 – 89 by alpha and gamma spectroscopy. For isotopes which could be identified via alpha decays we reached very low limit cross-sections of about 0.1 nb. This concerns most of the transfer products above Pb (Z > 82). In this region we performed extensive studies of the characteristics of DIT reactions like excitation functions, excitation energies, influence of shell effects and the beam energy dependence of isotopic yields. The respective results are discussed in [8].

In the following, we focus on the population of isotopes below Pb (Z < 82). Radioactive nuclei in this region are β^- - emitters or undergo electron capture decay. If the decay of an implanted mother nucleus populates an excited state in the daughter, the daughter nucleus will decay to the ground state by emission of gamma rays and/or conversion electrons. In our case, we used the emitted gamma rays for isotopic identification of the reaction products. The method is applicable for identification of isotopes which populate with sufficient branching an excited state in the daughter nucleus and which are not too long-lived. In the experiment discussed here, the upper limit for accessible halflives was less than one day, given by the measurement time and the production yield of the respective nucleus. The gamma spectrum recorded during beam-off periods in the focal plane of SHIP is shown in fig. 3 for the energy ranges (0 - 500) keV and (500 - 1000) keV. The identified transfer products with Z ≤ 82 are shown in the chart in fig. 4 (hatched squares). The velocity spectra of these isotopes

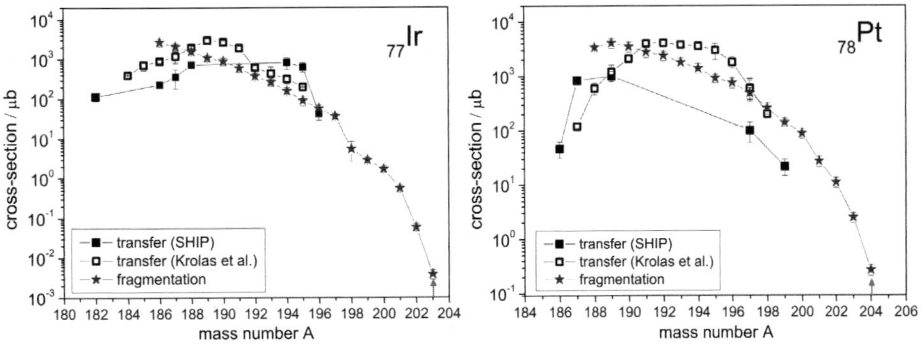

Figure 5. Measured cross-sections of Ir and Pt isotopes produced in deep inelastic transfer reactions in $^{64}Ni+^{207}Pb$ collisions at Coulomb barrier energies. Filled squares mark the cross-sections measured in our experiment and open squares show the data from [3]. For comparison, the cross-sections from fragmentation reactions of 1 GeV/u ^{208}Pb projectiles [11] are also shown (asterisks). The arrows mark N = 126 isotopes.

revealed a large energy dissipation during the reaction. The TKE values, which we deduced from the velocities, are independent of the beam energy and are located below the Viola energy [9] (the Viola energy usually denotes the TKE of fission fragments from an equilibrated compound nucleus). These are typical signatures of deep inelastic reactions where all kinetic energy is dissipated and the exit channel nuclei reseparate with their Coulomb barrier energy.

From the intensities of the measured gamma lines we deduced total production cross-sections for the respective isotopes taking into account the detection efficiency and the angular efficiency of SHIP for DIT products. Concerning the angular efficiency, experimental data are not available. Therefore, we used the efficiencies from simulations in the theoretical model [2] for $^{64}Ni+^{207}Pb$ [10]. Here, one has to stress that the transfer products discussed here result from deep inelastic reactions with large amount of energy dissipation and very broad angular distributions - in strong contrast to direct transfer reactions which peak at the grazing angle. As a consequence, the target-like transfer products are in our case emitted in a cone with an opening angle of about 50 degree in the laboratory system. The opening angle depends on A and Z of the transfer products and becomes smaller if A and Z increase. The fraction of the created nuclei which is accepted by SHIP for the transfer products shown in fig. 4 is in the range (0.1 - 0.5) % according to the simulations of [10].

The deduced total cross-sections for the examples of Ir (Z=77) and Pt (Z=78) isotopes are shown in fig. 5 (black squares). In the same figure, also the cross-sections for Ir and Pt transfer products from the other experiment with the very similar system $^{64}Ni+^{208}Pb$ [3] are shown (open squares).The cross-sections from both experiments agree surprisingly well, within less than a factor of ten, despite the uncertainties concerning angular efficiency in the case of SHIP and concerning beam energy in the thick target experiment [3]. The data from both experiments indicate a rather steep decrease of the cross-sections towards the neutron-rich side after the maximum is reached around nuclei located close to the stability line. The same holds for the data measured for other elements beside Ir and Pt which we are not going to show here. In both experiments no new isotopes were observed or identified, respectively, and the $N = 126$ shell was not reached for nuclei with $Z < 80$. Very interesting is the comparison of the isotopes and related cross-sections reached in DIT reactions with the ones from fragmentation reactions. In experiments on fragmentation reactions with ^{208}Pb projectiles on beryllium targets, rather neutron-rich isotopes were detected [11]. The measured fragmentation cross-sections are represented by the asterisks in fig. 5. One can notice that fragmentation and transfer

Table 1. Typical values for beam intensities N_{beam}, target thicknesses d_{target} and angular efficiencies in deep inelastic transfer reactions and in fragmentation reactions. The efficiency of < 5% for heavy deep inelastic transfer products is specifically given for SHIP.

	N_{beam}	d_{target}	angular efficiency
Deep inelastic transfer	5×10^{12} / s	500 μg/cm^2	< 5 %
Fragmentation	10^8 / s	5 g/cm^2	(50 - 100) %

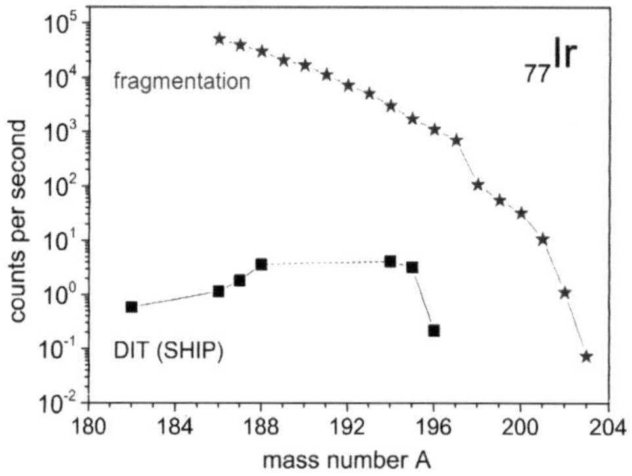

Figure 6. Count rates of Ir isotopes from deep inelastic transfer (DIT) and fragmentation which can be expected at the detection system. The count rates are related to the cross-sections in fig. 5, left. They were calculated with the beam intensities and target thicknesses given in table 1. Angular efficiencies of 50% for fragmentation products and 5% for transfer prodcuts were assumed.

cross-sections are quite similar (i.e. within the same order of magnitude) in the region where data from both, DIT and fragmentation, are available. In some cases the transfer cross-sections are even larger. This confirms theoretical predictions. Also the fragmentation cross-sections reveal a steep drop towards the neutron-rich side but the experiment was more sensitive and allowed to reach limit cross-sections below 10 nb. One has still to remark that the cross-sections which we measured in Ni+Pb reactions in our experiment and the ones given in [3] are on the same order of magnitude than the theoretical cross-sections predicted in [2] for reactions of Xe+Pb (theoretical calculations for the system Ni+Pb are presently not yet available for the nuclei discussed here).

Even more interesting is a comparison of the count rates which can be expected at the detector in DIT and fragmentation reactions because this is the decisive criterion for a successful application of a reaction for isotope production. To deduce the count rates from the cross-sections, one has to take into account the different conditions concerning beam intensity, target thickness and efficiencies for DIT and fragmentation reactions. Typical paramters are shown in table 1. The given angular efficiency range for DIT products is specifically for SHIP and for reactions with intermediate heavy projectiles and heavy target nuclei. Figure 6 shows count rates for Ir isotopes which can be expected at the detector from DIT and fragmentation reactions. The count rates have been calculated from the measured cross-sections given in fig. 5. For this, the beam intensities and target thicknesses given in table 1 were used. The angular efficiencies were assumed as 50% for fragmentation products and 0.1%

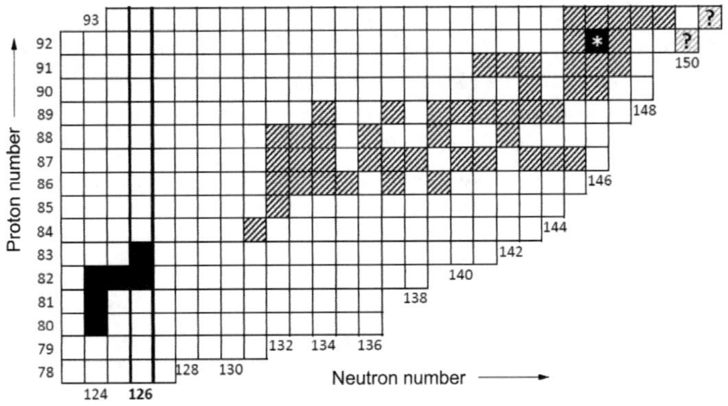

Figure 7. Deep inelastic transfer products identified in collisions of ^{48}Ca+^{238}U at 4.90 MeV/u (hatched squares). The question marks indicate that the correct identification of the two respective isotopes is not certain. The boarder of the chart on the neutron-rich side corresponds to the boarder of the present Karlsruhe chart of nuclides from 2012. The black squares mark stable isotopes. The target nucleus ^{238}U is marked by the asterisk.

for transfer products. The angular efficiency for transfer products was taken from the calculations in [10] for Ir nuclei. Detector efficiencies were assumed as 100% for this estimate. The count rates for the same Ir isotopes are typically at least 1000 times higher in fragmentation reactions. The same result we obtained also for isotopes of the other observed elements. This means, in order to become more profitable than fragmentation reactions, the count rates of transfer products would have to be increased by clearly more than three orders of magnitude. Therefore, also experimental setups with 100% angular efficiency for DIT would not allow to reach count rates larger than in fragmentation reactions.

Finally, another point has still to be mentioned: In fragmentation reactions, the isotopic identification of the reaction products can be performmed by measuring energy, energy loss, time-of-flight and magnetic rigidity of the nuclei. Due to the relativistic energies of the ions, A and Z can be resolved with one unit. This method cannot be applied for the identification of transfer products from reactions at Coulomb barrier energies because in the case of heavy and low-energetic reaction products, plasma effects and the resulting pulse height defect limits the A and Z resolution which is usually not better than three units in the region of Pb and above. Because of this, the isotope identification is presently performed via radioactive decays which limits the method to nuclei with appropriate decay properties while the method applied in fragmentation reactions gives access to all nuclei with halflives larger than several 100 ns. This time limit is given by the time of flight of the nuclei through the fragment separator.

4.2 Transfer Products from Collisions of ^{48}Ca+^{238}U

An option which was suggested by the dinuclear system model is the application of uranium targets and ^{48}Ca beams to populate transfer products along N = 126, below Pb [1]. In the following, we will briefly present our results on DIT reactions in this system at 4.90 MeV/u which we collected during a relatively short beam time of about one day. The isotope identification was performed via alpha decay spectroscopy or, for beta emitters, via gamma spectroscopy in the focal plane of SHIP. The limit cross-sections reached in this experiment were ~0.1 nb for alpha emitters and about 5 μb for beta emitters. We identified transfer products in the region of Z = 84 – 93 which are shown in the chart in fig. 7.

Nuclei with Z< 84 were not observed, most probably because their production cross-sections were already below the limit cross-section of 5 μb for beta emitters. This is consistent with the theoretical cross-sections given in [1] for these nuclei. It is also consistent with the cross-sections which we measured for isotopes with $Z \geq 84$. For these we found that the cross-sections decrease continuously with decreasing Z of the DIT product (i.e. with increasing distance from the target nucleus). For the isotopes with the lowest observed Z (Rn and Po) we measured cross-sections of several 10 μb which is already close to the limit cross-section.

In total, our experimental results show that reactions with Pb targets are more favourable for the production of neutron-rich transfer products in the region of $N = 126$, $Z < 80$. However, the transfer products which we observed in reactions of ^{48}Ca+^{238}U reach out far to the neutron-rich side in - and close to - the actinide region. For $Z = 87 - 93$ we detected nuclei which are only one or two neutrons away from the limits of the present chart of nuclides. The largest cross-sections were measured for nuclei which are located around the valley of beta stability. But also the most neutron-rich observed transfer products have still cross-sections up to several 100 μb. For example, the isotope $^{238}_{91}$Pa is populated with a cross-section of 180 μb and the isotope $^{232}_{87}$Fr with 15 μb. This indicates that also nuclei with still larger neutron numbers, and eventually also new isotopes, must have been populated in this experiment but could not be identified with our detection method due to their decay properties.

4.3 Summary, Conclusions and Outlook

Our studies of ^{64}Ni+^{207}Pb and ^{48}Ca+^{238}U collisions at Coulomb barrier energies showed that a vast region of nuclei is populated in deep inelastic transfer reactions with proton numbers around the target nucleus. In both systems, we identified relatively neutron-rich but not new isotopes. Concerning the population of nuclei in the region of $N = 126$, $Z < 82$, reactions with Pb targets turned out to be more favourable than reactions with U targets because they lead, for the same transfer products, to several orders of magnitude larger cross-sections. Our data and the ones from a previous experiment revealed that in transfer reactions cross-sections on the same order, or even larger, can be reached than in fragmentation reactions. This confirms the predictions of theoretical model calculations. However, a significant difference appears if one regards the expected count rates at the detector from both reactions which are about 1000 times larger in fragmentation reactions, resulting from the larger angular efficiency and effective target thickness. Even if an efficiency of 100% is assumed for a separation and detection setup for transfer products, fragmentation would still lead to comparable or higher count rates, respectively. Moreover, the isotope A and Z identification in fragmentation reactions can be performed by measuring energy, energy loss, time-of-flight and magnetic rigidity. A method which is not restricted to nuclei with appropriate decay properties like the identification of transfer products via alpha or beta decays. Concluding from the present situation, fragmentation reactions seem therefore superior to transfer reactions for producing as well as identifying neutron-rich nuclei along $N = 126$, $Z < 82$.

However, deep inelastic transfer reactions, in particular with the heaviest and most neutron-rich available targets like ^{238}U or ^{248}Cm, seem promising for the synthesis of new neutron-rich nuclei in the region $Z > 92$ - a region which cannot be accessed in fragmentation reactions and where transfer reactions would presently even be the only possibility. Already in our short beamtime of one day with the system ^{48}Ca+^{238}U, we observed nuclei in the region $Z > 84$ with neutron numbers up to the limits of the present chart of nuclides. The data indicate that still more neutron-rich nuclei and also nuclei with $Z > 93$ must have been populated but could not be identified in our experiment due to their decay properties. Therefore, the most urgent point and neccessary requirement for a successful application of transfer reactions for the synthesis of new isotopes is the development of detection techniques which allow the A and Z identification of very heavy and low-energetic nuclei, independent of their decay properties. A possible method is precision mass measurements where a resolving power of 10^5 is already sufficient for an isobaric identification of most of the heavy isotopes. A promising existing setup is the multiple reflection time-of-flight mass spectrometer MR-TOF-MS [12] which was developed for the Super-FRS project. It allows for a broadband detection of different isotopes and the complete setup, consisting of an ion catcher and the MR-TOF-MS, has an overall efficiency of about 20%, comparable to the efficiencies of present detection systems used in heavy element research.

References

[1] G.G. Adamian et al., Phys. Rev. C **81**, 057602 (2010)
[2] V. Zagrebaev and W. Greiner, Phys. Rev. Lett. **101**, 122701 (2008)
[3] W. Krolas et al., Nucl. Phys. A **724**, 289 (2003)
[4] W. Krolas et al., Nucl. Phys. A **832**, 170 (2010)
[5] M. Schädel et al., Phys. Rev. Lett. **48**, 852 (1982)
[6] H. Gäggeler et al., Phys. Rev. C **33**, 1983 (1986)
[7] A. Türler et al., Phys. Rev. C **46**, 1364 (1992)
[8] V. Comas et al., Eur. Phys. J. A **49**, 112 (2013)
[9] M. Wilpert et al., Phys. Rev. C **51**, 680 (1995).
[10] V.I. Zagrebaev, private communication.
[11] Teresa Kurtukian Nieto, *Production and β decay half-lives of heavy neutron-rich nuclei approaching the the stellar nucleosynthesis r-process path around A=196*, Ph.D. Thesis, University of Santiago de Compostela, 2006.
[12] W.R. Plaß et al., Nucl. Intrum. Methods B **266**, 4560 (2008).

Bimodal fission of Hs*

I M Itkis[1], M G Itkis[1], G N Knyazheva[1], E M Kozulin[1], T A Loktev[1], K V Novikov[1], F Hanappe[2] and E Vardaci[3]

[1] FLNR, Joint Institute for Nuclear Research, 141980 Dubna, Russia
[2] Université Libre de Bruxelles, CP229, B-1050 Belgique
[3] Istituto Nazionale di Fisica Nucleare and Dipartimento di Scienze Fisiche dell'Università di Napoli, Napoli, Italia

jitkis@nr.jinr.ru

Abstract. Mass and energy distributions of fission fragments obtained in the reactions ^{22}Ne + ^{249}Cf, ^{26}Mg + ^{248}Cm, and ^{22}Ne + ^{238}U have been measured. A special attention will be paid on the properties of mass-energy distribution of fission fragments obtained in the reaction ^{26}Mg + ^{248}Cm at an excitation energy of 35 MeV. At this energy shell effects should become more effective in fission, the TKE distribution of symmetric fragments obtained in the reaction ^{26}Mg + ^{248}Cm differs strongly from a Gaussian shape. Besides a low-energy component, a high-energy component, not foreseen in the LDM, arises. This is attributed to the fact that both fission fragments are close to the spherical neutron shell $N = 82$. It means that for the compound nucleus ^{274}Hs*, formed in the reaction ^{26}Mg + ^{248}Cm, the phenomenon of bimodal fission was observed for the first time. For the compound nucleus ^{260}No* formed in the reaction ^{22}Ne + ^{238}U at the initial excitation energy of 41 MeV the bimodal fission as well as superasymmetric fission were observed.

1. Intoduction

The question about symmetric and asymmetric modes in low energy nuclear fission arose immediately after the discovery of nuclear fission. An extensive set of experimental data on mass and energy distributions of fragments in the spontaneous and low energy fission of nuclei up to element with Z=104 have confirmed the validity of a hypothesis about the existence of independent fission modes stated by Turkevich and Niday in 1951[1]. The numerous theoretical works showed that multimodality is caused by the valley structure of the deformation potential energy surface of a fissioning nucleus. Following the Brosa model [2], the four fission modes can be distinguished: the Super Long symmetric mode; the Standard I mode caused by the influence of proton $Z = 50$ and neutron $N = 82$ shells; the standard II mode determined by deformed nuclear shells with $Z \approx 54$-56 and

$N \approx$ 86-88 and the Super Short mode, manifesting itself when both light and heavy fission fragments are close to the double magic tin with $A \approx 132$.

Bimodality means the co-existence of two fission modes with different total kinetic energies for the same symmetric mass division. The phenomenon of bimodal fission was discovered in the 1980's for the case of spontaneous and low energy fission of nuclei in the Fm-Rf region [3]. Some spontaneously fissioning isotopes of Fm, Md, and No were found to exhibit symmetric fragment mass distributions, whose widths are changing very rapidly from nucleus to nucleus. For example, for spontaneous fission of ^{258}Fm the mass distribution is unusually narrow, while for the neighboring nucleus ^{259}Md with only one more proton it is rather broad. Even more spectacular are the distributions of TKE, which are double humped. The high-energy mode is linked to a narrow mass distribution while the low-energy mode has a wide mass distribution.

This behavior of TKE distributions strongly differs from the Gaussian shape distributions found in fission of all other actinides. It is important to note that bimodal fission appears for Fm isotopes (Z=100) and more heavy elements when both fission fragments are close to spherical proton (Z=50) and/or neutron (N=82) shells.

By theory, bimodal fission is explained by the existence of two different paths on the potential energy surface leading to fission. One path is described by the liquid drop model (LDM), while for the other path shell effects play an important role. The LDM path leads to elongated scission configurations while the shell effect path corresponds to compact configurations. Both paths result in symmetric (or nearly so) mass distributions, but the high TKE valley is narrower while the low TKE valley is much wider. This explains the difference in width of the mass distributions.

The competition between various fission modes and diminishing of shell effects with increasing projectile energy is still an open question. To investigate the properties of bimodal and superasymmetric fission the mass and energy distributions of fission fragments of Hs and No isotopes were measured.

2. Experiment

The experiments were carried out at the Flerov Laboratory of Nuclear Reactions using beams of ^{22}Ne, ^{26}Mg ions extracted from the U-400 cyclotron at energies around the Coulomb barrier. The energy resolution was ~2%. Beam intensities on targets were 1-2 pnA. Layers of $^{238}UF_4$, ^{248}Cm, and ^{249}Cf, 120-200 μg/cm^2 thick, deposited on a 40-50 μg/cm^2 carbon backings, were used as targets.

Binary reaction products were detected in coincidence by the two-arm time-of-flight spectrometer CORSET [4]. Each arm of the spectrometer consists of a compact start detector and a position-sensitive stop detector, both based on microchannel plates. The distance between the start and stop detectors was 12-15 cm. A typical mass resolution of the spectrometer in these conditions is ~2-3 u.

The data processing assumes standard two-body kinematics. Primary masses, velocities, energies, and angles in the center-of-mass system of reaction products were calculated from measured velocities and angles in the laboratory system using the momentum and mass conservation laws with the assumption that the mass of the composite system is equal to $M_{\text{target}} + M_{\text{projectile}}$. Neutron evaporation before scission is not taken into account. This is justified by the fact that even at the highest reaction energies not more than 4 neutrons could be emitted. Hence, considering that the spectrometer resolution is 2-3 u, the neutron emission will not lead to visible effects on the mass-energy distributions. Fragment energy losses in the target, backing, and the start detector foils were taken into account.

The identification of the binary reaction channel with full momentum transfer and the removal of products of sequential and incomplete fission reactions, induced fission of target and targetlike nuclei, or reactions on impurity atoms in the target was based on the analysis of the kinematic diagram (the velocity vectors of two detected reaction products) in the center-of-mass system [5].

3. Analysis and results

4. Bimodal fission of Hs and No

The mass and energy distributions for the induced fission of $^{271,274}Hs^*$ formed in the reactions $^{22}Ne + ^{249}Cf$ and $^{26}Mg + ^{248}Cm$ at excitation energies below and above the Coulomb barrier are shown in figure 1.

For the both reactions at energies well above the Coulomb barrier the mass distributions exhibit a near Gaussian shape as predicted by LDM for the fission of relatively hot nuclei. In the framework of the LDM, the average kinetic energy has a parabolic dependence on the fragment mass and practically does not depend on the excitation energy and angular momentum of the compound nucleus. As one can see in figure 1, such behaviour of TKE is confirmed by experimental data.

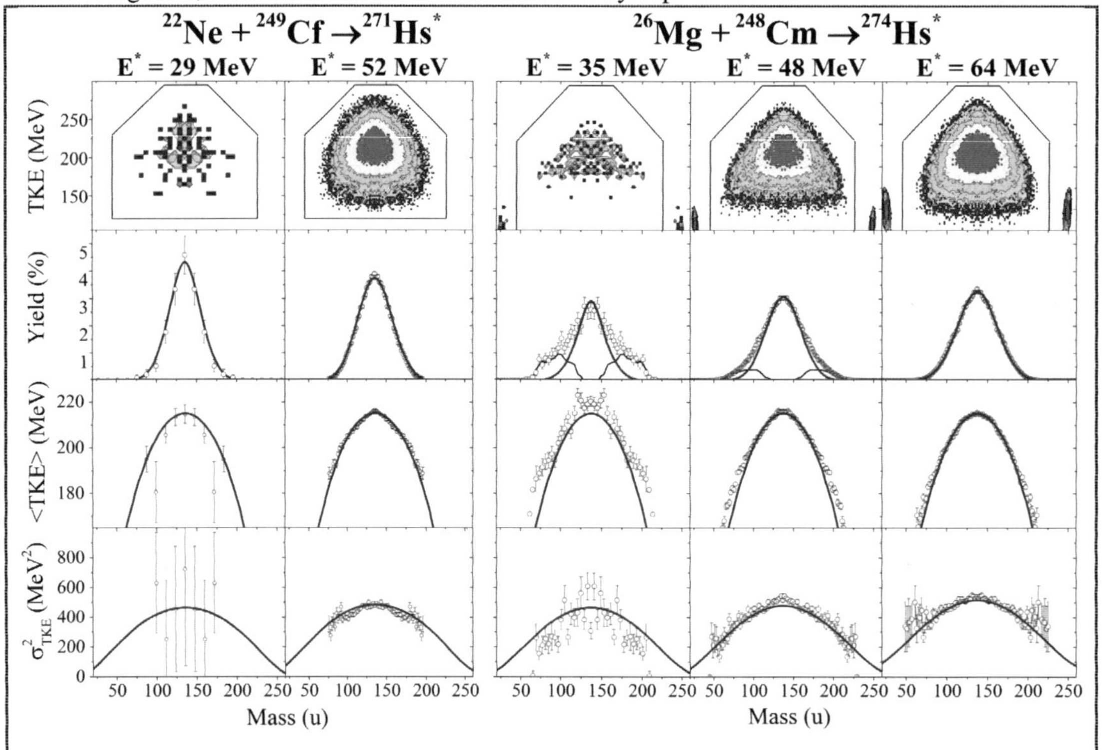

Figure 1. The mass-energy distributions of binary products for the $^{22}Ne + ^{249}Cf$ and $^{26}Mg + ^{248}Cm$ reactions. From top to bottom: the (TKE, *M*) matrices for binary products; the mass yields, the average total kinetic energy and its dispersion as a function of mass for fission-like fragments inside the outlined contour on the (TKE, *M*) matrices.

For the $^{26}Mg + ^{248}Cm$ reaction at an excitation energy of 35 MeV of the CN, some deviations from LDM predictions are observed for fragments, both symmetric and asymmetric. The increased yields of fragments with masses 70-100 u may be caused by the quasifission process since the interaction energy is ~10 MeV less than the Bass barrier, and the orientation effects on the reaction dynamics become apparent at this energy.

The TKE distributions for symmetric fragments with masses $A_{CN}/2\pm20$ u for both reactions are presented in figure 2. At the highest excitation energy the TKE distributions are well described by single Gaussian with the parameters coming from the LDM, whereas at subbarrier energies the TKE distributions of symmetric fragments have a complex structure. The mean TKEs and the dispersions of TKE are also higher than predicted by LDM. At the same time the mass distributions for both nuclei are symmetric. The structural features of the TKE distributions at low excitation may arise due to the fact that in symmetric fission both fragments are close to the spherical neutron shells with $N = 82$. A similar behaviour could be observed when both fission fragments are close to the spherical proton shell $Z = 50$. Thus the properties of mass and TKE distributions show the characteristic features for bimodal fission.

The TKE distributions of symmetric fragments of the ^{26}Mg + ^{248}Cm reaction at $E^* = 35$ MeV may be deconvoluted into two Gaussians, the constituent peaks lying near ~214 MeV and ~230 MeV. The lower energy corresponds to the linear dependency of the TKE on the Coulomb parameter established in the LDM, while the higher energy component is close to those found by Hulet for Fm, Md and No spontaneous fission [3]. Hence, for the reaction Mg + Cm leading to ^{274}Hs*, bimodal fission is observed. It should be noted that when the excitation energy of the compound nucleus is increased the high energy component decreases.

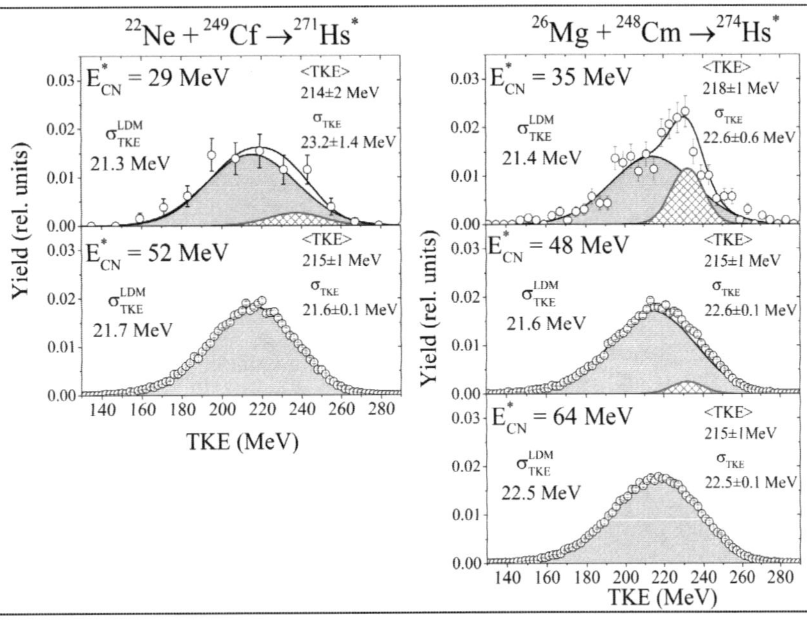

Figure 2. TKE distributions of fragments with masses $A_{CN}/2 \pm 20$ u for the reactions ^{22}Ne + ^{249}Cf (left panel) and ^{26}Mg + ^{248}Cm (right panel). High and low kinetic energy components are given as hatched and filled regions, respectively.

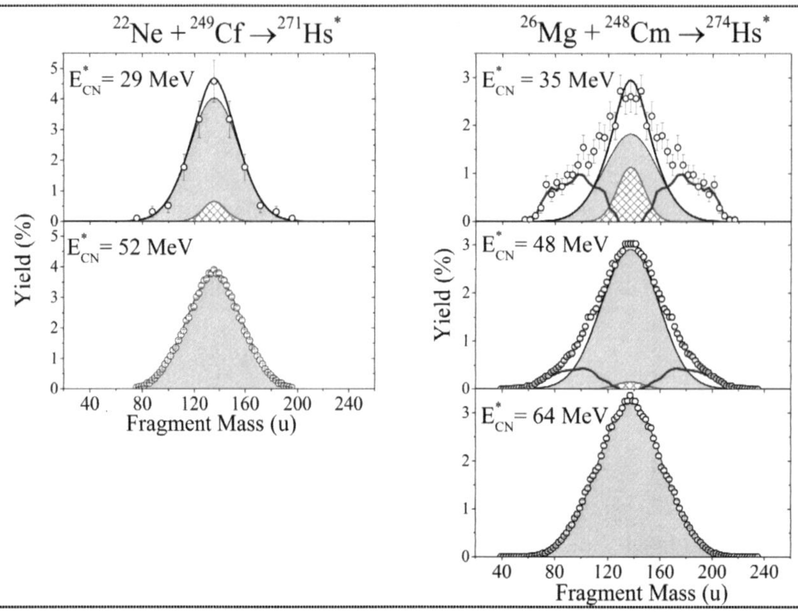

Figure 3. Mass distributions (symbols) of fission fragments. The filled region is associated with low energy component and the hatched region - with high energy component.

The yields of fission fragments that correspond to compact and elongated fission modes in the mass distribution being described by a Gaussian were weighted using the estimations that we have got from the TKE decomposition (for details see Ref. [6]). The results of this fitting procedure are shown in figure 3.

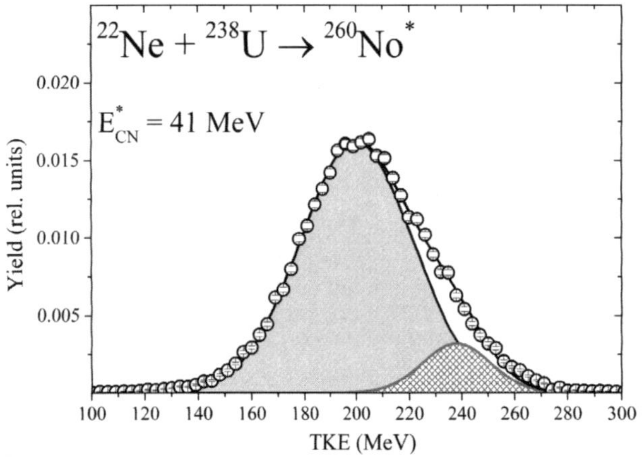

Figure 4. The TKE distribution of the fission fragments obtained in the reaction $^{22}Ne + ^{238}U$ and its' decomposition into high and low energy components.

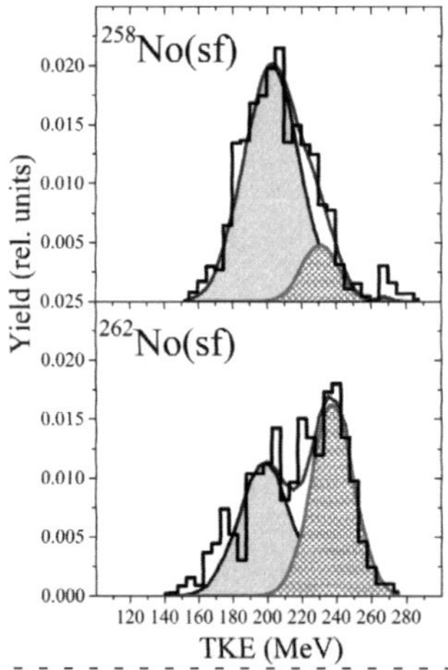

Figure 5. The TKE distributions of spontaneous fission fragments of ^{258}No and ^{262}No taken from [7].

The TKE distribution of symmetric fragments obtained in the $^{22}Ne + ^{238}U$ reaction at an excitation energy of $^{260}No^{*}$ of 41 MeV is shown in figure 4. For this reaction the contribution of high energy component is about 7%. In the case of spontaneous fission of ^{258}No and ^{262}No [7] the contributions of high energy component are about 14% and 54%, respectively (see figure 5).

5. *Superasymmetric fission of No*

For the first time the superasymmetric mode was observed in the compound nuclei fission in Pb region [8]. The enhancement of the mass yield in the region 65-75 u for the light fragment in the fission of ^{213}At and ^{210}Po compound nuclei is connected with the influence of double magic Ni ($Z = 28$, $N = 50$) and double magic Sn ($Z = 50$, $N = 82$). Notice that in this case the ratio $A_H/A_L \approx 2$, and the yield is around 0.01%. The superasymmetric fission mode with mass ratio of $A_H/A_L \approx 2.5$, caused by the closed shells $Z = 28$ and $N = 50$, was also found in thermal-neutron-induced fission of actinides nuclei [9]. In this case only the light fragment is close to the double magic Ni and the yield of superasymmetric mode does not exceed 10^{-4}%.

The question about the possibility of superasymmetric fission with mass ratio $A_H/A_L = 208/48 = 4.3$ when both fission fragments are close to the double magic ^{48}Ca and double magic ^{208}Pb arises. Such specific fission channel can be expected in the fission of ^{256}No ($^{48}Ca + ^{208}Pb$).

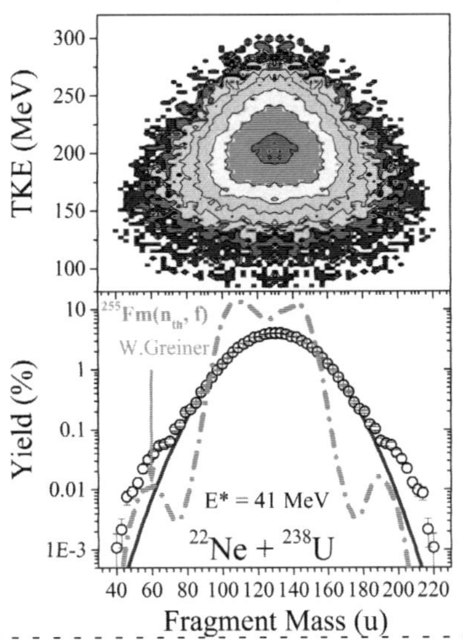

Figure 6. Double differential cross section and mass yield for the $^{22}Ne + ^{238}U$ reaction. Thick line represents the LDM calculation, dash-dot line is W. Greiner calculations for $^{255}Fm(n_{th},f)$

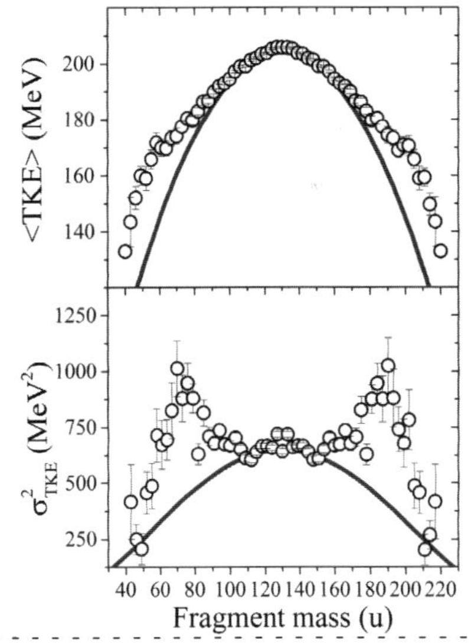

Figure 7. The mean TKE and its' dispersion as a function of fragment mass. The thick lines represent the LDM calculations.

To clarify this question the projectile-target combination $^{22}Ne + ^{238}U$ was chosen. The obtained mass and energy distributions of fragments are shown in figures 6 and 7. The mass distribution of the symmetric fragments has a nearly Gaussian shape and the average TKE shows a parabolic dependence on fragment mass typical for fission of excited compound nuclei as established by the LDM, whereas in the mass region around 52/208 u, that corresponds to the formation of fissioning pair of two magic nuclei Ca/Pb, an increase of fragment yields was observed. Moreover the total kinetic energy for these fragments is found to be about ~30 MeV higher than predicted by the LDM. The higher TKE indicates that the asymmetric fragments originate from more compact scission configuration as compared to normal fission, since the most part of final TKE is due to the Coulomb repulsion between fragments, and the shape elongation of the scission configuration determines the TKE.

In the low panel of figure 6 the experimental mass distribution obtained in the present investigation is compared with the predictions of Walter Greiner for thermal neutrons induced fission of ^{255}Fm. According to this calculation the yield of about 10^{-2}% is expected due to the influence of the closed shells, while for spontaneous fission of ^{255}Fm the yield of 10^{-4}% is expected for this superasymmetric mode [10]. In present work the superasymmetric mode with ratio $A_H/A_L \approx 4.3$ caused by the influence of double magic Ca (Z = 20, N = 28) and double magic Pb (Z = 82, N = 126) has been observed in fission of excited ^{260}No* compound nucleus. At excitation energy of 41 MeV the yield of these fragments is about 5×10^{-2}%.

6. Conclusion

The reactions ^{22}Ne + ^{249}Cf, ^{26}Mg + ^{248}Cm, and ^{22}Ne + ^{238}U have been studied at energies below and above the Coulomb barrier.

For the compound nucleus ^{274}Hs* formed in the reaction ^{26}Mg + ^{248}Cm at the initial excitation energy E^* = 35 MeV and ^{271}Hs* formed in the reaction ^{22}Ne + ^{249}Cf at the initial excitation energy E^* = 29 MeV the phenomenon of bimodal fission was observed.

For the compound nucleus ^{260}No* formed in the reaction ^{22}Ne + ^{238}U at the initial excitation energy E^* = 41 MeV the bimodal fission as well as superasymmetric fission with mass ratio $A_H/A_L \approx 4.3$ were observed for the first time.

7. Acknowledgments

The work was supported by the Russian Foundation for Basic Research (Grant No. 13-02-01282-a).

References

[1] Turkevich A and Niday J B 1951 *Phys. Rev.* **84** 52
[2] Brosa U, Grossmann S and Muller A 1990 *Phys. Rep.* **197** 167
[3] Hulet E K *et al.* 1986 *Phys. Rev. Lett.* **56** 313 770
[4] Kozulin E M *et al.* 2008 *Instrum. Exp. Tech.* **51** (1) 44
[5] Hinde D J *et al.* 1996 *Phys. Rev.* C **53** 1290
[6] Itkis I M *et al.* 2011 *Phys. Rev.* C **83** 064613
[7] Hulet E K *et al.* 1994 *Phys. Atom. Nucl.* **57** 1165
[8] Itkis M G, Okolovich V N, Russanov A Ya and Smirenkin G N 1985 *Z. Phys.* A **320** 433
[9] Rochmann D *et al.* 2004 *Nucl. Phys* A **735** 3
[10] Greiner W 2001 *Proc. Int. Workshop on Fusion Dynamics at the Extremes (Dubna, Russia, 25-27 May 2000)* ed Yu Ts Oganessian and V I Zagrebaev (Singapore: World Scientific) p 1

Journal of Physics: Conference Series **515** (2014) 012009 doi:10.1088/1742-6596/515/1/012009

The time scale of quasifission process in reactions with heavy ions

G.N. Knyazheva, I.M. Itkis, E.M.Kozulin
FLNR, Joint Institute for Nuclear Research, 141980 Dubna, Russia

E-mail: knyazheva@nr.jinr.ru

Abstract. The study of mass-energy distributions of binary fragments obtained in the reactions of ^{36}S, ^{48}Ca, ^{58}Fe and ^{64}Ni ions with the ^{232}Th, ^{238}U, ^{244}Pu and ^{248}Cm at energies below and above the Coulomb barrier is presented. .These data have been measured by two time-of-flight CORSET spectrometer. The mass resolution of the spectrometer for these measurements was about 3u. It allows to investigate the features of mass distributions with good accuracy. The properties of mass and TKE of QF fragments in dependence on interaction energy have been investigated and compared with characteristics of the fusion-fission process. To describe the quasifission mass distribution the simple method has been proposed. This method is based on the driving potential of the system and time dependent mass drift. This procedure allows to estimate QF time scale from the measured mass distributions. It has been found that the QF time exponentially decreases when the reaction Coulomb factor Z_1Z_2 increases.

1. Introduction

The quasifission (QF) is nonequilibrium process taking place in reactions with heavy ions. It is known that in superheavy composite systems QF is a dominant process and leads to the formation of the asymmetric fragments with mass asymmetry ā0.4 [1]. This type of QF process, so-called asymmetric quasifission (QF_{asym}), is characterized by asymmetric angular distributions in the centre-of-mass system and thus fast reaction times (10^{-21}s=1zs) [2, 3]. Besides the asymmetric also the symmetric component may be affected by the presence of the QF process (QF_{sym}). The angular distribution is symmetric for this type of QF that is similar to CN-fission. But in the case of QF, the anomalously large fragment anisotropies relative to those expected form CN-fission have been observed [4].

The QF is characterized by significant nucleon exchange and energy dissipation [2, 3] and the QF fragments exhibit similar characteristics to fragments of CN-fission process; therefore, distinguishing between CN-fission and QF fragments is complicated, especially for symmetric mass split.

Despite the fact that QF was discovered about 30 year ago, and a lot of studies on mass and angular distributions for different reactions have been done, at present there is no data on systematic behaviour of the properties of QF fragments, such as mass distribution, its dispersion, TKE, in dependence on "ion-target" combination and incident energy, since in previous investigations the main attention was paid to fusion probabilities and fusion-fission properties and QF has been considered as interfered process mainly.

2. The properties of QF fragments

Generally, in heavy-ion induced reactions the formation of QF_{asym} fragments is connected with the strong influence of the nuclear shell at Z = 82 and N = 126 (doubly magic lead). Indeed, the mass distribution of fission-like fragments formed in heavy ion induced reaction has a wide two humped shape. Fig. 1 shows the mass-energy distributions of binary fragments obtained in the reactions of ^{16}S, ^{48}Ca, ^{64}Ni ions with uranium target at energies close to the Coulomb barrier. These data have been measured by two time-of-flight CORSET spectrometer [5, 6]. The mass resolution of the spectrometer for these measurements was about 3u (FWHM) that allows to investigate the features of mass distributions with good accuracy. The Coulomb factors are strongly different for these systems: from 1472, 1840 to 2576 for the ^{16}S+^{238}U, ^{48}Ca+^{238}U and ^{64}Ni+^{238}U, respectively. Mass-energy distributions of all studied systems have typical wide two-humped shape caused by QF under the influence of closed shells with Z=82 and N=50, 126. In fact, as it was shown in Ref. [5], for the ^{48}Ca+^{238}U reaction the maximum yield corresponds to fragments with masses 208 u. However, in reactions with lighter projectiles on an uranium target, the asymmetric QF peak shifts to more symmetric masses [6]. By contrast, for the heavier projectile ^{64}Ni the maximum yield of QF_{asym} fragments corresponds to the heavy mass 215u [5]. In the bottom panel of Fig. 1 the driving potentials at scission point calculated with NRV code [7] using the proximity model together with experimental mass distributions are shown. It is clearly seen that the minimum of the driving potential corresponds to the maximum of the yield of asymmetric QF fragments. So, while the relative contribution of QF to the capture cross section mainly depends on the reaction entrance channel properties, the features of asymmetric QF are essentially determined by the driving potential of a composite system.

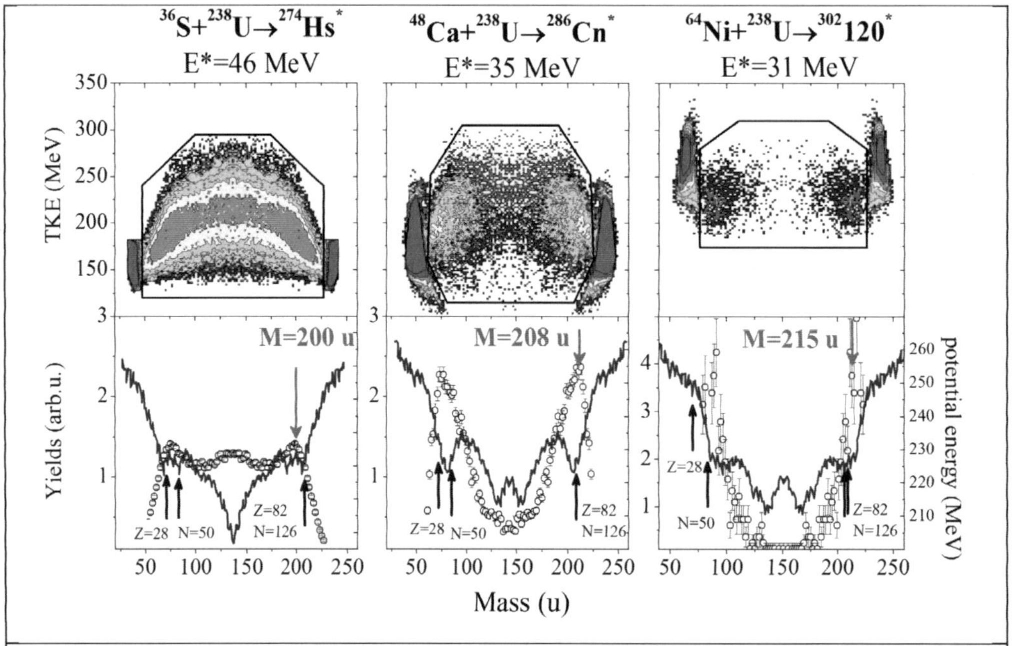

Figure 1. Top: Mass-energy distributions for the reactions ^{36}S, ^{48}Ca, ^{64}Ni +^{238}U at energies close to the Coulomb barrier; bottom: open circles are mass distributions for fission-like fragments inside the contour line on M-TKE matrices and solid lines are the driving potentials as a function of mass for studied systems.

Furthermore, the formation of QF_{asym} fragments depends on not only the driving potential of the composite system but also on the reaction entrance channel. As an example, the mass distributions of fission-like fragments obtained in the reactions $^{36}S+^{238}U$ and $^{26}Mg+^{248}Cm$ are shown in fig. 2. Even in the case of the $^{36}S+^{238}U$ and $^{26}Mg+^{248}Cm$ leading to the formation of the same composite nucleus $^{274}Hs^*$ the position of the asymmetric QF peaks changes [6] from 200 u to 185-190u, respectively. This trend is contrary to the multimodal fission [8] where the position of each mode determined by the nuclear shells is constant for specific compound nucleus and only the relative contribution of each mode varies in dependence of excitation energy. Thus, for more asymmetric reactions the Coulomb repulsion is expected to be smaller. For lighter projectiles this may lead to longer reaction times before separation for asymmetric QF and thus to larger numbers of exchange nucleons.

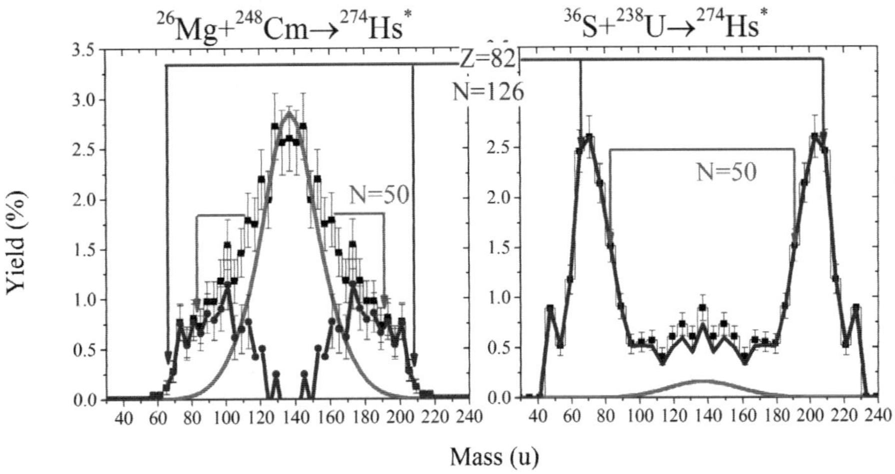

Figure 2. Mass distributions for fission-like fragments formed in the reactions $^{26}Mg+^{248}Cm$ and $^{36}S+^{238}U$ at energy below the Coulomb barrier. Red curves are fusion-fission components estimated from TKE analysis (see Ref.[6]), blue curves are QF components.

Another important characteristic of QF mass distribution is the width. Besides the position of peaks in the mass distributions of QF_{asym} fragments, also the widths of these peaks vary for different "ion-target" combinations. Fig. 3 presents the drift to the mass symmetry of QF_{asym} (estimated as a difference between the mass with maximum yield and more symmetric mass at half maximum yield) as a function of the energy above the Bass barrier for the reactions $^{48}Ca+^{232}Th, ^{238}U$, ^{244}Pu and ^{248}Cm. At energies below the Bass barrier the maximum drift is observed for the reaction with ^{248}Cm targets, the minimum – for ^{238}U and ^{244}Pu lies between these reactions. Notice that the deepest narrow minimum in driving potential at mass close to 208 u corresponds to the reaction $^{48}Ca+^{238}U$ that is the case of the narrowest QF_{asym} mass distribution, for the reaction with ^{244}Pu this minimum is wider and for ^{248}Cm is the widest. Consequently, for studied systems the drift to the mass symmetry for QF_{asym} is affected also by the features of the driving potential.

The drift forward mass symmetry increases with increasing collision energy up to the barrier energy. At energy above the barrier the width of QF_{asym} does not change virtually, while the estimation of the width of QF_{asym} impossible for higher energy due to the growing yield of symmetric fragments which contribution became dominant at these energies.

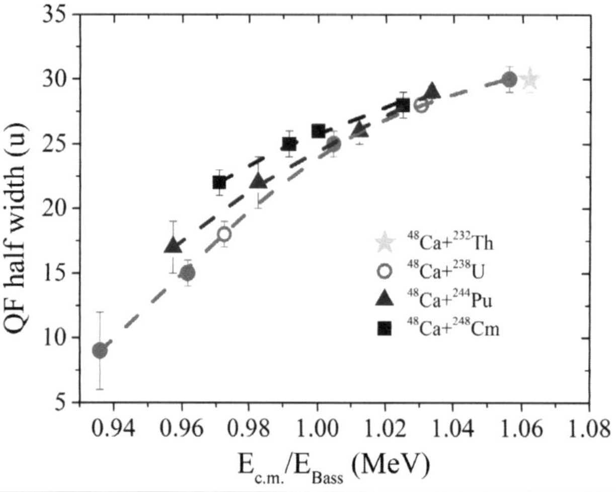

Figure 3. The width of QF_{asym} mass distributions for the reactions of ^{48}Ca with actinide targets as a function of interaction energy..

3. The method for estimation of QF time scale

In the reactions with ^{238}U-ions [2] the mass energy and angular distributions of QF fragments have been measured and the time scale for mass transfer in QF reactions has been derived from turning angles of the composite system. It has been found that the mass drift towards symmetry shows the characteristics of an overdamped motion with a universal time constant independent on scattering system and bombarding energy. It indicated that the drift forward mass symmetry occurs as overdamped motion given by:

$$\frac{\Delta A}{\Delta A_{\max}} = \frac{A_t - \langle A \rangle}{\frac{1}{2}\left(A_t - A_p\right)} = 1 - \exp\left[-(t - t_0)/\tau\right] \tag{1}$$

where τ =(5.3±1)zs is the time constant common to all systems, t_0ã 1zs is a time delay before mass drift sets in, At and Ap are the mass of target and projectile, respectively. The uncertainties in Eq. 1 are connected mainly with the calculation of moments of inertia for dinuclear systems. It follows from the equation that the key ingredient for the mass transfer is the distribution of the interaction time of the system.

Recently, the QF time scale has been derived from the measurements of mass-angular distributions for the reactions ^{186}W+^{34}S, ^{48}Ti and ^{64}Ni [8]. In this work the QF time distributions were parameterized using a half Gaussian followed by an exponential decay. The average (τ_0) and width of the Gaussian, and the decay time (τ_1), were individually adjusted to reproduce the measured mass-angular distributions.

Using the assumption for time distribution for QF process proposed in Ref. [9] together with the Eq. 1 it is possible to calculate the mass distribution for QF fragments. But, in the previous section it was shown that QF fragments are affected strongly by the shell effects. So, taken into account the shell effects in QF process, in the first approximation, the yield of QF fragment mass can be written as:

$$Y_{QF}(A) = p(U(A)) \times p(A(t)) \tag{2}$$

where p(U(A)) is the probability connected with the driving potential of the system (minimum of the driving potential corresponds to p(U(A))=1 and value of driving potential for the entrance channel corresponds to p(U(A))=0). The driving potentials were calculated as a sum of the LDM potential at distance of 15 fm between the mass centers of dinuclear system calculated with NRV code [7] and shell correction in ground state taken from [10]. It should be noted that the proximity model gives a very similar potential. The term p(A(t)) in Eq. 2 is the probability to exchange number of nucleons of $\Delta A=A_t-A$ at given fragment mass A and time *t* calculated with Eq. (1) in the assumption of the time distribution for QF from Ref. [6]. We have adjusted the parameters τ_0 and τ_1 (average of the Gaussian and decay time) of the QF time distribution to obtain the best agreement between experimental and calculated with Eq. 2 mass distributions. This procedure gives the possibility to extract the QF time scale from measured mass distributions with the accuracy of about 20% (mainly determined by the mass resolution of the experimental spectra).

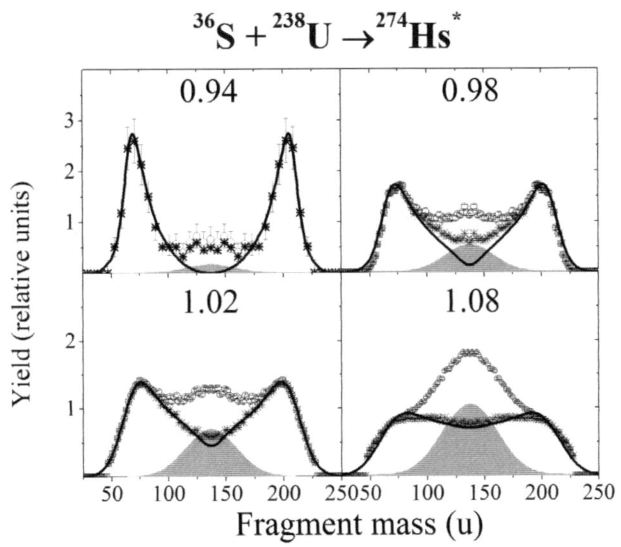

Figure 4. Mass distributions of fission-like fragments for the reaction $^{36}S+^{238}U$ at different energies ($E_{c.m.}/E_B$ is indicated for each plot); the filled regions and stars are the contribution of CN-fission and QF obtained from the analysis of TKE (see Ref.[6] for details), the lines are the fits of the experimental data with using Eq. 2.

When nuclei come to contact in the reactions with heavy ions they become excited, and fluctuations play an important role in the further evolution of the compound system. Due to these fluctuations fragments formed in the QF_{asym} process could have masses close to symmetry with some probability. To take into account these possible fluctuations in the present analysis the yield of obtained with Eq. 2 mass distribution was spread by Gaussian distribution for each fragment mass. The value of sigma of the Gaussian distribution was chosen the same for all masses and was obtained from the fitting procedure of the experimental mass distribution by the Eq. 2. It results in 13 u at energies above the barrier and monotonically decreases up to 2 u for energy $E_{c.m.}/E_B=0.94$.
Figure 4 show the calculated mass distribution for QF process using the proposed method in comparison with the experimental mass distribution of fission-like fragments for the reactions of ^{238}U with ^{36}S ions at energies below and above the Bass barrier. In the fitting procedure with Eq. 2 the values of τ_0 and τ_1 is determined by the outside and inside shape of asymmetric QF mass distribution respectively. The width of the experimental QF_{asym} distributions increases with increasing collision energies for the all studied reactions. For the case of the ^{36}S reaction decomposition into compound nucleus fission and QF obtained from the analysis of the total kinetic energy distributions of fission-like fragments (see Ref. [6] for details of this procedure) is also presented in Fig. 2. It is clearly seen that proposed method to describe the QF mass distribution gives good agreement with experimental data at energies above the barrier. At energies below the barrier the contribution of symmetric fragments formed in QF process estimated using this method is rather smaller than that obtained from the analysis of the TKE distribution. If one assumes that fluctuation of mass in QF process (sigma) is not constant for all fragment masses and dependent on the second derivative of the potential energy surface for given fragment mass, it is possible to reproduce the symmetric part at energies below the

barrier. But in this case the number of variable parameters increases and more experimental data are needed for reasonable choice of the parameters. As it was mentioned above, we supposed that the fluctuation (sigma) is constant for all fragment masses and depends only on the collision energy.

Nevertheless, the method describes well the asymmetric QF mass distribution and may be applied for estimation of the time scale for the asymmetric QF process. The obtained time scale of QF_{asym} process for different reactions a function of the Coulomb factor of the reactions Z_1Z_2 at energies close to the Coulomb barrier is presented in Fig. 5. It is clearly seen that the interaction time decreases exponentially with increasing the reaction Coulomb factor. At barrier energy the time scale for QF in the reaction with ^{36}S is about 6zs, ^{48}Ca – about 3zs and ^{64}Ni – 0.5zs.

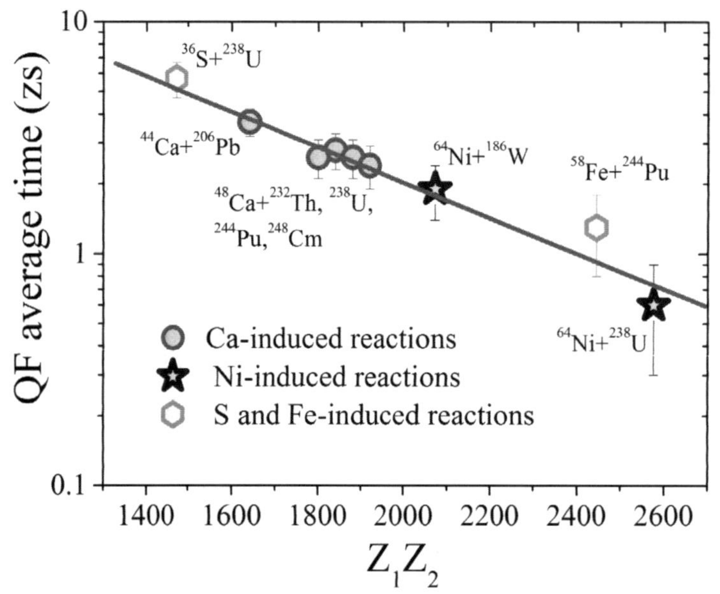

Figure 5. The estimated QF time as a function of the reaction Coulomb factor Z_1Z_2 at the Coulomb barrier energy.

4. Conclusion

In summary, from the analysis of mass-energy distributions of fission-like fragments formed in the of ^{36}S, ^{48}Ca, ^{58}Fe and ^{64}Ni ions with the ^{232}Th, ^{238}U, ^{244}Pu and ^{248}Cm at energies below and above the Coulomb barrier it was found:

1. While the relative contribution of QF to the capture cross section mainly depends on the reaction entrance channel properties, the features of asymmetric QF are determined essentially by the driving potential of a composite system.
2. The drift to the symmetry in the QF mass distributions increases with increasing of the projectile energy. This tendency may be explained by the increasing of interaction time for QF.
3. To describe the QF mass distribution the simple method has been proposed. This method is based on the driving potential of the system and time dependent mass drift. This procedure allows to estimate QF time scale from the measured mass distributions.

4. The interaction time for asymmetric QF process decreases with increasing the Coulomb factor Z_1Z_2.

Acknowledgements
This work was supported by the Russian Foundation for Basic Research (Grant No. 13-02-01282-a) and the JINR Grant No. 13-501-03.

References
[1] Itkis M G et al. 2007 *Nuclear Physics A* vol 787 p 105c
[2] Shen W Q et al 1987 *Physical Review C* vol 36 p 115
[3] Bock R et al 1982 *Nuclear Physics A* vol 388 p 334
[4] Back B B 1985 *Physical Review C* vol 31 p 2104
[5] Kozulin E M et al 2010 *Physics Letter B* vol 686 p 227
[6] Itkis I M 2011 *Physical Review C* vol 83 p 064613
[7] Zagrebaev V.I. et al *nrv.jinr.ru*
[8] Brosa U, Grossman S and Muller A 1990 *Physical Reports* vol 197 p 167
[9] Du Rietz R et al. 2011 *Physical Review Letters* vol 106 p 052701
[10] Moller P, Nix J P, Myers W D and Swiatecki W J 1995 *Atomic Date and Nuclear Data Tables* vol 59 p 185

Shell effects in fission, quasifission and multinucleon transfer reaction

E.M.Kozulin, G.N.Knyazheva, I.M.Itkis, N.I.Kozulina, T.A.Loktev, K.V.Novikov, I.Harca

FLNR, Joint Institute for Nuclear Research, 141980 Dubna, Russia

kozulin@jinr.ru

Abstract. Results of the study of mass-energy distributions of binary fragments for a wide range of nuclei with Z= 82-122 produced in reactions of ions located between ^{22}Ne and ^{136}Xe at energies close and below the Coulomb barrier are reported. The role of the shell effects, the influence of the entrance channel asymmetry and the deformations of colliding nuclei on the mechanism of the fusion-fission, quasifission and multinucleon transfer reactions are discussed. The observed peculiarities of the mass and energy distributions of reaction fragments are determined by the shell structure of the formed fragments. Special attention is paid on the symmetric fragment features in order to clarify the origin of these fragments (fission or quasifission). The influence of shell effects on the fragment yield in quasifission and multinucleon transfer reactions is considered. It is noted that the major part of the asymmetric quasifission fragments peaks around the region of the Z=82 and N=126 (double magic lead) and Z=28 and N=50 shells; moreover the maximum of the yield of the quasifission component is a mixing between all these shells. Hence, shell effects are everywhere present and determine the basic characteristics of fragment mass distributions.

1. Intoduction

In reactions with massive heavy ions at energies close to the Coulomb barrier the main competing processes are complete fusion, quasifission, and deep-inelastic collisions [1–4]. The relative strength of these processes strongly depends on entrance channel properties, such as mass asymmetry, deformation of the interacting nuclei, collision energy, and the Coulomb factor Z_1Z_2, but also on the entrance channel dynamics. The renewed interest in the study of heavy-ion collisions involving massive nuclei is driven by the search for new neutron-rich heavy nuclei. This, in turn, has rejuvenated the interest in the physics of mass transfer processes ranging from few to many nucleons. In earlier experiments the emphasis was placed on the investigation of heavy-ion transfer reactions with actinide targets. The energies were well above (20–30%) the Coulomb barrier and the aim was to produce superheavy nuclei [5, 6]. In these kinds of reactions the superheavy production cross sections decrease very rapidly with the increase of the atomic numbers of the colliding partners and binary reactions fully cover the reaction cross sections. Moreover, the amount of excitation energy available at bombarding energies well above the Coulomb barrier hinders the binding role of shell closures in the formation of fragments and reduces their survival probability against neutron evaporation or fission. Most of the reactions exploited to explore the production of superheavy nuclei with hot fusion reactions are, consequently, not suited to produce new neutron-rich nuclides. This brings us to the conclusion that one possible pathway to produce new neutron-rich nuclides is to count on the binding power of the shell closures in a condition of lowest excitation energy possible.

The aim of the present work has been experimental investigation of dynamics and properties fission, quasifission and multi-nucleon transfer reactions of heavy and superheavy composite systems formed in the reactions with heavy ions by measuring mass-energy and angular distributions of binary fragments. Velocity vectors of binary reaction products were measured using the two-arm time-of-flight spectrometer CORSET [7]. The extraction of the masses and Total Kinetic Energy (TKE) of the binary reaction products is based upon the analyses of the two-body velocity diagram. The mass resolution of the spectrometer for these measurements was about 3u (FWHM) that allows to investigate the features of mass distributions with good accuracy. Table 1 presents the reactions studied using the CORSET set-up.

Table 1. The reactions studied with Corset set-up.

Reaction	CN	E_{lab} (MeV)	$E_{c.m.}/E_B$	Ref.
$^{22}Ne+^{249}Cf$	^{271}Hs	102, 127	0.86, 1.08	Itkis [8]
$^{26}Mg+^{248}Cm$	^{274}Hs	125-160	0.90-1.15	Itkis [8]
$^{36}S+^{238}U$	^{274}Hs	168-198	0.92-1.09	Itkis [8]
$^{48}Ca+^{144}Sm$	^{192}Pb	178-245	0.95-1.31	Knyazheva [9]
$^{40}Ca+^{154}Sm$	^{194}Pb	175-210	0.98-1.18	Knyazheva [9]
$^{48}Ca+^{154}Sm$	^{202}Pb	168-260	0.92-1.43	Knyazheva [9]
$^{48}Ca+^{168,170}Er$	$^{216,218}Ra$	180-208	0.93-1.08	Chizhov [10], Sagaidak [11]
$^{44}Ca+^{206}Pb$	^{250}No	217, 227	1.00, 1.05	Knyazheva [12]
$^{48}Ca+^{208}Pb$	^{256}No	206-242	0.95-1.12	Prokhorova [13]
$^{48}Ca+^{232}Th$	^{280}Ds	244	1.06	Itkis [14]
$^{48}Ca+^{238}U$	^{286}Cn	212-258	0.91-1.11	Kozulin [15]
$^{48}Ca+^{244}Pu$	^{292}Fl	226-244	0.96-1.03	Itkis [14]
$^{48}Ca+^{248}Cm$	^{296}Lv	233-238	0.97-1.02	Itkis [14]
$^{58}Fe+^{208}Pb$	^{266}Hs	289-324	1.0-1.12	Itkis I M [8]
$^{58}Fe+^{244}Pu$	$^{302}120$	328	1.05	Itkis[14], Knyazheva [16]
$^{64}Ni+^{186}W$	^{250}No	300, 311	1.00, 1.05	Knyazheva [12]
$^{64}Ni+^{238}U$	$^{302}120$	330-382	0.98-1.13	Kozulin [15]
$^{86}Kr+^{208}Pb$	$^{294}118$	453, 466	1.06, 1.09	Itkis [17]
$^{88}Sr+^{176}Yb$	^{264}Hs	435	1.03	Kozulin [18]
$^{136}Xe+^{208}Pb$	$^{344}136$	700-1020	0.98-1.43	Kozulin [19]

2. Fission and quasifission of superheavy systems

In reactions with heavy ions complete fusion and quasifission are competing processes. It is known that in superheavy composite systems quasifission mainly leads to the formation of asymmetric fragments with mass asymmetry ~0.4 [2]. This type of quasifission process, so-called asymmetric quasifission, is characterized by asymmetric angular distributions in the center-of-mass system and thus fast reaction times (~10^{-21} s) [9, 12]. The TKE for these fragments is observed to be higher than that for CN fission [12, 14] and hence this process is colder than fission. Due to this reason shell effects in quasifission are more pronounced [14]. Besides the asymmetric component, also the symmetric component may be affected by the presence of the quasifission process. Consequently, the question of whether the symmetric fragments originate from fission or quasifission processes arises. On the one hand, the angular distribution for all these mass-symmetric fragments is symmetric with respect to 90° in the center-of mass system and the estimated reaction time is ~10^{-20} s, typical for CN fission processes [3, 9]. On the other hand, the calculations of potential energy surfaces for heavy-ion-induced reactions along with Langevin-type dynamic equations of motion show that one of the possible reaction channels for such systems is a process occurring without a CN stage, but with

Journal of Physics: Conference Series **515** (2014) 012010 doi:10.1088/1742-6596/515/1/012010

fragment properties close to those known from fission. This process is characterized by long reaction times sufficient for mass equilibration and resulting in the formation of symmetric fragments [symmetric quasifission].

A guideline for the interpretation of the pattern following from mass-energy, angular distributions and cross sections comes from dymanical models. At present there are several theoretical approaches to describe the dynamics of the reactions with heavy ions (for example [20-24]).

Mass-energy distributions of binary fragments formed in the reactions $^{48}Ca+^{238}U$, $^{58}Fe+^{244}Pu$ and $^{64}Ni+^{238}U$ at excitation energies of formed CN of 45MeV are shown in Fig.1. The reaction products having masses close to those of projectile and target are identified as quasielastic and deep-inelastic events in the TKE-M matrix, and we will not consider them. Reaction products lying between elastic peaks can be identified as totally relaxed events, i.e., as fission (or fission-like) fragments. We have outlined them by solid lines in the panels. Henceforth we consider the properties of these events only.

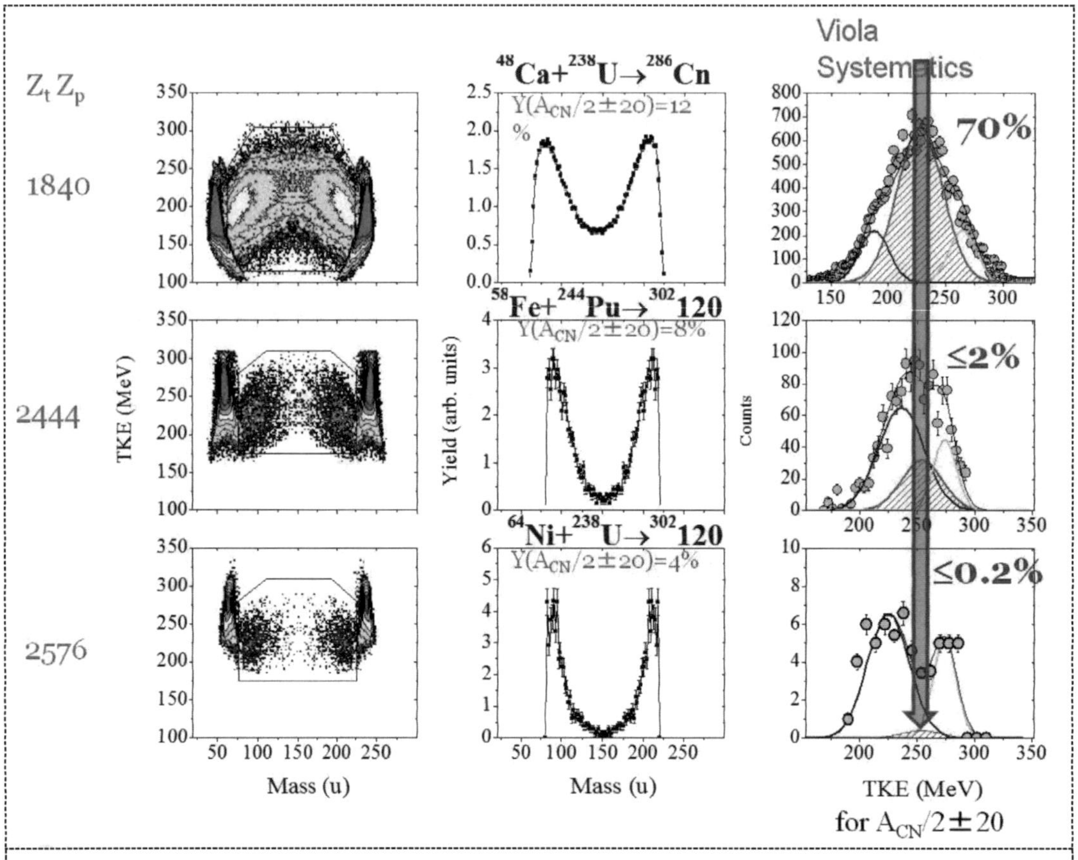

Figure 1. Mass-energy distributions of binary fragments formed in the reactions $^{48}Ca+^{238}U$, $^{58}Fe+^{244}Pu$ and $^{64}Ni+^{238}U$ at excitation energies of formed CN of 45MeV (left panel);mass distributions for event inside the contour lines on mass-energy distributions (middle panel); TKE distribution for symmetric fragments (right panel).

Mass-energy distributions for the reaction $^{64}Ni+^{238}U$ have the wide two-humped shape caused by quasifission process mainly determined by the influence of spherical closed shells with Z = 82 and N = 50, 126. The maximum yield corresponds to the fragments with masses about 215 a.m.u. and complementary light ones. Based on simple assumption on N/Z equilibration the nuclear shells with Z=82 and N=126 correspond to the heavy fragment mass 207-209 a.m.u., neutron shell at N=50 results in light fragment mass 82-83 a.m.u. and the complementary heavy masses for this nuclear shell is 219 a.m.u. So, the major part of the asymmetric quasifission peak fits into the region of the Z=82 and N=126 (double magic lead) and N = 50 shells and maximum of yield of asymmetric QF component is a compromise between all these shells. In the formation of the quasifission asymmetric component the closed shell at N=50 seem to be effective on a par with shells Z=82 and N=126, and it leads to the shift of asymmetric quasifission peak from mass 208 a.m.u, observed in the reaction ^{48}Ca+^{238}U, to 215 a.m.u. at the transition from 112 to 120 superheavy nucleus.

At first sight the mass-energy distributions for the reactions ^{58}Fe+^{244}Pu and ^{64}Ni+^{238}U at the CN excitation energies about 45 MeV are similar: the wide two-humped shape with large quasifission component peaked around the mass 215 a.m.u. However, at the same CN excitation energy the mass drift to the symmetry (estimated as a width between masses corresponding to the maximum and half maximum of quasifission yields) is 22 nucleons in the case of ^{58}Fe reaction and only 11 nucleons in the case of ^{64}Ni-ions. It is significant that the mass drift to the symmetry is about 34 a.m.u. for the ^{48}Ca+^{238}U at the same CN excitation energy. At the symmetric mass region $A_{CN}/2\pm20$ the contribution of the fragments is about 12%, 8% and 4% for Ca, Fe and Ni-ions, respectively.

The TKE distributions for symmetric fragments with masses $A_{CN}/2\pm20$ u for all reactions are presented in figure 1. It is readily seen that both TKE distributions have a complex structure which is not consistent with only CN fission. In fact, it is known that in such a case the average TKE of the partner fragments is substantially independent on the excitation energy and shows a typical Gaussian-like shape. The TKE distributions of symmetric fragments of the ^{48}Ca+^{238}U, ^{58}Fe+^{244}Pu and ^{64}Ni+^{238}U may be deconvoluted into three Gaussians. We use the Viola systematics as a starting point to evaluate mean and variance of the CN fission mode. After a 3-Gaussian fitting procedure we can evaluate the cross-sections due to each of the three components: CN-fission, asymmetric and symmetric quasifissions. In contrast to ^{58}Fe+^{244}Pu, for the reaction ^{64}Ni+^{238}U the TKE distribution has more pronounced low and high energy components (see fig. 1 right column), while the component with average value of 252 MeV (corresponding to the Viola systematics) is highly hindered. Because of the low statistics, only an upper value for the relative yield of the CN-fission component can be reasonably given.

On the basis of the reasonably good success of the analysis method proposed, we can draw some main conclusion. The capture cross-sections are about a few hundred millibarns for Ca and Ni induced reactions, whereas the formation of symmetric fragments is one order of magnitude less for the reaction ^{64}Ni+^{238}U. Yet, in the case of the Ca + U at the highest energy, approximately 70% of the events have the TKE expected for the CN fission process, whereas in the case of the ^{64}Ni + ^{238}U only a few percent of symmetric fragments have the TKE compatible with the Viola prediction for the ^{302}120 CN fission. While the ^{64}Ni + ^{238}U reaction has lower excitation energy at center of mass energies close to the Bass barrier, the CN fission cross-section is suppressed by stronger symmetric and asymmetric quasifission processes and the expected gain in CN survival probability was not observed. The CN fission cross-section in the ^{64}Ni + ^{238}U → ^{302}120 case drops three orders of magnitude with respect to the ^{48}Ca + ^{238}U →^{286}112 case. This is unfortunately a limiting factor. Furthermore, the relative contribution of the CNF from ^{64}Ni + ^{238}U is much lower than in the case of ^{58}Fe + ^{244}Pu → ^{302}120. Recently the experiments aimed at the synthesis of isotopes of element Z = 120 have been performed using the ^{244}Pu(^{58}Fe, xn)$^{302-x}$120 reaction [25] and ^{238}U(^{64}Ni, xn)$^{302-x}$120 reaction [26]. A cross-section limit of 0.4 pb at $E* = 44.7$ MeV for the former reaction and 0.09 pb at $E* = 36.4$ MeV for the latter reaction were obtained. In the case of ^{48}Ca + ^{238}U reaction the evaporation residue cross-section for 3n, 4n channels is about a few pb. Thereby in the transition from Ca to Fe and Ni ions, the evaporation residue cross-section drops down at least one and two orders of magnitude, respectively. Thus, we conclude that the reaction ^{64}Ni + ^{238}U is less favorable compared to ^{58}Fe + ^{244}Pu for production of the superheavy element with atomic number 120.

In the symmetric region of fragment masses with $A_{CN/2}\pm20$ the fusion-fission process may coexists with quasifission and nuclear shells with Z=50 and N=82 are now involved. In Fig 2 the mass-energy distribution for the reaction ^{48}Ca+^{238}U at an excitation energy of 35MeV for the compound nucleus ^{286}112 is presented. On the bottom panel of this figure the mass distribution (normalized to 200%) for fragments inside the contour line on the M-TKE matrix is shown as solid circles. The solid curve is a description of the mass distribution by the sum of Gaussians with average masses corresponding to nuclear shells Z =28 and 82, and N =50 and 126. Open squares depict the symmetric component (normalized to 200 %) obtained as the difference between the experimental mass distribution and the fit to the quasi-fission contribution. One can see that the mass distribution for symmetric fragments is asymmetric in shape with the light fragments mass at about 132-134 a.m.u.

The same trends were observed for mass distributions of reaction fragments measured in the reactions ^{48}Ca+^{244}Pu and ^{48}Ca+^{248}Cm [14]. With the assumption that the fusion-fission process prevails in the symmetric mass region, the induced fission reactions for the compound nuclei ^{286}112, ^{292}114 and ^{296}116 have asymmetric mass distribution like in the familiar cases of fission of actinide nuclei.

The question about the possibility of the formation of superasymmetric fragments in fusion-fission, quasifission and multinucleon transfer reactions when both fission fragments are close to the double magic nuclei arises. The mass distribution of fission fragments of ^{260}No formed in the reaction of ^{22}Ne+^{238}U is presented in fig. 3. The increase of fragment yields in the mass region around 52/208 u that corresponds to the formation of fissioning pair of two magic nuclei Ca/Pb was observed.

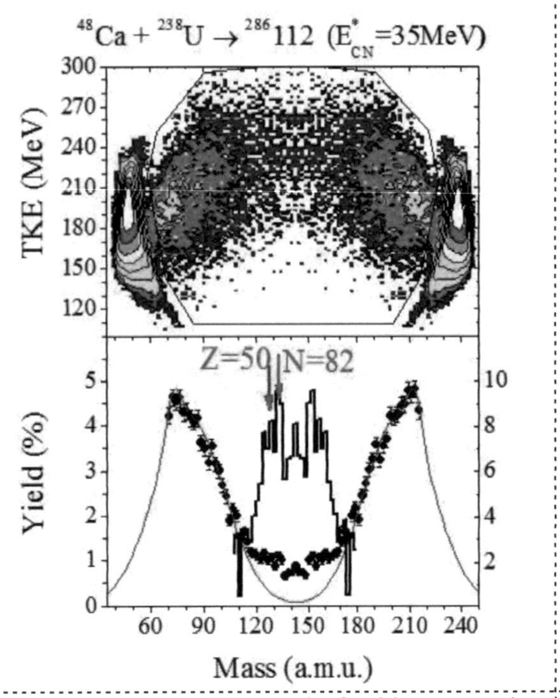

Figure 2. Mass-TKE matrix for binary reaction fragments formed in the reaction ^{48}Ca+^{238}U at 35 MeV excitation energy and mass distribution for all fission-like events (solid circles and left axis) and for symmetric fragments only (open squares and right axis).

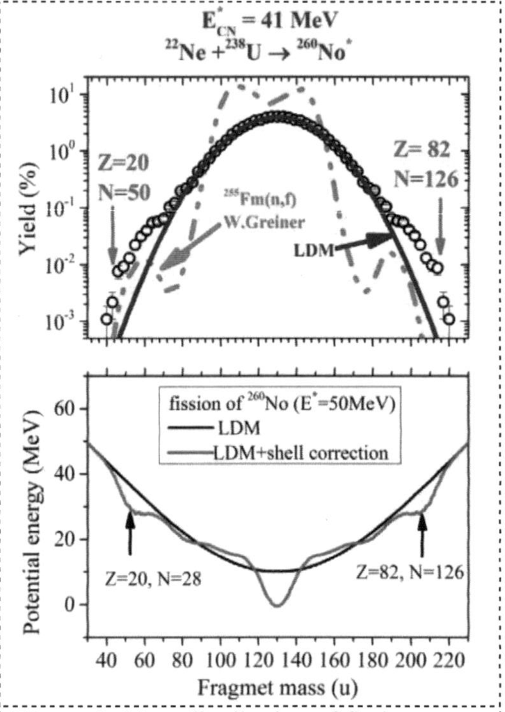

Figure 3. Mass distribution of fission fragments of ^{260}No formed in the reaction ^{22}Ne+^{238}U (open circles); blue line – mass distribution, calculated in the framework of the LDM; dash-dot line – the calculations of W. Greiner for ^{256}Fm (nth, f) taking into account the shell corrections.

3. Transition from fission to multinucleon transfer reactions

In order to investigate on the role of shell effects on the fragment productions in colliding systems with total charge Z = 108 we have used reactions of ^{22}Ne, ^{26}Mg, ^{36}S, and ^{58}Fe beams on ^{249}Cf, ^{248}Cm, ^{238}U, and ^{208}Pb targets at energies below and above the Coulomb barrier [8,18]. The mass and energy distributions of binary fragments formed in these reactions at energies below and above the Coulomb barrier are shown in figure 2. The entrance channel properties of these systems vary strongly: the entrance channel mass asymmetry is η= 0.571for the reaction ^{58}Fe+^{208}Pb, 0.737 for ^{36}S+^{238}U, 0.810 for ^{26}Mg+^{248}Cm, and 0.838 for ^{22}Ne+^{249}Cf. It is important to note that all reaction partners, except ^{208}Pb, are well deformed nuclei. In the reactions with deformed nuclei the potential energy surface strongly depends on the relative orientation of the reaction partners. Except for reactions with strong mass-asymmetry in the entrance channel the dominance of tip configurations at energies below the barrier leads to the increase of quasifission contributions. As demonstrated in Fig. 4 the mass-energy distributions change with decreasing the asymmetry η in the entrance channel from typical for fusion – fission triangular shape for incoming ^{22}Ne-ions to strongly two humped asymmetric shapes for incoming ^{58}Fe and ^{88}Sr ions.These changes are understood as reflecting the relative contributions of different processes depending on the reaction studied.

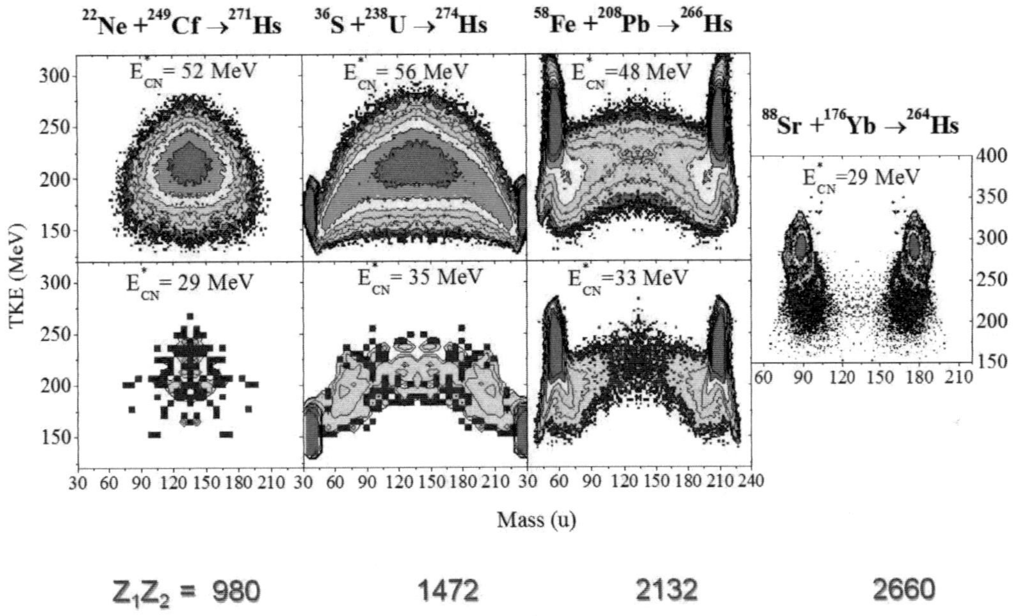

Figure 4. The mass-energy distributions of binary products of Hs (Z=108) composite system formed in the reactions ^{22}Ne+^{249}Cf, ^{36}S+^{238}U, ^{58}Fe+^{208}Pb and ^{88}Sr+^{176}Yb at energies below (bottom panels) and above the Coulomb barrier (top panels).

Previously to distinguish between quasifission and CN-fission the angular and mass distributions of fission-like fragments were used. The present analysis of the TKE distributions of fragments with masses $A_{CN}/2\pm20$ u for different reactions studied shows that the variance of the TKE distribution is sensitive to the presence of the quasifission process. At an excitation of CN larger than 40 MeV (when the shell in CN-fission is practically disappeared) the broadening of the TKE distribution points out to the presence of the both quasifission and CN-fission processes. The narrowing of the TKE distributions indicates that quasifission is a dominant process. This tendency is illustrated in Fig.5 for the case of Hs-composite systems formed in the reactions ^{22}Ne+^{249}Cf, ^{36}S+^{238}U and ^{58}Fe+^{208}Pb.

Our measurements [8] have shown that in the case of the reactions induced by ^{36}S (at energy below the Coulomb barrier) and ^{58}Fe (for energies below and above the Coulomb barrier) asymmetric quasifission is the dominant process. This is caused by the influence of the closed shells at Z = 28, 82 and N = 50,126. The fragments formed in such asymmetric quasifission processes have masses around

200 u. However, the entrance channel asymmetry for both projectiles [$\eta = (A_{projectile} - A_{target})/(A_{projectile} + A_{target}) = 0.73$ in the case of ^{36}S and 0.56 for ^{58}Fe] is larger than the mass asymmetry of asymmetric quasifission fragments (0.45–0.50). This means that nucleons flow mainly from target to projectile in the above mass range as a consequence of the entrance channel asymmetry and the shape of the potential energy surface for heavy nuclei [8]. These results pave the way toward the search for entrance channel conditions which favor the flow of nucleons in the opposite direction, which is a necessary condition for the neutron-rich nuclei to be produced.

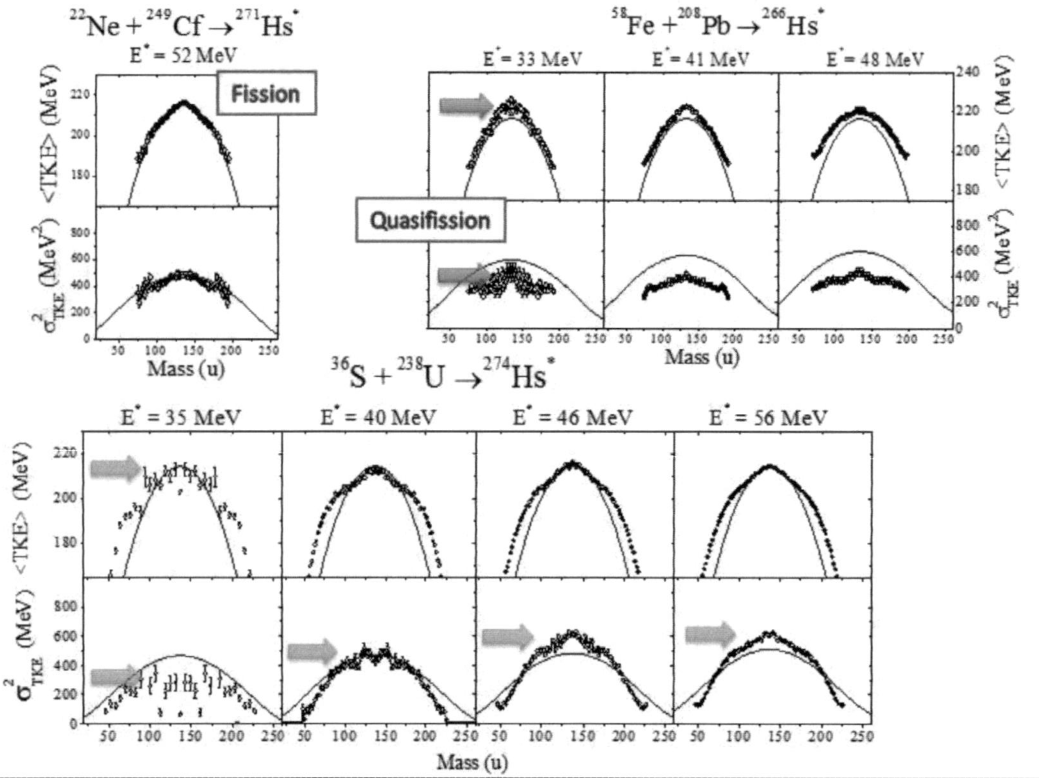

Figure 5. The average TKE and dispersion of TKE of fission-like fragments formed in the reactions ^{22}Ne+^{249}Cf, ^{36}S+^{238}U, ^{58}Fe+^{208}Pb at energies below and above the Coulomb barrier.

To explore the influence of shell effects on the formation of neutron-rich binary fragments in damped collision, even in the case of a large mass transfer, we have investigated binary reaction channels in the reaction ^{88}Sr+^{176}Yb at an energy slightly above the Coulomb barrier ($E_{c.m.}/E_{Bass}$ = 1.03). The total charge of such system is Z = 108, as in the case of the reactions ^{36}S+^{238}U (Z_1Z_2 = 1472) and ^{58}Fe+^{208}Pb (Z_1Z_2 = 2132). What is different in the reaction ^{88}Sr+^{176}Yb is that, in contrast to the reactions with ^{36}S and ^{58}Fe ions, the projectile nucleus has to transfer about 25 nucleons to the target to form fragments with mass of about 200 u. The mass-energy distribution of binary fragments formed in the reaction ^{88}Sr+^{176}Yb (Fig 4 last column) suggests that besides the elastic and quasielastic components, a significant part of the events has a large dissipation of the entrance channel kinetic energy $E_{c.m.}$, which indicates the occurrence of strongly damped collisions. The fragments with TKE dissipation larger than 20 MeV are located mainly in the region 85–115 u for projectile-like fragments and 150–180 u for target-like fragments. The contribution of symmetric fragments with masses ($A_{target} + A_{projectile})/2 \pm 20$ u to all damped collision events with TKE losses >20 MeV is about 1.6%. It is very unlikely that these fragments were formed in fusion-fission processes due to the large value of the Coulomb factor Z_1Z_2 for this reaction. Even in the case of the more asymmetric reaction ^{58}Fe+^{208}Pb the contribution of the fusion-fission component to the symmetric fragments is less than a few percent [8]. As was shown in Ref. [16], the contribution of fusion-fission rapidly decreases with the increasing of the Coulomb factor of a reaction. The presence of symmetric fragments in damped collisions is caused by the driving potential of the system. The minimum of the potential energy of the system favors the creation of symmetric fragments. In particular, the minimum at symmetric masses is strengthened by the nuclear shells at Z = 50 and N = 82. Furthermore, the yield of the symmetric component strongly depends on the reaction time and the nucleon transfer rate. The yield of target-like fragments with mass larger than 170 u at laboratory angles from 25° up to 35° is shown in Fig. 6. This laboratory angles correspond to the angle of grazing collisions for the recoil nucleus and we may expect the maximum yield for the production of target-like fragments at this condition. We observe heavy fragments with mass up to 200 u. Considering the mass resolution of the CORSET spectrometer, this remarkably means that a net mass transfer from projectile to target of about 20–25 nucleons occurs in this reaction. Such a large net mass transfer has also been observed in the reaction $^{136}Xe+^{208}Pb$ (up to 16 nucleons from Xe to Pb) at the energy of $1.23E_{Bass}$ with a cross section of the order of 200 μb, for the lower mass transfer, and a few μb for the larger mass transfer [19].

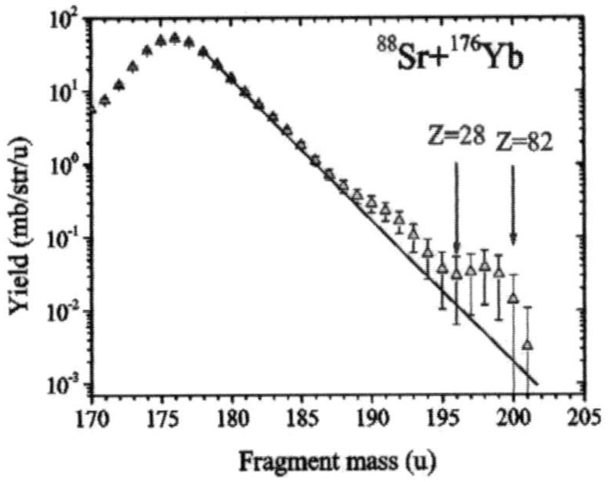

Figure 6. Double differential cross sections of heavy fragments formed in the reaction $^{88}Sr+^{176}Yb$ at $E_{c.m.}$= 290 MeV and detected at laboratory angles from 25° up to 35°.

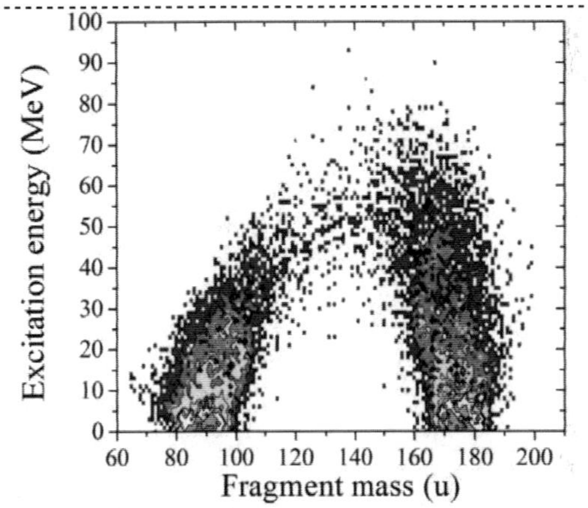

Figure 7. Excitation energy of primary fragments formed in the reaction $^{88}Sr+^{176}Yb$ at $E_{c.m.}$= 290 MeV.

For target-like fragments heavier than the target nucleus, the production cross section of the primary fragments, starting from the region of no-shell closures (maximum at mass around 176), is compatible with an exponential decrease, which is outlined by the solid line in Fig. 6. Because of the absence of shell closures in the mass degree of freedom. For target-like fragments heavier than 190 u, namely, in the mass region progressively closer to shell closures, an enhancement of the yields, with respect to the extrapolation of the exponential decrease to the shell closure region, is quite evident. This trend makes us to suspect that the proton shells at Z = 28 and 82 play an important role and ignite the increase by half an order of magnitude of the yield of the reaction products even for the transfer of twenty nucleons. As was mentioned above, in the previous study of the mass-energy distributions of binary fragments obtained in the reactions of $^{36}S+^{238}U$ and $^{58}Fe+^{208}Pb$ leading to composite systems with the same Z = 108 it was found that the maximum yield of asymmetric quasifission fragments corresponds to the heavy mass of about 200 u, but the transfer of nucleons occurs from the target to projectile. The difference with the present case here is in the entrance channel mass asymmetry, which translates in a different entry point in the potential surface. This difference plays a role on deciding where the main flow of nucleons might be directed. Additional characteristic features of the reaction under study come from the following considerations about the survival probability of the primary fragments. All fragments formed in damped collisions are excited, and de-excited by neutron evaporation mainly. Since the interest to study this type of reactions is connected first of all with the possibility to produce new heavy isotopes, the cross sections of the fragments after the de-excitation process are important. In the present case, the bombarding energy was chosen, using the potential energy surface as a guideline, to maximize the production cross section for large mass transfers (persistence of shell closure) and the survival probability of the primary fragments with respect to neutron evaporation or fission. We estimate the available excitation energy of both fragments as $E_f^*= E_{c.m.}-TKE+Q_{gg}$ and assume that this excitation is divided between the two primary fragments according to their mass ratio. The obtained excitation energy for each fragment is shown in Fig. 7. The particular shape of the distribution in Fig. 7 is due to the (hypothetical) method used to split the excitation energy between the two primary fragments and to the distribution of Q_{gg} values. The excitation energy is largest for symmetric fragments and reaches values up to 90 MeV. For target-like fragment heavier than the target the excitation is about 30–50 MeV. Each neutron takes away on the average 10 MeV (sum of binding energy of one neutron and its kinetic energy). Hence, fragments with mass around 200 u evaporate 3–5 neutrons on average during their de-excitation.

The enhancement found in the yield of products with masses heavier than the target mass confirms that low-energy multinucleon transfer reactions are a possible pathway for producing new neutron-rich isotopes. This result is particularly promising because such mechanism was proposed in Ref. [27] for the synthesis of neutron-rich superheavy elements (SHE), which are not reachable in fusion reactions.

4. Production of heavy neutron-rich nuclei in multi-nucleon transfer reactions

The transfer of many nucleons has been pointed out in several works as a feasible route to synthesize heavy nuclei on the neuron-rich side of the stability line. The choice of the reaction $^{136}Xe + ^{208}Pb$ is based on two qualifying points: 1) the stabilizing effects of the neutron closed shell N = 82 for ^{136}Xe and N = 126 for ^{208}Pb; 2) the specific trend of the Q_{gg} values of all possible mass transfer channels: Q_{gg} values are close to zero for mass transfers that push the primary target-like fragments toward symmetry and become more and more negative for target-like fragments heavier than ^{208}Pb. The two points above mentioned offer three important advantages. First, protons may experience a higher mobility because of the neutron closed shell in the projectile and target, and the transfer of several protons from Xe to Pb may lead to very neutron-rich nuclides, i.e., in the region of osmium and platinum. Second, for the transfers that produce target-like fragments with masses around 200 u or lower, the Q_{gg} are about zero, and windows of total kinetic energy lost select in turn windows of available excitation energy to the primary fragments. In this way, it is possible to gate on target-like fragments that experience lower excitation energy.

Mass-energy matrices of binary fragments for the reaction $^{136}Xe + ^{208}Pb$ at c.m. energies of 526 MeV that were measured by the CORSET setup are presented in Fig. 8. Figure shows the measured TKE-mass distributions integrated over the center-of mass angle 40°-140° of the primary fragments in the reaction $^{136}Xe+^{208}Pb$. Projectile-like and target-like fragments can be identify as elastic and quasi-elastic events at mass-energy matrices. Besides the elastic and quasi-elastic components, a significant part of events has a large dissipation of the initial kinetic energy which indicates the presence of strongly damped collisions. The primary mass distributions of fragments with energy lost greater than 40 MeV are shown in Fig. 8 (bottom panel). Due to such selection of TKEL, most of the quasi-elastic events have been removed. The mass distribution has a two-humped shape. The yield of the fragments in the mass region 136 ± 8 u is more than 60% of all damped events at all measured energies. Furthermore, we observe fragments with mass up to 238 with cross section of the order of 0.1 mb in the reaction at $E_{c.m.}$ = 526 MeV. Considering the mass resolution of the CORSET spectrometer, this remarkably means that a net mass transfer from projectile to target of about 20 nucleons occurs in this reaction.

The theoretical primary mass distribution is given. Also in this case, the calculated yield curve has been convoluted with the experimental mass resolution. One can see a good agreement between theoretical and experimental yields only in the nearly symmetric region. Indeed, the artificial normalization of the calculated yield curve, by a factor of about 0.4, reveals that the shape of the experimental curve can be very well reproduced in the full mass range except for the superasymmetric region, where the underestimation will increase even more. This kind of normalization problem is quite frequent and this result constitutes a feedback for the present model. This allows us to conclude that the model is, roughly within a factor two, in good agreement with the experimental data in the region where Q_{gg} is nearly zero (the region of major interest for this work) but it underestimates by a large extent the primary mass distribution in the superasymmetric fragment mass region. Clearly, the lack of knowledge of the atomic number distribution in this mass region is an unwanted drawback, but this result may give impulse for a more detailed experimental work with respect to the present exploration of the features of the reaction products.

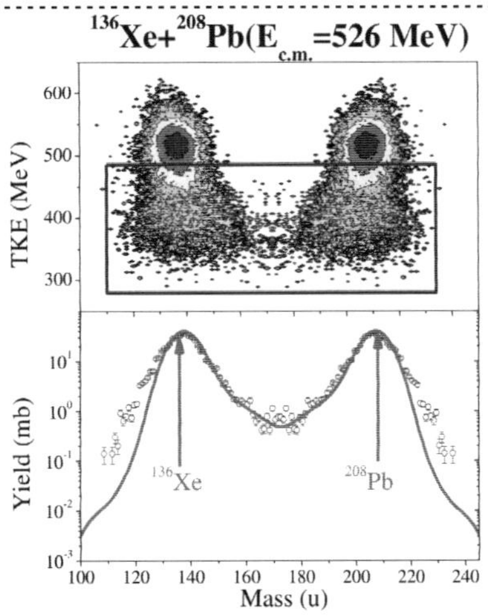

Figure 8. Mass distribution for the fragments with large energy losses (blue contour on matrix); Red line is the calculation in the framework of dynamical Langevin equations by V.Zagrebaev.

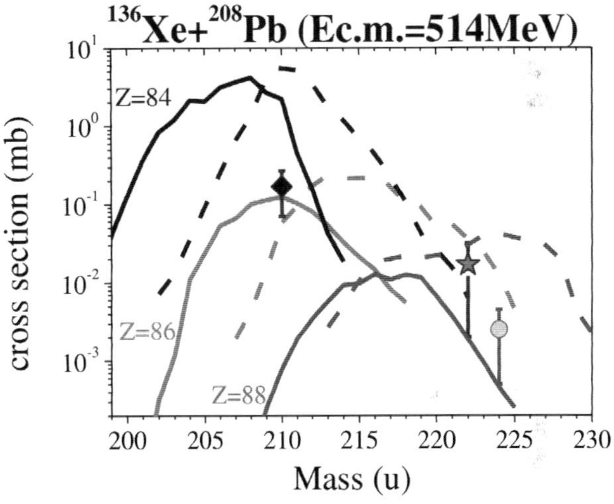

Figure 9. The calculated cross sections for the primary (dash line) and survived (solid line) fragments compared with the values obtained by activation analysis for ^{210}Po, ^{222}Rn and ^{224}Ra.

To provide a firm identification of some target-like fragments with mass greater than 210 u and also to reject any artifact due to the limited resolution of the CORSET setup, we performed a second experiment on the same system at the bombarding energy of 850 MeV by using a catcher-foil activity analysis. Since the major part of nuclides with masses larger than 210 u undergoes α-decay, a careful analysis of the α-decay spectra from nuclides implanted in a catcher-foil can indeed allow us to reconstruct the cross section of the α-emitter isotopes produced in the reaction. This method also calls for a more direct comparison of the cross sections with the expectations of the model.

The cross-section of 200 ± 100 μb, 17 ± 14 μb, and 2.5 ± 2 μb were obtained, respectively, for ^{210}Po, ^{222}Rn, and ^{224}Ra nuclides. In Fig. 9 these values are compared to the new calculation of the model in Ref. [28]. For ^{210}Po the experimental cross-section is overestimated by one order of magnitude; contrarily, the cross section of ^{222}Rn and ^{224}Ra are underestimated by about one order of magnitude, even though at the limit of the experimental error. Both results are quite unexpected because the model is known to well reproduce cross section for transfer of few nucleons; the same trend as for ^{222}Rn and

^{224}Ra in Fig. 9 was indeed recently observed in the reaction ^{160}Gd + ^{186}W at $E_{c.m.}$ = 461.9MeV [29] for the transtarget reaction products near Z = 79.
Yet, the direct measurement of reaction product through their α activity confirms that a mass transfer up to 16 nucleons can occur with a cross section of the order of 200 μb for the lower mass transfer and few μb for the larger mass transfers. These cross sections are certainly of the order of magnitude that make a more detailed search for neutron-rich nuclei accessible with this reaction.

5. Conclusion

From these series of the experiments it was shown that the quasifission process leads not only to formation of asymmetric fragments, but also contributes to the symmetric mass region of fragments.

It was established that in the case of the quasi-fission process the influence of the shell effects on the observed characteristics is much stronger than in the case of classical fission of heavy compound nuclei.

In the reactions of inverse quasifission and multinucleon transfer it has been found that the yield of nuclei with masses heavier than target mass is relatively large. This makes even more promising the production of new neutron-rich heavy and superheavy nuclei in such kind of reactions.

6. Acknowledgments

The work was supported by the Russian Foundation for Basic Research (Grant No. 13-02-01282-a). The authors are grateful to Prof. Giorgio Giardina for the interesting and fruitful discussions. The authors are grateful to the Organizing Committee of the International Symposium "Entrance Channel Effect on the Reaction Mechanism in Heavy Ion Collisions" and Prof. Giorgio Gardina for the support and warm hospitality during stay at the University of Messina.

References

[1] Volkov V V 1978 *Phys. Rep.* **44** 93
[2] Itkis M G *et al.* 2007 *Nucl.Phys. A* **787** 150c
[3] Bock *et al.* 1982 *Nucl. Phys. A* **388** 334
[4] Itkis M G, Itkis I M, Knyazheva G N, Kozulin E M 2013 *Proc. Symposium on Exciting Physics (Makutsi, Safari Farm, South Africa 2011)* ed W Grainer (Springer, Heidelberg) p 33
[5] Schadel M *et al.* 1978 *Phys. Rev. Lett.* **41** 469; 1982 *Phys. Rev. Lett.* **48** 852
[6] Moody K J *et al.* 1986 *Phys. Rev. C* **33** 1315
[7] Kozulin E M *et al.* 2008 *Inst. Exper. Tech.* **51** 44
[8] Itkis I M *et al.* 2011 *Phys. Rev. C* **83** 064613
[9] Knyazheva G N *et al.* 2007 *Phys. Rev. C* **75** 064602
[10] Chizhov A Yu *et al.* 2003 *Phys. Rev. C* **67** 011603(R)
[11] Sagaidak R N *et al.* 2003 *Phys. Rev. C* **58** 014603
[12] Knyazheva G N *et al.* 2008 *Phys. Part Nucl. Lett.* **5** 21
[13] Prokhorova E V *et al.* 2008 *Nucl. Phys. A* **802** 45
[14] Itkis M G *et al.* 2004 *Nucl. Phys. A* **734** 136
[15] Kozulin E M *et al.* 2010 *Phys. Lett. B* **686** 227
[16] Knyazheva G N et al 2010 *Proc. Int. Symposium on Exotic Nuclei EXON2009 (Sochi, Russia 2009)* eds Yu E Penionzhkevich, S M Lukyanov (Melville, New Yourk) AIP 1224 p 377
[17] Itkis M G, Knyazheva G N, Kozulin E M 2008 *Inter. J. Mod. Phys. J* **17** 2208
[18] Kozulin E M *et al.* 2014 *Phys. Rev. C* **89** 014614
[19] Kozulin E M *et al.* 2012 *Phys. Rev. C* **86** 044611
[20] Aritomo Y, HaginoK, Nishio K and Chiba S 2012Phys Rev C85 044614

[21] Nasirov A *et al.* 2013 *Eur. Phys. J. A* **49** 147
[22] Simenel C *et al.* 2013 *EPJ Web of Conferences* **63** 02001
[23] Zagrebaev V and Greiner W 2008 Phys. Rev. C 78 034610
[24] Giardina G et al., 2011, J. Phys.: Conf. Series 282 012006
[25] Oganessian Yu Ts *et al.* 2009 *Phys. Rev. C* **79** 024603
[26] Hofmann S *et al.* 2008 *GSI Scientific Reports* 131
[27] Zagrebaev V I and Greiner W 2011 *Phys. Rev. C* **83** 044618
[28] Zagrebaev V I and Greiner W 2008 *Phys. Rev. Lett.* **101** 122701
[29] Loveland W, Vinodkumar A M, Peterson D, and Greene J P 2011 *Phys. Rev. C* **83** 044610

Study of angular momentum variation due to entrance channel effect in heavy ion fusion reactions

Ajay Kumar
Department of Physics, Banaras Hindu University, Varanasi-221005, India

Abstract. A systematic investigation of the properties of hot nuclei may be studied by detecting the evaporated particles. These emissions reflect the behavior of the nucleus at various stages of the deexcitation cascade. When the nucleus is formed by the collision of a heavy nucleus with a light particle, the statistical model has done a good job of predicting the distribution of evaporated particles when reasonable choices were made for the level densities and yrast lines. Comparison to more specific measurements could, of course, provide a more severe test of the model and enable one to identify the deviations from the statistical model as the signature of other effects not included in the model. Some papers have claimed that experimental evaporation spectra from heavy-ion fusion reactions at higher excitation energies and angular momenta are no longer consistent with the predictions of the standard statistical model.

In order to confirm this prediction we have employed two systems, a mass-symmetric ($^{31}P+^{45}Sc$) and a mass-asymmetric channel ($^{12}C+^{64}Zn$), leading to the same compound nucleus $^{76}Kr^*$ at the excitation energy of 75 MeV. Neutron energy spectra of the asymmetric system ($^{12}C+^{64}Zn$) at different angles are well described by the statistical model predictions using the normal value of the level density parameter **a** = A/8 MeV^{-1}. However, in the case of the symmetric system ($^{31}P+^{45}Sc$), the statistical model interpretation of the data requires the change in the value of **a** = A/10 MeV^{-1}. The delayed evolution of the compound system in case of the symmetric $^{31}P+^{45}Sc$ system may lead to the formation of a temperature equilibrated dinuclear complex, which may be responsible for the neutron emission at higher temperature, while the protons and alpha particles are evaporated after neutron emission when the system is sufficiently cooled down and the higher ℓ-values do not contribute in the formation of the compound nucleus for the symmetric entrance channel in case of charged particle emission.

1. Introduction

Heavy-ion induced fusion reactions are capable of producing compound nuclei with high angular momentum, high excitation energy and large deformations. High excitation energy implies that the nucleus de-excites by emitting several particles and γ-rays so that the decay pattern involves a number of different paths. While the statistical model has been used for many decades to analyze a variety of observables related to the compound nucleus decay, the successful description of the particle emission remains essential for evaluating the validity of the model and the choice of the parameters within it. The emission of the light particles in heavy ion fusion reactions has been used very frequently to understand the behavior of the hot rotating nuclei in several investigations. Detailed experimental data and different model calculations allow us to probe whether the foundation of the statistical model holds for the compound nuclei populated in these reactions. In the case of the compound nuclei at moderate energies and angular momenta, such as those produced with light ion projectiles, the experimental charged particle spectra are well explained in terms of the statistical models employing the optical model transmission coefficients. However, in the case of heavy ion induced fusion reactions, there have been several claims of serious discrepancies between the predictions of the standard statistical model and the measured light charged particle energy spectra [1-9]. Measured light charged particle spectra have been characterized as having lower energy than predicted. Several papers reported that these nuclei are subjected to the lower emission barriers as compared to the inverse absorption channels due to the large deformations at higher excitation energy and the angular momentum [2-7]. Some other authors claim that these spectra may be well explained in terms of the statistical model incorporating only a spin dependent level density and without lowering the emission barriers [10-14].

Assuming that the statistical nature of the compound nucleus decay is experimentally ascertained, some open questions are still related to the description of the average shapes of the highly excited, rapidly rotating nuclei and their influences on the basic nuclear properties like level densities, yrast line position and emission barrier heights [15-16]. Several papers have been devoted to these topics and the field is not yet free from controversies. The study of the neutron emission is advantageous because of the absence of Coulomb effects which complicate the interpretation of the charged-particle spectra in terms of a uniform nuclear temperature. Recently some papers have claimed [17-18] that experimental neutron evaporation spectra from heavy-ion fusion reactions at higher excitation energies and angular momenta are no longer consistent with the predictions of the standard statistical model. Specially, it has been observed that in such cases measured neutrons have been characterized as having higher average energies than predicted [17-18]. This is interpreted as neutron emission from the temperature-equilibrated intermediate di-nuclear complex during the time of its evolution towards compound nucleus formation. In order to confirm these observations, we have studied two systems, a mass-symmetric channel ^{31}P + ^{45}Sc and a mass-asymmetric channel ^{12}C + ^{64}Zn, leading to the same compound nucleus $^{76}Kr^*$. We compared the neutron evaporation spectra as well charged particle spectra with the statistical model calculations and observed that for the asymmetric entrance channel there is no deviation from the statistical model calculations. However, for the symmetric entrance channel the experimental neutron spectra is harder than the statistical model calculations, indicating the entrance channel effects in the neutron evaporation spectra, while the charged particle spectra shows opposite kind of trends w.r.t the statistical model calculations

2. Experiment

The data were obtained using 15 UD Pelletron at Inter University Accelerator Centre (IUAC), New Delhi, India. The ^{12}C pulsed beam of 85 MeV on a ^{64}Zn target was used to form the compound nucleus $^{76}Kr^*$ with excitation energy of 75 MeV. In the other experiment ^{31}P beam at 120 MeV on ^{45}Sc target was used so that the compound nucleus ($^{76}Kr^*$) is formed to match the excitation energy of 75 MeV of the asymmetric system. All the beam energies mentioned are the mid target energies. The self-supporting isotopically enriched (99.93%) targets of 1mg/cm^2 thickness were used in both cases. The experiment was done using the 1.5m diameter stainless steel general purpose scattering chamber (GPSC) available at IUAC. Neutron detectors having liquid scintillator cells of BC501 of 12.5 cm diameter and thicknesses of 12.5 cm were used at an angle of 30°, 60°, 90° and 120°, respectively, with respect to the beam direction and were placed at a distance of 1 m apart from the target. The n-γ pulse-shape discrimination was employed to reduce the g background. The neutron energy was determined by the time-of-flight technique. The pulse from the neutron detectors was used as the start while the stop pulse to the time-to-analog converter (TAC) was provided by the pulsed beam. The γ-ray peak with the neutron spectrum allowed us to calibrate the time spectrum and the overall time resolution of <1 ns. The time-of-flight spectra thus obtained were converted into neutron energy and the intensity was normalized using the neutron detection efficiency code MODEFF [19]. A neutron energy threshold of 0.5 MeV was selected for all the detectors using standard γ-ray sources with proper electron to neutron energy conversion. We have also measured the charged-particle spectra during this experiment using ΔE–E (40 μm–5 mm) detector telescopes .

3. Analysis

(a) Statistical model calculations:: The statistical computer code CASCADE [20] was used to perform theoretical calculations, which assumes the reaction to occur in two steps: first the formation of a compound nucleus and second the statistical decay of the equilibrated system. The maximum value of the angular momentum l_{max} is calculated by the Bass model [21] and the diffuseness (d) is assumed to be 2 ħ.

There are two aspects of the physical processes which govern the flow of an evaporation cascade: the spin dependent level density defining the available phase space and the transmission coefficients that control the access to this phase space. The transmission coefficients mainly affect the lower-energy part of the particle spectrum. In the standard application of CASCADE, the transmission coefficients are derived for neutrons using optical model parameters [22] for the inverse fusion reactions. In heavy-ion fusion reactions at high excitation and in particular the levels with high angular momentum have a meaningful influence on the de-excitation cascade.

The dependence of the level density on deformation caused by the periodic changes in the shell structure is well known for low-spin systems [23-24]. In the high-energy limit, the shell effect on the level density can be described in terms of a constant correction to the intrinsic excitation energy at which this density is to be derived using the Fermi gas formula. The dependence of the level density on the excitation energy and the spin is a crucial quantity in statistical model calculations for the heavy ion induced reactions. However, very little is known experimentally about the spin dependence of level densities for the large spins and high excitation energies.

(b) Dynamical trajectory model calculations (HICOL):: T In the model developed byFeldmeier [25], various aspects of the dissipative heavy ion collisions are brought out for the center of mass energies ranging from the Coulomb barrier up to several MeV per nucleon above the barrier. The lower limit is for treating the classical trajectories and the upper limit is to ensure that the mean field approximation is valid. The macroscopic properties of a large scale nuclear motion are obtained, where the coupling between the intrinsic and collective degrees of freedom is treated in a microscopic picture of particle exchange [26], which provide the friction and the diffusion tensor. The dynamical evolution of the two colliding nuclei is described by a sequence of shapes which basically consists of two spheres connected by a conical neck. Throughout the collision the volume of the shape is conserved so that the uniform mass and charge densities remain the same. The macroscopic shapes of the nuclear system are represented by axially symmetric configurations with sharp surfaces. These shapes are uniquely determined by three macroscopic degrees of freedom, the distance between the nuclei, the neck coordinate , and the asymmetry co- ordinate .

The potential is obtained as a double volume integral of a Yukawa plus exponential folding function and the Coulomb potential is calculated assuming a uniform charge distribution with a sharp surface. The motion of the system is governed by strong dissipative force $X(t)$, which is related to the friction and the diffusion terms obtained from the particle exchange model. One-body dissipation is assumed to be predominant as it has been found to be more relevant for these types of reactions [27]. This model gives a realistic macroscopic description of the nucleus-nucleus collision, based on the concept of one body dissipation. It does not contain free parameters and consistently describes the dynamical evolution of the various composite systems formed in nucleus-nucleus collisions in a wide range of impact parameters.

4. Results and Discussion

The neutron spectra of the composite system $^{76}Kr^*$ formed through the asymmetric reaction ($^{12}C+^{64}Zn$) at a maximum angular momentum of 41 ħ and excitation energy of 75 MeV are shown in Fig. 1(a) for angles at 30°, 60°, 90°, and 120° with respect to the beam direction. The angles (≥30°) are selected so that the contribution, if any, from inelastic, transfer, deep inelastic, and pre-equilibrium processes are negligible as these are focused in the forward direction [9]. The neutron spectra are in good agreement with the statistical model calculations using the normal-level density parameter $a=A/8$ MeV^{-1}, the rotating liquid drop model moment of inertia, and the optical model transmission coefficients for the respective inverse absorption channels.

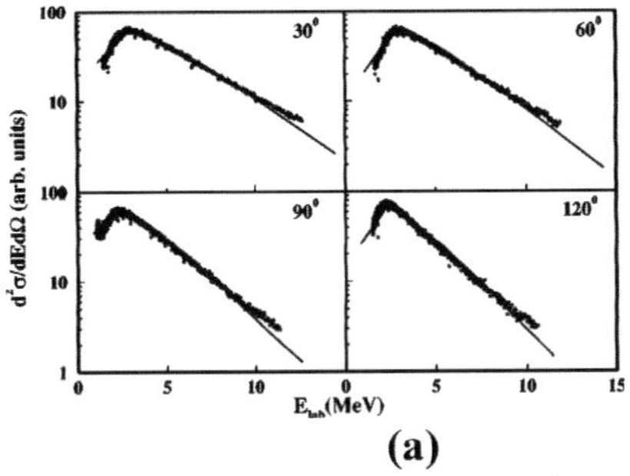

(a)

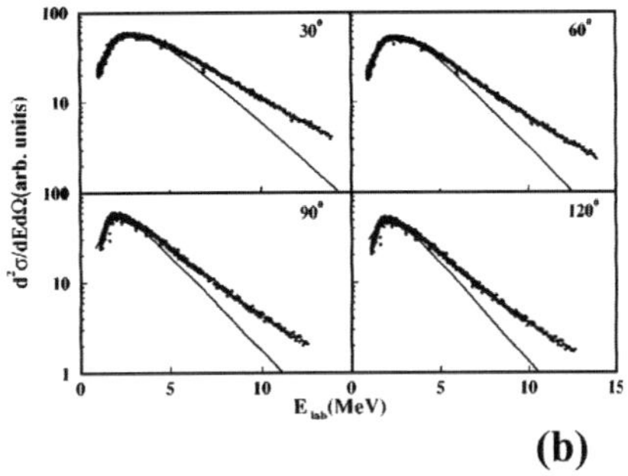

(b)

FIG. 1. Comparison of the experimental neutron spectra (dots)with the statistical model (solid line) (a) using r_0=1.25 and $a=A/8$ for the asymmetric reaction $^{12}C+^{64}Zn$ with ℓ_{max}=41 ħ and E^*=75MeV at E_{lab}=85 MeV (b) r_0=1.25 and $a=A/8$ for the symmetric reaction $^{31}P+^{45}Sc$ with ℓ_{max}=43ħ and E^*=75MeV at E_{lab}=120 MeV

The neutron emission for the comparatively mass-symmetric ($^{31}P+^{45}Sc$) system at different angles for the same excitation energy (75 MeV) is shown in Fig. 1(b). As is clear from the figure these spectra are not in agreement with the statistical model predictions using normal parameters as used for the asymmetric system. It is also clear that the high energy part of the neutron spectra in the case of a symmetric system are harder than the statistical model predictions, indicating neutron evaporation at a higher temperature. The slope of the high-energy part of the neutron spectra is very sensitive to the level density ($\rho \propto \exp[2(aE)^{1/2}]$) and thus on the level density parameter *a*.

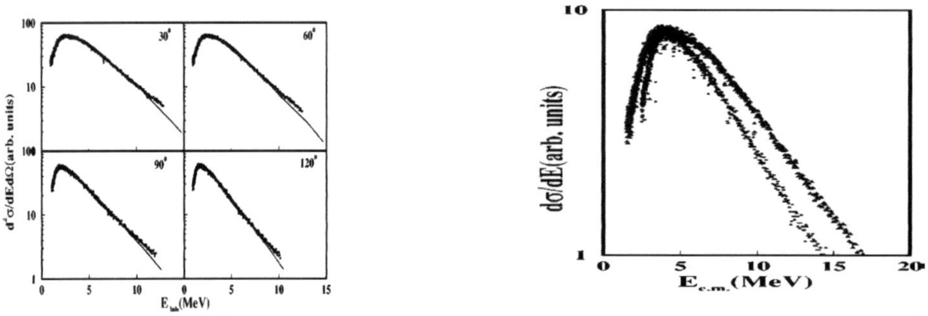

(a) **(b)**

FIG. 2. (a) Comparison of the experimental neutron spectra (dots) with the statistical model (solid line), using r_0=1.25 and $a = A/10$ for the symmetric $^{31}P+^{45}Sc$ with $\ell_{max}=43\hbar$ and E^*=75MeV at E_{lab}=120 MeV. (b) Comparison of the experimental neutron spectra for the symmetric reaction ^{31}P + ^{45}Sc (triangle) and for the asymmetric system ^{12}C+ ^{64}Zn (dots), in the center-of-mass system.

In order to verify quantitatively the experimental trends, the statistical model calculation was performed by changing the value of the level density parameter $a=A/10$ MeV^{-1}. Results of these calculations are shown in Fig. 2(a) for the symmetric $^{31}P+^{45}Sc$ reaction at 120 MeV. This provides a reasonable description of the data and reproduces the shape of the spectra very well. The lower value of the level density parameter $a=A/10$ MeV^{-1} manifests an effective higher nuclear temperature [$T=\sqrt{(E/a)}$] for the neutron evaporation in the case of the mass-symmetric system. This change of level density parameter *a* in the case of neutron evaporation from $^{28}Si+^{118}Sn$ and $^{28}Si+^{124}Sn$ has also been reported by Wile *et al.* [15]. We have also compared the experimental neutron spectra for the two entrance channels in the center-of-mass system in Fig.2(b) in order to avoid any kinematic bias. The spectrum for the symmetric system was found to be harder than the asymmetric system.

In order to understand the behavior of nuclear reaction dynamics more clearly, we have also compared the experimental charged-particle spectra with statistical model calculations using the normal-level density parameter $a=A/8$ for alpha and protons in Figs. 3(a) and 3(b) for

Journal of Physics: Conference Series **515** (2014) 012011 doi:10.1088/1742-6596/515/1/012011

asymmetric and symmetric systems. E^*

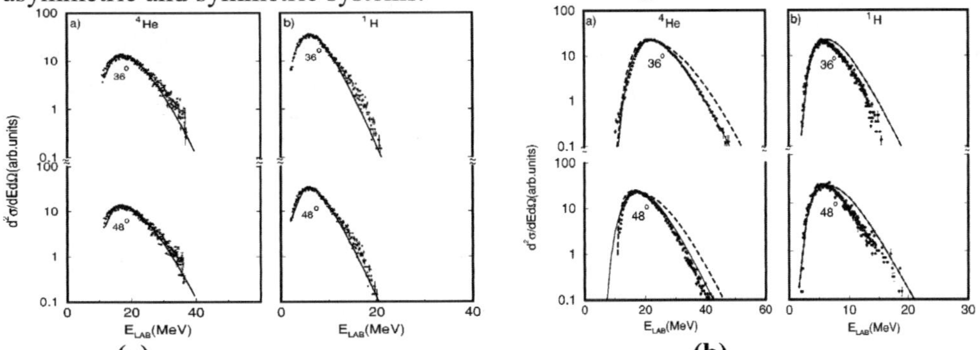

(a) **(b)**

FIG. 3. Comparison of the experimental alpha and proton spectra with the statistical model (solid line) using the transmission coefficients for the spherical nuclei and the RLDM moment of inertia for the (a) asymmetric reaction ^{12}C+^{64}Zn with ℓ_{max} = 41ħ and E*=75 MeV at E_{lab}=85 MeV and (b) symmetric reaction ^{31}P+^{45}Sc with ℓ_{max} = 43ħ and =75 MeV at E_{lab}=120 MeV.

In the case of the charged particles we find that the experimental data is very consistent with statistical model calculations for asymmetric system (^{12}C+^{64}Zn), while the high energy part of evaporation spectra found to be softer as compared to the theoretical spectra in case of symmetric system (^{31}P+^{45}Sc). This is contrary to the neutron spectra, which are found to be harder as compared to the statistical model calculation.

In order to explain the difference in behavior of the proton and alpha spectra from the symmetric system, we carried out dynamical model calculations using the HICOL code. The excitation energies as well as the elongations of the fusing nuclei are plotted as a function of time in Figs. 4 and 5, respectively.

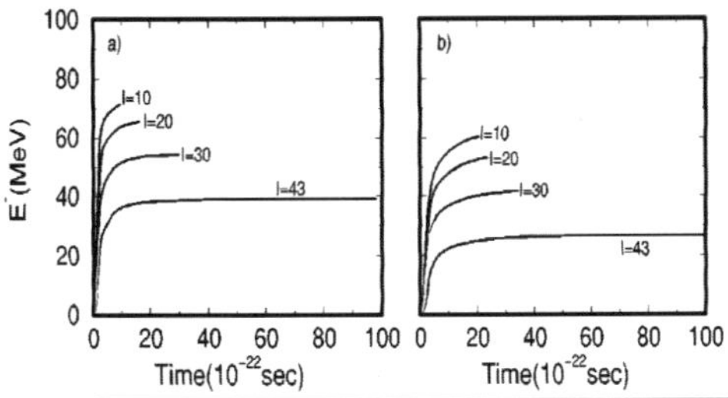

FIG.4. Calculated evolution of the excitation energy (E*) of the colliding nuclei as a function of time for the reactions (a) ^{12}C+^{64}Zn at 85 Mev and (b) ^{31}P+ ^{45}Sc at 120 MeV.

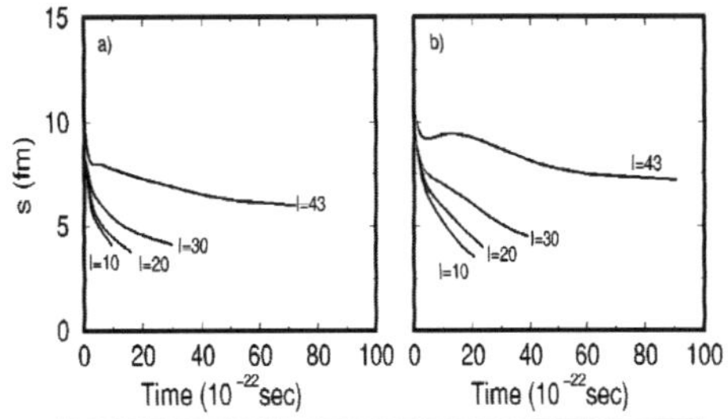

FIG. 5. Calculated evolution of the separation(s) of the colliding nuclei as a function of time for the reactions (a) ^{12}C+ ^{64}Zn at 85 Mev and (b) ^{31}P+^{45}Sc at 120 MeV.

One finds that the higher ℓ values fuse slowly and are saturated at lower excitation energies. In the case of the asymmetric system, the ℓ_{max} contribution to the fusion process is equal to 41 ħ, the same as suggested by the Bass model [21], but in the case of the symmetric system the maximum value of angular momentum which fuses is 30 ħ as opposed to 43 ħ obtained from the Bass model. The time evolution of the two reactions for an angular momentum of 30 ħ is shown in Fig. 6. One finds that the symmetric system $^{31}P+ ^{45}Sc$ evolves more slowly compared to the asymmetric $^{12}C+ ^{64}Zn$ system and the formation time of the asymmetric system is significantly smaller than the decay time while the formation time in the case of the symmetric system is closer to the decay time as shown in Table I.

S. No.	System	E_{Lab} (MeV)	ℓ_{max} (Bass Model)	Excitation Energy (MeV)	Formation Time(10^{-22} s)	Decay Time (10^{-22} s)	HICOL predicted ℓ_{max}
1.	$^{12}C+^{64}Zn \rightarrow ^{76}Kr^*$	85	41ħ	75	29	37	41ħ
2.	$^{31}P+^{45}Sc \rightarrow ^{76}Kr^*$	120	43ħ	75	37	40	30ħ

TABLE I. Comparison of the various parameters used for the two systems.

The statistical model calculations with the HICOL predicted value of ℓ = 30 ħ for the symmetric system is compared with experimental alpha and proton spectra in Fig. 7. The excellent agreement with theory suggests that the deviation from the statistical model results from dynamical effects arising from the mass symmetry of the entrance channel during the formation of the compound nucleus.

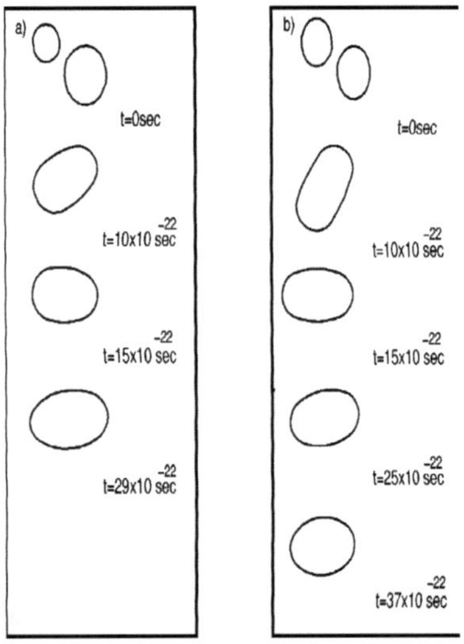

FIG.6. Dynamical model time evolution of the reactions (a) $^{12}C+ ^{64}Zn$ and (b) $^{31}P+ ^{45}Sc$.

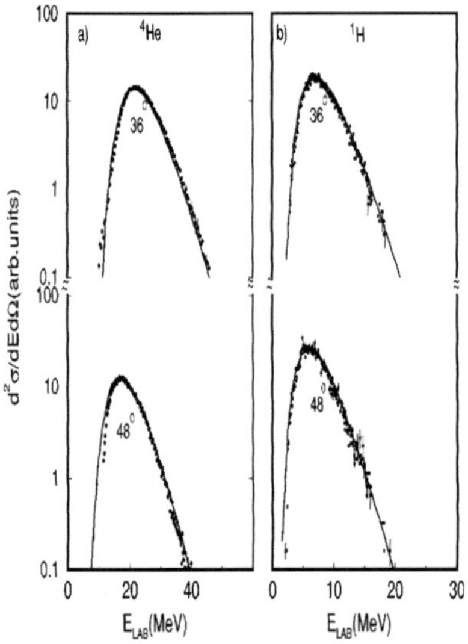

FIG.7.Comparison of experimental spectra with the statistical model for the reaction $^{31}P+^{45}Sc$ using the HICOL predicted ℓ_{max}=30 ħ (a) α-spectra (b) Same for proton spectra.

5. Summary

The compound nucleus $^{76}Kr^*$ is formed in heavy-ion fusion reactions by an asymmetric entrance channel $^{12}C+^{64}Zn$ and the symmetric entrance channel $^{31}P+^{45}Sc$ at the excitation energy of 75 MeV. Neutron energy spectra of the asymmetric system ($^{12}C+^{64}Zn$) at different angles are well described by statistical model predictions using the normal value of the level density parameter $a = A/8$ MeV^{-1}. However, in the case of the symmetric system ($^{31}P+^{45}Sc$), the statistical model interpretation of the data requires a change in the value of $a = A/10$ MeV^{-1}. However the charged particle spectra for the symmetric system is not consistent with the statistical model calculation by using the angular momentum obtained by Bass model and the experimental charged particle data for symmetric system is very consistent with statistical model, when the HICOL predicted angular momentum value is used.

But again if the HICOL predicted angular momentum is used for symmetric system particle spectra (charged particle and neutrons), the neutron spectra becomes more harder comparative to statistical model calculation. So to get more clear information about the variation of angular momentum, we need to do more experiments for determination of the value of fused angular momentum in both type of systems i.e. symmetric as well as asymmetric system.

More likely the delayed evolution of the compound system in the case of the symmetric system may lead to the formation of a temperature-equilibrated system, which may be responsible for neutron emission at higher temperature, while the protons and alpha particles are evaporated after neutron emission when the system is sufficiently cooled down. However, this needs to be explored further with other systems.

Acknowledgments

The authors acknowledge with thanks the Accelerator crew of Inter University Accelerator Centre, New Delhi, for providing a high-quality beam during the course of this experiment. The author also acknowledges financial support from SERB, DST, New Delhi, India, through project grant SR/FTP/PS-068/2010.

[1] Fornal B *et al.* 1998 *Phys. Rev. C* **37** 2624.
[2] B. Fornal *et al.* 1990 *Phys. Rev. C* **41** 127.
[3] Choudhury R K *et al.* 1984*Phys. Lett.B* **143** 74 .
[4] Viesti G *et al.* 1988 *Phys. Rev. C* **38** 2640.
[5] Rana G L *et al.* 1988. *Phys. Rev. C* **37** 1920.
[6] Majka Z *et al.* 1987 *Phys. Rev. C* **35** 2125.
[7] W. E. Parker *et al.* 1991 *Phys. Rev. C* **44** 774.
[8] Agnihotri D K et al. 1993 *Phys. Lett. B* **307**, 283.
[9] Govil I M et al 1998 *Phys. Rev. C* **57** 1269.
[10] Govil I M et al. 1987 *Phys. Lett. B* **197**, 51..
[11] Fornal B et al. 1989 *Phys. Rev. C* **40**, 668.
[12] Huizenga J R et al 1989 *Phys. Rev. C* **40** 668.
[13] Cohen S et al 1974 *Ann. Phys. (N.Y.)* **82**, 557.

[14] Thoennessen M et al. 1993 *Phys. Rev. Lett.* **70**, 4055.
[15] Wile J L et al. 1993 *Phys. Rev. C* **47** 2135.
[16] Kolata J J et al.1976 *Phys. Lett. B* **65**, 333.
[17] Kumar A et al. 2003 *Phys. Rev. C* **68**, 034603.
[18] Kumar A et al. 2008 *Nucl. Phys. A* **68**, 034603.
[19] Cecil R A et al. 1979 *Nucl. Instrum. Methods* **161**, 439.
[20] Puhlhofer F 1997 *Nucl. Phys. A* **280**, 267.
[21] Bass R 1973 *Phys . Lett. B* **47** 139.
[22] Hodgson P E 1967 *Annu. Rev. Nucl. Sci.* **17**, 1.
[23] Huizenga JR and Moretto LG 1972 *Annu. Rev. Nucl. Sci.* **22**, 427.
[24] Williams F C Jr., Chan G and Huizenga JR 1972 *Nucl. Phys. A* **187** 225 .
[25] Feldmeier H 1987 *Rep. Prog. Phys.* **50**, 915.
[26] Feldmeier H and Spangenberger H 1985 *Nucl. Phys. A* **435**, 229.
[27] Blocki J, Planeta R, Brzychczyk J and Grotowski K 1992 *Z. Phys. A* **341** 307.

The upgrade programme of the major experiments at the Large Hadron Collider

P La Rocca[1] **and F Riggi**[2]

[1] Research Fellow, Department of Physics of Catania and INFN Section of Catania, Italy
[2] Full Professor, Department of Physics of Catania and INFN Section of Catania, Italy

E-mail: paola.larocca@ct.infn.it

Abstract. After a successful data taking period at the CERN LHC by the major physics experiments (ALICE, ATLAS, CMS and LHCb) since 2009, a long-term plan is already envisaged to fully exploit the vast physics potential of the Large Hadron Collider (LHC) within the next two decades. The CERN accelerator complex will undergo a series of upgrades leading ultimately to increase both the collision energy and the luminosity, thus maximizing the amount of data delivered to all experiments. As a consequence, the experiments have also to cope with very high detector occupancies and operate in the hard radiation environment caused by a huge multiplicity of particles produced in each beam crossing. In parallel to the accelerator upgrades, the LHC experiments are planning various upgrades to their detector, trigger, and data acquisition systems. The main motivation for the upgrades is to extend and to improve their physics programme also in the increasingly challenging LHC environment. In this paper a general overview of the upgrade programme of the major experiments at LHC will be given, with some additional details concerning specifications and physics programme of new detector subsystems.

1. Introduction

The Large Hadron Collider (LHC) is the most powerful particle accelerator built to date. Since 2009 it has successfully provided collisions to the four large experiments installed along its circumference: ALICE, ATLAS, CMS and LHCb. The ATLAS and CMS experiments are general-purpose detectors, designed to see a wide range of particles produced in LHC collisions and to search for new phenomena, including the Higgs boson, supersymmetry and extra dimensions. The ALICE experiment is mainly devoted to research in heavy-ion physics and Quark Gluon Plasma (QGP) formation, whereas LHCb is primarily designed to investigate the decays of B-particles and so provide an insight into the phenomenon of CP-violation.

During its first 4 years of operation, the LHC ramped up its energy from the start-up energy of 900 GeV up to 8 TeV, which is just over half of the design energy of the machine for proton-proton collisions. In November 2012 the LHC reached a peak instantaneous luminosity of 7.7×10^{33} cm^{-2}s^{-1}, close to the design luminosity of 10^{34} cm^{-2}s^{-1} (even though at half the design energy and twice the beam crossing time). The first LHC heavy-ion PbPb collisions took place in 2010 at a center-of-mass energy of 2.76 TeV. The LHC continued providing more PbPb data at the same energy during the consecutive years. Finally, in 2012 the LHC successfully demonstrated its capability to deliver asymmetric collisions by providing a short, 4-hour long "pilot" run of pPb collisions at $\sqrt{s_{NN}} = 5.02$ TeV, which was replicated and extended during 2013.

These data-taking periods comprise the so-called LHC Run 1, which has successfully ended in February 2013, bringing the LHC to the first long shutdown necessary for its energy upgrade to the design specifications. This is the first step of a more ambitious upgrade programme that the LHC will undergo in the next long-term future. The exciting discoveries achieved during the first years of LHC operation have lead the high energy physics community to foresee a challenging upgrade of the accelerator, called High Luminosity LHC (HL-LHC). In order to further increase its discovery potential beyond 2022, LHC needs an upgrade to increase its luminosity by a factor of 10 beyond its design value. As a highly complex and optimized machine, such an upgrade of the LHC must be carefully studied and requires about 10 years to be implemented. A more powerful LHC would provide more accurate measurements of new particles and enable observation of rare processes that occur below the current sensitivity level.
To meet with the foreseen higher luminosity at the High Luminosity LHC, also the four big LHC experiments will take the opportunity to perform upgrades and routine repairs.

A detailed description of the LHC upgrade programme is discussed in Section 2, whereas the future plan of the main LHC experiments is described in Section 3.

2. The High Luminosity LHC Project

According to its envisaged schedule, the LHC is expected to continue providing collisions over the next 7 years. In this period, the LHC will achieve its design energy and luminosity, with a 25 ns beam crossing time (instead of the current 50 ns). Two major shutdowns (the so-called LS1 and LS2) of longer than a year each will be needed to accomplish these objectives. At that point many elements of the machine, as well as many sub-systems of the LHC experiments, will be damaged by radiation and will need to be replaced. This situation has lead to a foregone question about the future of LHC beyond 2022.

Thanks to the spectacular performance of the accelerator and to the interesting discoveries obtained by the experiments, the upgrade of the machine has become one the highest priority in the European Strategy for Particle Physics [1] - recently adopted by the CERN Council at Brussels - to extend and improve the physics program of the LHC experiments. The novel machine configuration will provide a considerably higher annual integrated luminosity, perhaps by a factor of 5, over that achieved in the past. That would allow the LHC to deliver ~3000 fb^{-1} to the experiments by 2036, allowing them to complete their exploration and study of physics at the Terascale. At the same time as the accelerator is upgraded, the experiments will also undergo major transformations to handle the higher luminosity: indeed, most of the detectors, including the main tracking systems, must be rebuilt to deal with the extreme radiation levels and large numbers of interactions per beam crossing.

This twenty year period of LHC operation, from its start-up to 2030, will be characterized by three long shutdown periods, that correspond to three evolution phases of the machine performance:

Phase-0 In this stage, which started with the first long shutdown in February 2013, the LHC will increase the center of mass energy of pp collisions to $\sqrt{s}$ = 13 - 14 TeV with a peak luminosity of $L_{peak} = 10^{34}$ $cm^{-2}s^{-1}$ delivering about 100 fb^{-1} of data. During the 18 months of LS1, the magnet interconnections will be consolidated and a long list of other improvements will be carried out to bring all the equipments to the level needed for 7 TeV beam energy. Most of the LHC experiments will also undergo several consolidation works in order to cope with the new running conditions.

Phase-1 In this period, which will start with the second long shutdown in 2018 and will extend until 2020, the LHC will achieve its design energy and luminosity through a series of improvements, including the installation of a new injector. Towards the end of this period,

Journal of Physics: Conference Series **515** (2014) 012012 doi:10.1088/1742-6596/515/1/012012

the LHC will reach a peak luminosity of 2 - 3 $\times$ 10^{34} $cm^{-2}s^{-1}$, beyond the original design value, and deliver about 300 fb^{-1} of pp collision data. Improvements and upgrades to the LHC experiments will be necessary to fully exploit the luminosity, especially towards the end of this phase.

Phase-2 In its ultimate running conditions (beyond 2022, after the third long shutdown) the LHC is expected to reach a peak luminosity of 5 - 7 $\times$ 10^{34} $cm^{-2}s^{-1}$ (depending on the leveling option). By exchanging aged parts with improved components (performance-improving consolidation) the upgrade will be done gradually. An example for this is the replacement of the new focusing magnets. The expected delivered luminosity at $\sqrt{s} = 14$ TeV of about 3000 fb^{-1} will further increase the physics reach of the LHC experiments.

A schematic timeline of the LHC roadmap is shown in Figure 1.

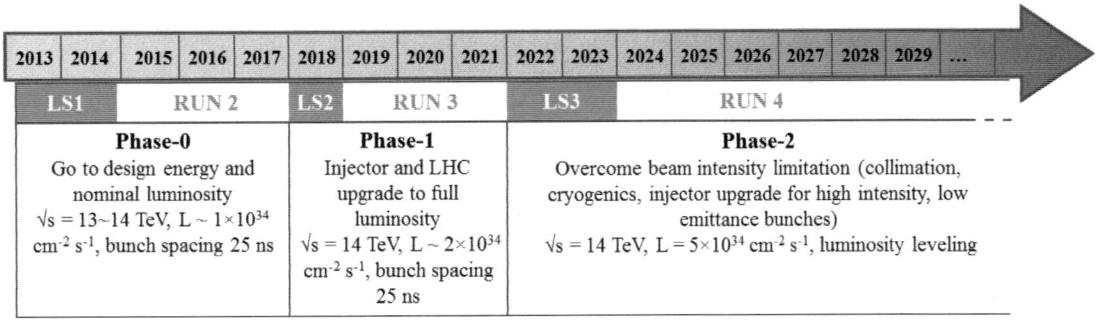

Figure 1. LHC roadmap to achieve its full potential.

At the same time, the experiments will have to upgrade their detectors significantly to cope with the higher luminosity and the foreseen long running time. Some of the upgrades and replacements of detectors become necessary at around 2022, independent of the future increase in luminosity. Especially to be mentioned here is the replacement of the large inner tracking systems of ATLAS and CMS, which will reach the end of their lifetime by that time. The detailed upgrade programme of the major LHC experiments is described in the next Section.

3. The upgrade of the major LHC experiments

The four major LHC experiments, ALICE, ATLAS, CMS and LHCb, have been taking excellent data with high efficiencies. Today, after the initial phase of LHC operations at roughly half of its design energy, some new milestones have already been achieved. Hundreds of scientific papers have been published by the four LHC experiments in only three years of activity. Some of the results are listed below:

- First, the highlight of physics achievements is clearly the discovery of a new particle, compatible with the Standard Model Higgs boson within the present experimental errors and with a mass near 125 GeV. After the discovery, both ATLAS and CMS worked hardly to measure various properties of the new particle: today the Higgs boson mass has been measured to a remarkable precision (0.43% in ATLAS and 0.34% in CMS), as well as its production cross section relative to the SM prediction (σ/σ_{SM}).
- In flavour physics, the LHCb experiment has now overtaken the remarkable achievements of the B factories and Tevatron experiments. In particular, the first evidence of the rare $B_s^0(\mu\mu)$ decay seems to finally hint a deviation from the SM predictions, but still awaits

for a confirmation from ATLAS or CMS. Moreover, there are important new results on CP-violation in the heavy-flavor sector, as observed in B_s^0 decays.

- Physics studies of ultra relativistic heavy ion collisions by ALICE, ATLAS and CMS made a new step, opening a new horizon. Unprecedented and very exciting new results were obtained for example on jet quenching, dijet production, jet-photon correlations, elliptic flow, multiparticle correlations and charm suppression.

A complete list of results and technical information can be found on the web-sites of the LHC experiments [2, 3, 4, 5].

Investigation of new physics has just started and further studies are needed to conclude the exploration of the TeV scale. Concrete upgrade plans have already been proposed, requiring time and resources for at least the next 10 years. The installation of the upgraded detector components presents a serious challenge, due to many factors, such as difficulties to access some detectors, risk of physical damage to the detectors, radiation exposure and accident risk in the pits, limited time windows to perform upgrade operations. All the experiments have produced detailed documents about their upgrade strategy, whose activity will be distributed over the three long shutdowns foreseen by the machine.

In the following we briefly describe the major LHC experiments and their upgrade plans in view of the realization of the HL-LHC.

3.1. ALICE

ALICE [6] is a general-purpose detector, especially suited for the investigation of heavy-ion collisions at the LHC energetic regime. The ALICE detector has a central part, which is made of several subdetector systems placed inside a solenoidal magnet providing a field up to 0.5 T, and a forward muon spectrometer, together with additional small detectors located at small angles and at large distances from the interaction point. The central barrel includes a set of detectors with full azimuthal coverage: the Inner Tracking System (ITS), a large Time-Projection-Chamber (TPC), a Time-of-Flight Detector (TOF), and a Transition Radiation Detector (TRD). Electromagnetic calorimetry is provided by two e.m. calorimeters with different segmentations, covering a large range in ϕ and located such as to be able to detect back-to-back jet correlations. A high momentum particle identificator, based on a Ring Imaging Cherenkov Detector, is also embedded in the central region.

The overall organization of the ALICE detector allows a detailed study of several observables, both in pp and heavy ion collisions, which arise from the detection of hadrons, electrons, muons and photons. The physics results achieved by ALICE during the first phase of LHC data taking, with pp, pPb and PbPb collisions, have demonstrated the feasibility of most of the physics program planned since the beginning. The observation of a hot hadronic matter at extreme values of temperature, density and volume, together with precision measurements of all QGP probes, has confirmed the initial picture of a QGP as an almost perfect liquid. High precision measurements of rare probes will benefit of the LHC upgrade, thus extending the potential for a detailed characterization of the nuclear matter under extreme conditions. The detailed characterization of the QGP requires the investigation of various properties, with measurements of different probes. In-medium parton energy loss mechanisms, jet-jet and photon-jet correlations, quarkonium dissociation, study of charm and beauty are only a few of the planned studies in this respect.

To address such topics, the ALICE detector would require high statistics measurements. The experimental setup is then being modified to allow the readout of all interactions, and become able to accumulate 10 nb^{-1} of Pb-Pb collisions. The upgrade strategy of the ALICE experiment [7] follows the LHC plans to increase progressively, after the long shutdown LS2, the luminosity of heavy ion beams in order to reach an interaction rate of about 50 kHz, i.e.

instantaneous luminosities of L = 6 × 10^{27} cm^2s^{-1}.

The planned ALICE upgrades include at the moment several items: a new beampipe with smaller diameter; a new, high-resolution and low-material budget Inner Tracking System possibly entirely based on small size pixel detectors; the upgrade of the TPC with GEM detectors and new readout electronics; the upgrade of the readout electronics of TRD, TOF, PHOS and Muon Spectrometer for high rate operation; the upgrade of the forward trigger detectors, online systems and reconstruction. The inclusion of additional detectors or modification of the existing ones is also being discussed in view of long term decisions by the involved funding agencies. Concerning in particular the role of the Inner Tracking System (ITS), its upgrade is fundamental to improve its resolution and readout rate capabilities. The recent progress in Si detector technology, an improved integration to reduce the distance between the interaction region and the first layer of the ITS and a minimized material budget will make it possible for ALICE to address such goals.

3.2. ATLAS

ATLAS [8] is the largest detector at the Large Hadron Collider, with overall dimensions of 44 m length and 25 m diameter. ATLAS is a general-purpose experiment, especially designed to measure pp collisions at the presently envisaged frontiers of LHC, i.e. center-of-mass energies up to $\sqrt{s}$ = 14 TeV and peak luminosity of 10^{34} $cm^{-2}s^{-1}$.

The main components of the ATLAS setup include the Inner Detector, operating inside a solenoidal field of up to 2 T, a Calorimeter and the Muon Spectrometer with a toroidal magnet and a field up to 0.5 T.

The physics topics of interest for the ATLAS Collaboration include most of the physics potential discovery at LHC: measuring the Higgs properties, searching for new physics beyond the Standard Model, especially supersymmetry studies and new gauge bosons.

In response to the new challenges offered by the LHC upgrade, in particular by the High Luminosity improvement of the machine, major changes are required to improve the detector performance in an environment providing higher detection occupancy and radiation level. While the barrel calorimeters and muon chambers are able to cope with the new scenario, significant improvements are needed for the Inner Detector, forward calorimeters and muon spectrometer.

Following the planned program of the LHC machine towards the HL-LHC, with its three long shutdown periods, the ATLAS Collaboration has split its upgrade program in three phases [9, 10]. Apart from a series of small modifications and improvements, Phase-0 (taking place during the 2013-2014 shutdown), will focus on the installation of a new barrel layer in the present Pixel Detector, with a new Beryllium beam pipe in the central region. This will improve the vertex capabilities of the tracker, extending the potential for physics analyses.

For the Phase-1 in 2018, after which the LHC luminosity will be increased by a factor 2, thus reaching 2 × $10^{34}cm^{-2}s^{-1}$, a partial replacement of the Muon Spectrometer is envisaged, in order to ensure efficient tracking at high particle rate and large pseudorapidity. Due to the increased collision rates, also better trigger strategies will be implemented, to allow on-line selection algorithms for b-tagging and lepton identification.

Finally, a further Phase-2 upgrade is scheduled for 2022-2023, to fully exploit the high luminosity feature of HL-LHC. In this final upgrade strategy a new Inner Detector is planned, together with trigger and calorimeter upgrades. This is partly due to the need to replace the previous silicon detector and associated electronics due to cumulated radiation damage, and at the same time to provide a new system able to handle a detector occupancy between 5 and 10 times higher than before. Even though most of these new facilities are still in the R&D status, the new silicon Inner Tracker should consist of several pixel and Si-strip layers in the barrel part and in the endcap regions, with an overall higher granularity, smaller material budget and increased radiation tolerance for the readout electronics.

3.3. CMS
The CMS detector [11] is built making use of a large (13 m long, 6 m diameter) superconducting solenoid providing a magnetic field up to 4 T. This is large enough to accommodate with a modular strategy the tracking detector, based on Silicon microstrip and pixel devices, the electromagnetic calorimeter (ECAL) which uses $PbWO_4$ crystals and the lead/scintillator sampling hadron calorimeter (HCAL). Additional calorimeters extend the coverage of the CMS detector at larger rapidities. The iron yoke is equipped with four stations for muon tracking, covering most of the 4π solid angle.

The physics interests of the CMS Collaboration include all the potential which can be addressed by the study of pp and heavy-ion collisions at LHC, due to the setup capability to detect electrons, muons, τ, W and Z-bosons, top quarks, and other probes. During the first period of LHC data taking, CMS has already studied a wide amount of physics topics, from the Standard Model precision measurements to Higgs physics, Supersymmetry, top and B physics, quarkonia, as well as heavy ion physics.

Going towards the increase of the LHC luminosity within the year 2022, the machine peak luminosity will exceed the value for which CMS was originally designed, and a series of modifications/improvements are needed to successfully operate in the new conditions, from the point of view of instantaneous and cumulated luminosity and non collisional background. Other issues of concern for an upgrade deal with the optimization of data taking (minimization of the periods when the detector is not ready to take data), as well as with the replacement of obsolete equipments after long operational periods, as it is usual for experiments running for a long time.

The main changes that are proposed by the CMS Collaboration to handle the new LHC conditions after the LS2 period will be concentrated on the upgrade of the Muon System, the Hadron Calorimeters, the Pixel System, the Trigger and Data Acquisition [12].

For the Muon System it is mainly planned to add a new layer of Cathode Strip Chambers, improving also the associated readout electronics and trigger, and an additional layer of Multigap Resistive Plate Chambers to extend pseudorapidity coverage. In the hadron calorimeters, the upgrade activity will address the problem to handle an increased instantaneous and integrated luminosity, as well as to increase the robustness and efficiency. This will require to replace the HPD photosensors in the central calorimeters with the new Silicon photomultiplier devices, and the existing photomultipliers in the Forward Hadron Calorimeter with new devices having thinner glass windows and larger radiation tolerance. Since the present Pixel System cannot sustain the future high luminosity conditions still maintaining a high tracking efficiency, significant changes are planned to replace several of the existing layers in the barrel and in the end-cap. The upgraded detector will have a reduced innermost radius and material budget, thus improving tracking capabilities, vertexing and b-jet identification. The trigger will be improved switching to a new, more flexible, technology no more based on the existing VME standard, while the Data Acquisition System will have an increased band width, by a factor 2 to 5, in order to follow the expected increase in data collection.

3.4. LHCb
The LHCb experiment [13] is a single-arm spectrometer that covers the angular region from approximately 10 mrad to 300 (250) mrad in bending (non-bending) plane, corresponding to the pseudo-rapidity interval $1.9 < \eta < 4.9$. The spectrometer consists of: the vertex locator (VELO) system (including the pile-up veto stations), for a precise measurements of track coordinates close to the interaction region; the tracking system made of a Trigger Tracker (a silicon microstrip detector TT) in front of the spectrometer magnet, and three tracking stations behind the magnet, made of silicon microstrips in the inner parts (IT) and of Kapton/Al straws for the outer parts (OT); two Ring Imaging Cherenkov counters, the first positioned directly

behind the VELO (RICH1), the second located behind the magnet and the tracking system (RICH2), both aimed at identifying charged particles over the momentum range 1-150 GeV/c; the calorimeter system composed of an electromagnetic (shashlik type) calorimeter (ECAL) and a hadron (Fe and scintillator tiles) calorimeter (HCAL), designed to stop particles as they pass through the detector, measuring the amount of energy lost as each one grinds to a halt; the muon detection system, that comprises five rectangular stations composed of MWPCs and GEM detectors, designed for the detection of muons of B meson decays; the spectrometer magnet, a warm dipole magnet providing an integrated field of 4 Tm.

The main purpose of the LHCb experiment is to study indirect evidence of new physics (beyond the Standard Model) in the beauty and charm sector. This is performed through the precise measurement of CP violation and rare decays of beauty and charm hadrons. LHCb runs at a lower luminosity than ATLAS and CMS, to avoid excessive pile-up of pp interactions, since it is designed to study the vertex structure of events. Nevertheless the experiment has now accumulated $\sim$3 fb^{-1}, about a factor of ten less than ATLAS and CMS, but corresponding to an enormous sample of the order of 10^{12} produced B decays.

The operation and the results obtained from the data collected in LHC Run1 demonstrate that the detector is robust and functioning very well. However, following the improved performance of the LHC, the expected luminosity could in theory double the amount of beauty and charm decays generated, while reducing the complexity of events by a factor two. For this reason the prospect to increase the physics yield in the LHCb dataset seems very attractive, but it can not be pursued without improving the detector. The strategy for the upgrade of the LHCb experiment consists of ultimately removing the first-level hardware trigger entirely [14]. The direct consequences of this approach are that all LHCb sub-detectors with silicon sensors will need to be redesigned to cope with an average luminosity of up to 2×10^{33} cm^{-2}s^{-1} and to be equipped with new trigger-less Front-End electronics, and the entire read-out architecture must be redesigned in order to cope with a multi-Tb/s readout network and the full 40 MHz dataflow.

The installation and commissioning of the upgraded detector and readout system is planned for the LHC LS2. After that, the LHCb will perform measurements beyond Flavor Physics: many of the current exploration studies will become precision studies, enabling LHCb to reach a sensitivity comparable to the theory's one.

4. Concluding remarks

The LHC began operation in late 2009 and the first three years of excellent performance of the machine and detectors brought in the first major discovery and a whole new program of precision measurements and searches. However, a series of improvements and upgrades to the machine are foreseen in order to extend the detectors research program. This will represent a great challenge for the LHC experiments, which will have to cope with unprecedent track densities, from the high instantaneous luminosity, and with extremely high radiation levels. In order to keep the performance roughly at present levels in the harsher HL-LHC conditions, the major LHC experiments are planning a series of detector upgrades synchronized to the LHC upgrade schedule, with a large number of detector installations foreseen during the three long shutdowns.

Such upgrade plans require detailed schedules and a long phase of R&D activity, that is already started. The upgrade of an operating detector presents a serious challenge, and many practical issues have to be taken into proper account. Indeed, even if most of the detectors can be quickly opened at the beginning of a shutdown, some components are not readily accessible and upgrading them within the planned shutdowns requires an elaborate planning. At the same time, the risk of physical damage to the detector due to upgrade activity must be minimized, as well as the exposure to ionizing radiation and contamination of the workers maintaining the detector.

For this reason the upgrade of the accelerator and detectors is a long process, that will cover a period of at least 10 years. The official approval of the upgrade projects of the LHC experiments is almost completed and part of the activity have already started during the ongoing long shutdown.

References

[1] Aleksan R *et al* 2013 Physics Briefing Book : Input for the Strategy Group to draft the update of the European Strategy for Particle Physics *Report* CERN-ESG-005
[2] The ALICE Experiment **http://aliceinfo.cern.ch/**
[3] The ATLAS Experiment **http://atlas.web.cern.ch/Atlas/Collaboration/**
[4] The CMS Experiment **http://cms.web.cern.ch/**
[5] The LHCb Experiment **http://lhcb.web.cern.ch/lhcb/**
[6] Aamodt K *et al* (The ALICE Collaboration) 2008 *JINST* **3** S08002
[7] Abelev B *et al* (The ALICE Collaboration) 2012 Letter of Intent for the Upgrade of the ALICE Experiment *Report* CERN-LHCC-2012-012/LHCC-I-022
[8] Aad G *et al* (The ATLAS Collaboration) 2008 *JINST* **3** S08003
[9] The ATLAS Collaboration 2011 Letter of Intent for the Phase-I Upgrade of the ATLAS Experiment *Report* CERN-LHCC-2011-012/LHCC-I-020
[10] The ATLAS Collaboration 2012 Letter of Intent for the Phase-II Upgrade of the ATLAS Experiment *Report* CERN-LHCC-2012-022/LHCC-I-023
[11] Chartrchyan S *et al* (The CMS Collaboration) 2008 *JINST* **3** S08004
[12] The CMS Collaboration 2011 Technical Proposal for the Upgrade of the CMS Detector through 2020 *Report* CERN-LHCC-2011-006/LHCC-P-004
[13] Alves A *et al* (The LHCb Collaboration) 2008 *JINST* **3** S08005
[14] Aaij R *et al* (The LHCb Collaboration) 2011 Letter of Intent for the LHCb Upgrade *Report* CERN-LHCC-2011-001/LHCC-I-018

Entrance channel effects in medium mass nuclear systems at 25 MeV/nucleon

I. Lombardo[1], L. Acosta[2], C. Agodi[2], F. Amorini[2], A. Anzalone[2], L. Auditore[5], I. Berceanu[8], G. Cardella[3], M.B. Chatterjee[7], E. De Filippo[3], E. Geraci[3,4], G. Giuliani[9], L. Grassi[10], J. Han[2], E. La Guidara[3], G. Lanzalone[2,6], D. Loria[5], C. Maiolino[2], T. Minniti[5], A. Pagano[3], M. Papa[3], S. Pirrone[3], G. Politi[3,4], F. Porto[2,4], F. Rizzo[2,4], E. Rosato[1], P. Russotto[2], A. Trifirò[5], M. Trimarchi[5], G. Verde[3], M. Vigilante[1]

[1] Dipartimento di Fisica, Università di Napoli Federico II, and INFN-Sezione di Napoli, Via Cintia, I-80126, Napoli, Italy
[2] INFN - Laboratori Nazionali del Sud, Via S. Sofia, Catania, Italy
[3] INFN - Sezione di Catania, Via S. Sofia, Catania, Italy
[4] Dipartimento di Fisica e Astronomia, Università di Catania, Via S. Sofia, Catania, Italy
[5] Dipartimento di Fisica, Università di Messina, and INFN-Gr.Coll. Messina, Italy
[6] Facoltà di Ingegneria ed Architettura, Università Kore di Enna, Enna, Italy
[7] Saha Institute of Nuclear Physics, Kolkata, India
[8] Institute for Physics and Nuclear Engineering, Bucharest, Romania
[9] Texas A&M University, College Station, TX, United States
[10] Ruder Boskovic Insitute, Zagreb, Croatia

E-mail: ivlombardo@na.infn.it

Abstract. We discuss experimental data concerning ^{48}Ca+^{48}Ca and ^{42}Ca+^{54}Fe reactions at 25 MeV/nucleon obtained by using the 4π multi-detector Chimera. Effects due due to the different neutron to proton ratios (N/Z) of the two systems have been investigated. Charge distributions of forwardly emitted fragments show even-odd staggering whose amplitude is larger for the neutron poor system. Moreover, the neutron excess of the total system influence the evolution of fusion-like sources formed in semi-central collisions. In particular, for the neutron rich system, the relative emission yield of heavy residues is larger, while multi-fragmentation and binary-like events prevail in the neutron poor system.

1. Introduction

The possibility of performing nuclear reactions between heavy ions with different charge and neuron number allows to investigate effects related to the different entrance channels [1]. For example, entrance channels can differ in total mass, projectile/target mass asymmetry, total neutron (N) to proton (Z) number ratio, and also projectile/target neutron to proton (N/Z) asymmetry. The availability of a large range of (stable) isotopes is elements such as Calcium, Nickel, Tin and Xenon allow to perform nuclear reactions aimed to study different projectile/target combinations. In this way, effects related to the various degree of freedom before mentioned can be investigated. As an example, effects related to the mass asymmetry in incomplete fusion events at low energies ($<$ 20 MeV/nucleon) have been investigated [2]. In more recent times, data concerning systems with noticeable differences in N/Z have been increasingly reported in the literature [3, 4]. In this way, it was possible to investigate in details the effects played by the isospin degree of freedom on various aspects of reaction dynamics, such as the co-existence of various competitive mechanisms [5, 6], the neutron exchange between reaction partners (isospin diffusion) [7, 8, 9, 10] and the neutron enrichment of low-density regions formed in peripheral collisions (isospin drift) [5, 10, 11]. The analysis of these phenomena is useful also to obtain details about the behavior of the Symmetry Energy of the Nuclear Equation of State, a topic of relevant interest also in the Nuclear Astrophysics and Nuclear Structure fields [12, 13].

Interesting findings can be obtained by analyzing the distributions of light fragments emitted at forward angles. Z-distributions are characterized by even-odd oscillations that could be attributed to the last steps of the de-excitation phase of excited pre-fragments, as discussed in various paper reported in the literature [14, 15, 16, 17, 18, 19, 20, 22]. It has been observed moreover that, by increasing the N/Z of total system, the amplitude of the even-odd staggering of the Z-distributions is more an more smoothed. As expected from symmetry considerations, even-odd staggering effect has been recently found also for the N-distributions, as reported in [18, 20]. As suggested in Refs. [16, 21], also the reaction mechanism involved could have a small influence on the amplitude of the even-odd effect.

In the case of medium-mass systems, at low values of impact parameter, we can observe the co-existence of incomplete fusion phenomena in competition with binary-like and multi-fragmentation events [23, 24, 25]. In this case the correlation between the mass and velocity of the two or three biggest emitted fragments can represent an useful way to discriminate between the various reaction mechanisms [3, 17]. Recently, by investigating ^{40,48}Ca+^{40,48}Ca reactions at 25 MeV/nucleon we observed that the N/Z of the total system plays an important role in the competition between these mechanisms [6, 26], as predicted by many theoretical models [27, 28, 29]. Moreover, comparisons of our data with the CoMD-II dynamical model calculations has allowed to extract information about the density dependence of the Symmetry Energy in the Nuclear Equation of state at near-saturation densities [6, 26].

In this paper we focus our attention to the analysis of the two isobaric systems ^{48}Ca+^{48}Ca (N/Z=1.4) and ^{42}Ca+^{54}Fe (N/Z$\simeq$1.09, preliminary results) at 25 MeV/nucleon. The use of isobaric systems allow to reduce the influence of spurious effects due to the bare difference in their total mass. Charge distributions of light fragments emitted at forward angles show even-odd staggering effects with a large amplitude in the case of the more neutron poor system, while for the neutron rich system the relative probability of heavy residue emission is favored. The results of these analysis will be briefly discussed in the following sections.

2. Experimental apparatus

The experiment was performed at the INFN-Laboratori Nazionali del Sud in Catania. ^{42,48}Ca beams, accelerated at 25 MeV/nucleon by the Super-Conducting Cyclotron, were driven on isotopically enriched self-supporting target of ^{48}Ca (2.7 mg/cm^2 thick) and ^{54}Fe (2.0 mg/cm^2 thick).

The reaction ejectiles were detected by means of the 4π multi-detector Chimera [30, 31]. It covers practically the whole solid angle (about 94% of 4π) and it is constituted by 1192 Si-CsI(Tl) telescopes arranged with azimuthal symmetry around the beam axis. The nominal thickness of silicon detectors is $300\mu m$, while caesium iodide thickness varies as a function of the polar angle from 12 cm (forward angles) down to 3 cm (backward angles). The coupling of different identification techniques offers the opportunity to identify in charge and/or in mass the reaction products. These methods are respectively: (1) the ΔE-E method, useful to discriminate the Z of ejectiles punching through the Si detector, and also to give isotopic identification of fragments with Z$\approx$10 or more; (2) the Time of Flight (ToF) method, that allows to obtain the mass of slow

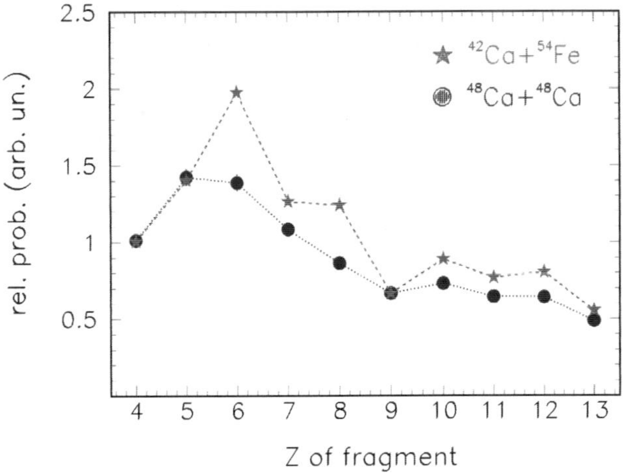

Figure 1. Charge distributions of light fragments emitted at $\theta_{Lab} = 11.5°$ and identified by means of the telescope technique in ^{48}Ca+^{48}Ca (N/Z=1.4, blue full dots), ^{42}Ca+^{54}Fe (N/Z=1.09, red stars) reactions at 25 MeV/nucleon. In both distributions the beryllium yields have been normalized to unit.

fragments stopped in Si detectors; (3) the fast-slow analysis of CsI(Tl) light output, allowing the isotopic identification of light particles, such as hydrogen and helium nuclei; (4) the study of the rise time of signals coming from silicon detectors, useful to determine the Z of stopped fragments. Further details about the Chimera array and its detection and identification capabilities can be found in Refs. [30, 32, 33]. Time resolution obtained in the performed experiment is $\approx 800ps$; it is due mostly to timing characteristics of the beam. The obtained mass resolution ($\frac{\Delta m}{m}$) is around 5% [6, 26] for nuclei having mass $A \approx 50$.

Profiting of the performances of the experimental apparatus, we consider in the data analysis only complete events, having at least the 80% the total charge and the 70% of the total parallel momentum. We impose moreover an upper constraint on the total detected charge: it must be less than the 100% of the total entrance charge. These selections allow to discard spurious events of reaction on oxygen contaminant in the target and pile-up events. Very peripheral quasi-elastic events were removed by the on-line electronic trigger by requiring the detection of at least 3 charged particles.

3. Even-odd staggering effects on Z distributions of light fragments

Charge distributions of light fragments emitted in heavy ion collisions often show the presence of even-odd staggering. This effect has been recognized in a large variety of bombarding energy domains, going from few MeV/nucleon (a regime dominated by Deep Inelastic Collisions and Compound nucleus phenomena) [16, 17, 34] up to 1000 MeV/nucleon (as seen in projectile fragmentation studies) [35]. In recent times, due to the increasingly improved identification power of the detectors, this subject has obtained a renewed interest, as seen in the literature [16, 18, 19, 20, 25].

Even-odd effects can be qualitatively explained by considering the last steps of de-excitation cascade involving excited pre-fragments formed during the dynamical phase of the reaction. The de-excitation cascade is correlated to the one-particle separation energy threshold, and this quantity shows strong even-odd oscillations due to pairing effects [15, 35, 16, 18].

It seems moreover that the amplitude of the even-odd effect is influenced by the N/Z of entrance channels [14, 16, 17]. In particular, by increasing the N/Z of the entrance channel, the amplitude of the even-odd oscillations are more and more softened. In Figure 1 we show the experimental Z distributions for light fragments emitted at $\theta_{Lab} = 11.5°$ and identified by means

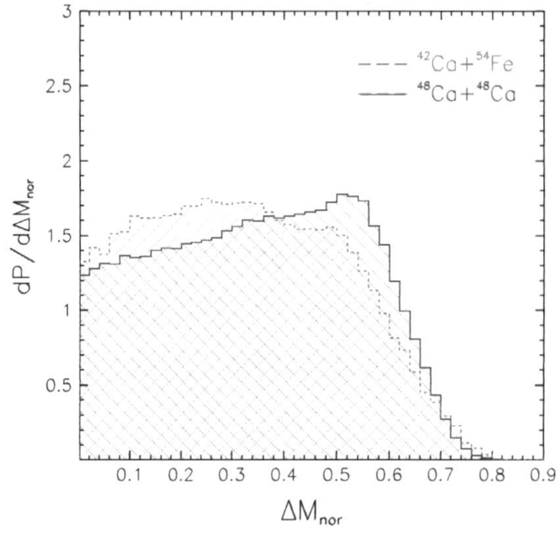

Figure 2. ΔM_{nor} distributions for selected central events of reaction in ^{48}Ca+^{48}Ca (blue solid line) and ^{42}Ca+^{54}Fe (red dashed line, *preliminary results*) systems at 25 MeV/nucleon.

of the telescope technique in ^{48}Ca+^{48}Ca (N/Z=1.4) and ^{42}Ca+^{54}Fe (N/Z=1.09) reactions. In both data sets, data have been matched at Z=4 (beryllium). We can observe that, even if the total mass of the two systems is the same, the amplitude of the even-odd oscillations are more softened for the neutron rich system. This effect can be qualitatively explained by considering that, for neutron rich systems, neutron rich pre-fragments would be formed in the first stage of reaction. These fragments will de-excite preferably by emitting neutrons, and therefore the proton number distributions will be less sensitive to the even-odd staggering of the particle emission threshold distribution. Moreover, for the neutron poor system (^{42}Ca+^{54}Fe) a quite large emission of carbon isotopes is seen, while the yields of boron and fluorine isotopes relatively to beryllium are quite similar for the two systems.

4. Semi-central events of reactions: heavy residue emission

The centrality of reaction events can be selected by using constraints on the charged particle multiplicity m_{cp} [36].

In order to investigate the competition between various reaction mechanisms that can be observed in central events of reaction, we firstly select the most central collisions by imposing a proper selection on the charged particle multiplicity m_{cp} [36]. As extensively discussed in [6, 26], two slightly different cut on m_{cp} should be used for systems with very large differences in the total N/Z to adequately take into account the presence of undetected neutrons. In our case we choose $m_{cp} \geq 5$ for the neutron rich system ^{48}Ca+^{48}Ca and $m_{cp} \geq 6$ for the neutron poor ^{42}Ca+^{54}Fe system. Moreover, we focused our attention on events where the second or third biggest fragment has a velocity larger than the center of mass one (v_2 or $v_3 \geq 0.13$ c). With this constraint, we focus our attention mainly on massive transfer phenomena where a part of the projectile fuses with the target.

For these events, we can discriminate the principal reaction mechanisms by investigating the correlation between the masses of the two biggest emitted fragments, m_1 and m_2. As discussed in details in Refs. [6, 26], we can define the ΔM_{nor} quantity as $\frac{m_1 - m_2}{m_{tot}}$, where m_{tot} is the total mass of the system. Large value of ΔM_{nor} are associated to heavy residue emission in fusion-like events, while low value of ΔM_{nor} are linked to binary-like, fusion-fission and multi-fragmentation events. In Figure 2 we show the experimental ΔM_{nor} distributions for the ^{48}Ca+^{48}Ca and ^{42}Ca+^{54}Fe systems here investigated. As seen in Figure 2, for the neutron rich system, ΔM_{nor} distributions are pushed towards higher values, while the opposite happens for the neutron poor one. It is interesting to note that the shape of the ΔM_{nor} distribution for the ^{42}Ca+^{54}Fe (N/Z=1.09) is similar to the one of ^{40}Ca+^{46}Ti (N/Z=1.05) as reported in [6]. Therefore, combining the results of the present analysis and of Refs. [6, 26] it seems that the relative yield of events leading to the formation of an heavy residue is larger for the more neutron rich systems, and the present effect can be mainly attributed to the N/Z degree of freedom and not to the mass asymmetry or the total mass of entrance channels. As discussed elsewhere [6, 26, 27], these effects can be explained by considering the interplay between the Coulomb and Symmetry terms of the Nuclear Equation of State. We plan to perform CoMD-II dynamical model calculations also for the ^{42}Ca+^{54}Fe data with the aim of refining the estimate of the stiffness of the Symmetry Energy for the Nuclear Equation of State at near-saturation densities [6, 26].

5. Conclusions

In this paper we report experimental results obtained by investigating ^{48}Ca+^{48}Ca (N/Z=1.4) and ^{42}Ca+^{54}Fe (N/Z=1.09) systems at 25 MeV/nucleon. Effects due to the isospin degree of freedom can be studied profiting of the good events reconstruction and isotopic identification given by the Chimera 4π array, that is used as detection device.

The analysis of Z-distributions of light fragments emitted at forward angles shows even-odd staggering effect that can be attributed to the last steps of the de-excitation cascade of excited pre-fragments. In this case, the amplitude of the even-odd effect is enhanced for the neutron poor system ^{42}Ca+^{54}Fe, while it is strongly suppressed in the very neutron rich ^{48}Ca+^{48}Ca system.

We find that, at semi-central impact parameters, the relative emission yield of heavy residues in fusion-like phenomena is enhanced for the neutron-rich system, while binary-like and multi-fragmentation events prevail for the neutron poor system. This effect has been recognized by studying the correlation between the masses of the two biggest emitted fragments.

As discussed in various theoretical papers in the literature, this effect can represent a useful probe to extract information about the density dependence of the Symmetry Energy of the Nuclear Equation od State at near-saturation densities. All these results stimulate moreover to continue the investigation on isospin effects in medium mass nuclear reactions by using radioactive beams or targets, in order to enlarge the range of N/Z of the formed systems.

References

[1] Bromley DA 1988, *Treatise on Heavy Ion Science*, (Plenum Press, NY), Vols. 2,3,8
[2] Morgenstern H et al 1984, *Phys. Rev. Lett.* 52 1104
[3] Lautesse P et al 2006, *Eur. Phys. Jour.* A 27 349
[4] Eudes P et al 2013, *J. Phys. Conf. Ser.* 420 012133
[5] De Filippo E et al 2012, *Phys. Rev.* C 86 014610
[6] Amorini F et al 2009, *Phys. Rev. Lett.* 102 112701
[7] Tsang MB et al 2004, *Phys. Rev. Lett.* 92 062701
[8] Galichet E et al 2009, *Phys. Rev.* C 79 064614
[9] Keksis AL et al 2010, *Phys. Rev.* C 81 054602
[10] Lombardo I et al 2010, *Phys. Rev.* C 82 014608
[11] Barlini S et al 2013, *Phys. Rev.* C 87 054607
[12] Botvina AS and Mishustin IN 2004, *Phys. Lett.* B 584 2332
[13] Carbone A et al 2010, *Phys. Rev.* C 81 041301(R)
[14] Yang LB et al 1999, *Phys. Rev.* C 60 041602
[15] Ricciardi MV et al 2004, *Nucl. Phys.* A 733 299
[16] D'Agostino M et al 2011, *Nucl. Phys.* A 861 47
[17] Ademard G et al 2011, *Phys. Rev.* C 83 054619
[18] Lombardo I et al 2011, *Phys. Rev.* C 84 024613
[19] Casini G et al 2012, *Phys. Rev.* C 86 011602
[20] Piantelli S et al 2013, *Phys. Rev.* C 88 064607

[21] Lombardo I 2012, *Eur. Phys. J. Plus* 127 88
[22] Pirrone S et al 2012, *this conference proceedings*
[23] Rivet MF et al 1986, *Phys. Rev.* C 34 1282
[24] Bootsma TMV et al 1997, *Zeit. Phys.* A 359 391
[25] Pirrone S et al 2001, *Phys. Rev.* C 64 024610
[26] Cardella G et al 2012, *Phys. Rev.* C 85 064609
[27] Papa M 2013, *Phys. Rev.* CC 87 014001
[28] Farine M et al 1989, *Zeit. Phys.* A 339 363
[29] Colonna M et al 1998, *Phys. Rev.* C 57 1410
[30] Pagano A et al 2004, *Nucl. Phys.* A 734 504
[31] Pagano A 2012, *Nucl. Phys. News* 22 1
[32] Le Neindre N et al 2002, *Nucl. Instrum. Methods* A 490 251
[33] Alderighi M et al 2002, *Nucl. Instrum. Methods* A 489 257
[34] Papa M et al 2005, *Phys. Rev.* C 72 064608
[35] Napolitani P et al 2007, *Phys. Rev.* C 76 064609
[36] Cavata C et al 1990, *Phys. Rev.* C 42 1760

Journal of Physics: Conference Series **515** (2014) 012014 doi:10.1088/1742-6596/515/1/012014

Bifurcations in dissipative fermionic dynamics

Paolo Napolitani[1], Maria Colonna[2] and Mariangela Di Prima[2]

[1]IPN, CNRS/IN2P3, Université Paris-Sud 11, 91406 Orsay cedex, France
[2]INFN-LNS, Laboratori Nazionali del Sud, 95123 Catania, Italy

E-mail: napolita@ipno.in2p3.fr

Abstract. The Boltzmann-Langevin One-Body model (BLOB), is a novel one-body transport approach, based on the solution of the Boltzmann-Langevin equation in three dimensions; it is used to handle large-amplitude phase-space fluctuations and has a broad applicability for dissipative fermionic dynamics. We study the occurrence of bifurcations in the dynamical trajectories describing heavy-ion collisions at Fermi energies.

The model, applied to dilute systems formed in such collisions, reveals to be closer to the observation than previous attempts to include a Langevin term in Boltzmann theories. The onset of bifurcations and bimodal behaviour in dynamical trajectories, determines the fragment-formation mechanism. In particular, in the proximity of a threshold, fluctuations between two energetically favourable mechanisms stand out, so that when evolving from the same entrance channel, a variety of exit channels is accessible.

This description gives quantitative indications about two threshold situations which characterise heavy-ion collisions at Fermi energies. First, the fusion-to-multifragmentation threshold in central collisions, where the system either reverts to a compact shape, or splits into several pieces of similar sizes. Second, the transition from binary mechanisms to neck fragmentation (in general, ternary channels), in peripheral collisions.

1. Introduction

The general context of this study is describing the dynamics of fermionic systems in presence of instabilities which produce fluctuations of so large amplitude to produce bifurcations in the dynamical evolution. In the case of dissipative heavy-ion collisions, one given projectile-target system evolving from a given entrance channel, defined by an impact parameter and an incident energy, may produce a variety of exit channels. For instance, when approaching Fermi energies, heavy-ion collisions may oscillate between fusion and multifragmentation for small impact parameters, or between binary and ternary mechanisms for large impact parameters (fig. 1). To describe such a chaotic behaviour, dominated by fluctuations and bifurcations, pure mean-field equations should be replaced by more adapted theories, like stochastic approaches [1].

Fluctuations are the effect of the correlations which a many-body system generates. To take them into account there exist two main lanes. Either molecular-dynamics approaches, developed from a two-body Hamiltonian. Or a hierarchy of N-body contributions, which can be reduced to a mean-field description supplemented by a residual interaction. In order to keep an efficient description of effects like spinodal instabilities (relevant in central collisions) and isospin transport [2] (like migration toward a neck region, relevant in peripheral collisions), we follow this second approach. In this framework, and in a general quantum description, large-

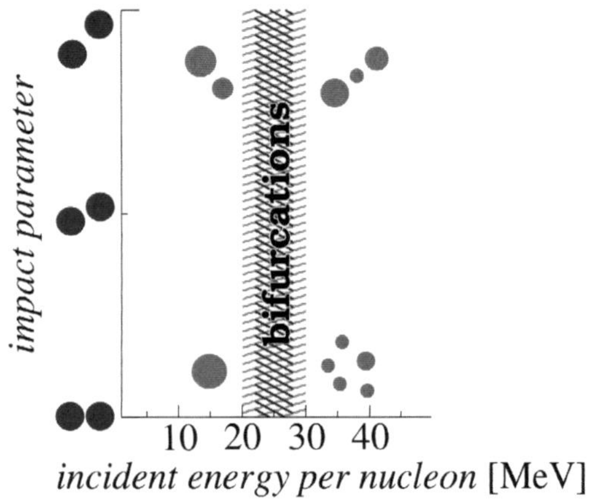

Figure 1. Schematic illustration of a variety of exit channels as a function of different entrance channels (impact parameter and incident energy per nucleon): there are regions where the system oscillates between more possible configurations for one given entrance channel.

amplitude fluctuations in dynamical trajectories may be obtained by defining subensembles of Slater states, each one presenting small fluctuations about its corresponding mean-field [3]. We adopt a semiclassical analogue of this picture, the Boltzmann-Langevin (BL) transport equation, where the residual interaction carries the unknown N-body correlations and is written in terms of the one-body distribution function as

$$\partial_t f - \{H[f], f\} = \bar{I}[f] + \delta I[f] \,, \tag{1}$$

so that the left-hand side gives the Vlasov evolution for the distribution function f in its own self-consistent mean field, and the right-hand side introduces the residual interaction, containing the average Boltzmann hard two-body collision integral $\bar{I}[f]$ and the fluctuating term $\delta I[f]$. This form indicates that the residual interaction, which carries the unknown N-body correlations is written in terms of the one-body distribution function. The fluctuation term is of Markovian type and it acts as a dissipating force while preserving single-particle energies; it is related to the collision integral through the fluctuation-dissipation theorem in analogy to the Langevin description of the Brownian motion. The effect of the fluctuation term on the dynamical trajectories is the possible appearing at any time, whenever the system presents instabilities, of bifurcation branches which propagate in phase space [4] (fig. 2).

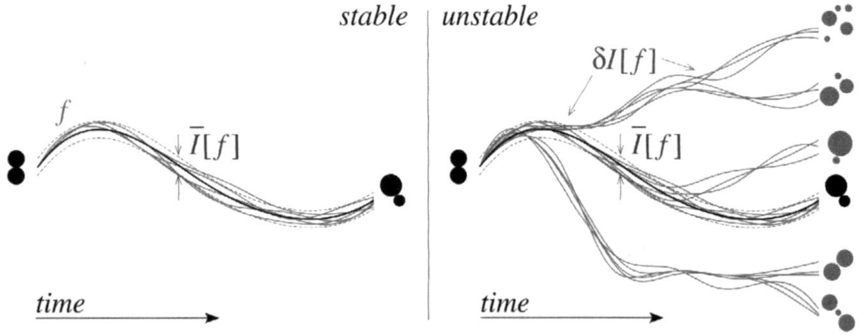

Figure 2. Schematic illustration of the temporal evolution of the one-body distribution function in absence (left) and in presence (right) of fluctuation.

2. Solution of the Boltzmann-Langevin equation for fermionic systems

The solution of the BL equation in full phase space is the aim of the Boltzmann-Langevin One-Body model (BLOB) [5]. It is obtained by replacing the conventional Uehling-Uhlenbeck average collision integral by a similar form where one binary collision does not act on two test particles a, b but it rather involves extended phase-space agglomerates of test particles of equal isospin A$= a_1, a_2, \ldots$, B$= b_1, b_2, \ldots$ (leading to the final states C$= c_1, c_2, \ldots$, D$= d_1, d_2, \ldots$) to simulate wave packets:

$$\begin{aligned} \bar{I}[f] + \delta I[f] &= g \int \frac{\mathrm{d}\mathbf{p}_b}{h^3} \int \mathrm{d}\Omega \; W(\mathrm{AB}\leftrightarrow\mathrm{CD}) \; F(\mathrm{AB}\rightarrow\mathrm{CD}) \qquad (2) \\ &= g \int \frac{\mathrm{d}\mathbf{p}_b}{h^3} \int \mathrm{d}\Omega \; \left\langle |v_a - v_b| \frac{\mathrm{d}\sigma}{\mathrm{d}\Omega} \right\rangle_\Sigma \left[(1-f_\mathrm{A})(1-f_\mathrm{B}) f_\mathrm{C} f_\mathrm{D} - f_\mathrm{A} f_\mathrm{B} (1-f_\mathrm{C})(1-f_\mathrm{D}) \right], \end{aligned}$$

At each interval of time all phase space is scanned for collisions and all test-particle agglomerates are redefined in phase-space cells of volume h^3. The above procedure introduces correlations which are then exploited through a stochastic procedure, where the effective collision probability $W \times F$ is confronted with a random number. As a consequence, fluctuations develop spontaneously in the phase-space cells of volume h^3 with the correct fluctuation amplitude, determined by a variance which at equilibrium is equal to $f(1-f)$ if the agglomerates contain a number of test particles N_{test} equal to the total number of test particles divided by the number of nucleons constituting the system [6]. Since N_{test} test particles are involved in one collision, and since those test particles could be sorted again in new agglomerates to attempt new collisions in the same interval of time as far as the collision is not successful, the transition rate W in eq.1 should contain a cross section equal to the nucleon-nucleon cross section divided by N_{test}: $\sigma = \sigma_{NN}/N_{\mathrm{test}}$. The transition rate $W(\mathrm{AB}\leftrightarrow\mathrm{CD})$ is the average of the elementary transition rates $W(\mathrm{ab}\leftrightarrow\mathrm{cd})$ over the ensemble Σ of all the couples of test particles belonging to the agglomerates A and B.

The BLOB model applies a precise shape-modulation technique [7] which ensures that the occupancy distribution does not exceed unity in any phase-space point in the final states; this leads to a correct Fermi statistics for the distribution function f, in term of mean value and variance. The main constraint of the above procedure is to impose a phase-space metric characterised by the phase-space cells of volume h^3, but the metric in momentum and coordinate space are unconstrained in the present approach except for imposing the maximum compactness for the agglomerates in momentum space which does neither violate Pauli blocking nor energy conservation. Such strong localisation in momentum space makes the collisions more effective in agitating the phase space. However, beyond the present application to Fermi energies, further attention should be paid to the compactness of the wave packets also in coordinate space when dealing with effects like collective flow and stopping, which at intermediate energy become relevant.

3. Mean-field response in presence of instabilities

If we look for instabilities, we find regions of the equation of state (EoS), like the spinodal region, where the incompressibility $\chi^{-1} = \rho\,\partial P/\partial\rho$ is negative. In such a situation we can explore unstable modes of wave number k growing with a characteristic time τ_k, in relation with the form of the mean-field potential [8].

In a linear-response regime we can write, for a three-dimensional system, an analytical expression of the dispersion relation [9] which connects the growth rate τ_k^{-1} to the form of the mean-field potential as $\tau_k^{-1} = f(k, \chi, \rho', T, \sigma)$, where a temperature of $T = 3$MeV and a density of $\rho' = 0.05$ fm $^{-3}$ (to be compared to the saturation density of 0.16 fm^{-3}) are chosen in correspondence with a mechanically unstable situation; the parameter χ is related to the finite range of the interaction (i.e. the width of the gaussian smearing of the mean-field potential is let range from 0.8 to 0.9 fm, which is consistent with the width of the triangular functions used in the numerical calculation). Such analytical calculation is represented by the band in fig. 3, and it is closely comparable to the result of a transport calculation corresponding to the same mechanically unstable situation.

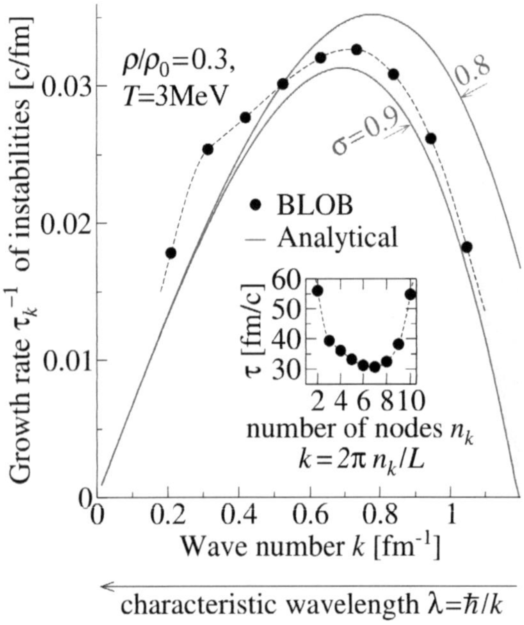

Figure 3. Dispersion relation: comparison between an analytical calculation in a linear-response regime (the band) and a corresponding numerical calculation (dots) in a periodic box of size L, for a mechanically unstable situation like $T = 3$MeV and $\rho/\rho_0 = 0.05\text{fm}^{-3}/0.16\text{fm}^{-3}$. The inset gives the evolution of the characteristic time as a function of the number of nodes.

The transport calculation is done in a periodic cubic box of edge size $L = 60$fm, letting the different modes k develop spontaneously. The linear response is calculated tracking the evolution of the density profile along one axis and averaging along the other two transverse coordinates. The collision term is let agitate the density profile in an initially uniform system over several wavelengths: from the analysis of this evolution it was deduced how the mean field amplifies each k wave. The calculation gives the set of points illustrated in the inset of fig. 3, where the characteristic time is shown as a function of the number of nodes of the wave. From these points, the evolution of the growth rate of the instabilities as a function of the wave number is extracted. As a consequence of the fact that this numerical approach introduces a coupling among different k values, a shoulder is produced in the spectrum for small k values from the combination of small wavelengths (large k) into larger wavelengths (small k). On the other hand, the analytical calculation does not show any shoulder because, of course, all wavelengths are decoupled.

Around this efficient description of the dispersion relation, the BLOB model is built and it can therefore be applied to the description of the spinodal behaviour. In this respect, when a nuclear system is brought to small densities (and, correspondingly, large excitation), if spinodal conditions are attained, the system tends to break into fragments with a size $A_{\text{fragm}} \approx \lambda_0^3\rho_0$, determined by the maximum unstable mode k_0 and the corresponding characteristic wavelength $\lambda_0 = \hbar/k_0$.

4. Instability growth versus mean-field resilience

We apply the above phenomenology to the situation of dissipative heavy-ion collisions. For convenience (because experimental data exist, measured by the INDRA collaboration [10]), we simulate the reaction ^{136}Xe+^{124}Sn at 25 and 32AMeV with a central impact parameter: in this situation the system is close to a threshold between fusion and multifragmentation (below for 25AMeV, above for 32AMeV). The BLOB transport model is employed with the prescription of a soft EoS (with $K_{\rm inf}$ = 200MeV), a linear asy-EoS, an in-medium cross section as prescribed in ref. [11], with $N_{\rm test}$ = 40, and with a statistics of 600 events. The dynamics is followed for a time of 300 fm/c.

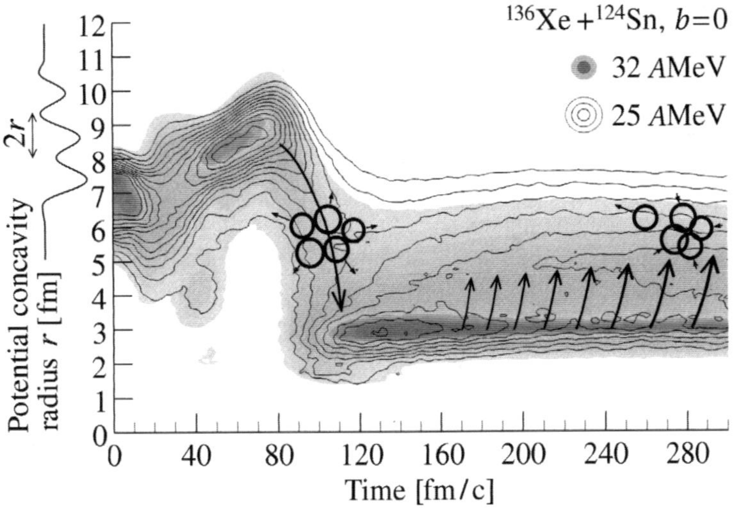

Figure 4. Calculation (BLOB) of the evolution of the size of inhomogeneities developing in the potential landscape as a function of time for the reaction ^{136}Xe+^{124}Sn at 32AMeV (colours) and 25AMeV (contours) for a central impact parameter.

Fig. 4 shows the evolution of the size of inhomogeneities which develop in the potential landscape as a function of time for the two incident energies 25 and 32AMeV. These inhomogeneities correspond to blobs of matter which, if the dynamics is sufficiently explosive, may eventually leave the system as emitted fragments. At short times, as discussed in the previous section, spinodal instabilities tend to split the system into several fragments of comparable size $A_{\rm fragm} \approx \lambda_0^3 \rho_0$, corresponding approximately to the region of Neon. Later on, this process enters in competition with a process of partial coalescence determined by the mean-field resilience, which tends to revert the system to a compact shape, and which is more effective for the lowest incident energy.

This picture results in a large variety of exit channels for one given macroscopic initial condition (entrance channel), ranging from fusion to multifragmentation and passing through very asymmetric configurations with a small fragment multiplicity, eventually binary, and a large size asymmetry. Such intermediate configurations may recall asymmetric fission with the difference that the distribution of momentum transfer could reach large values more compatible with multifragmentation.

5. Fusion-multifragmentation competition in central collisions

The distribution of different configurations for the exit channel of a given system can be represented by the charge asymmetry $(Z_1 - Z_2)(Z_1 + Z_2)$ between the fragment of largest charge Z_1 and the fragment of second largest charge Z_2. If the same calculation of fig. 4 is repeated for a range of different incident energies we obtain the distribution of exit channels illustrated in fig. 5. We observe that for small incident energies the fusion channel is favoured, while for larger incident energies multifragmentation dominates. In between, a region exist where dynamical trajectories show bifurcations between these two competing channels.

This dynamical picture, where energy fluctuations result from the same macroscopic initial conditions and lead to oscillations between two energetically favoured configurations, provides

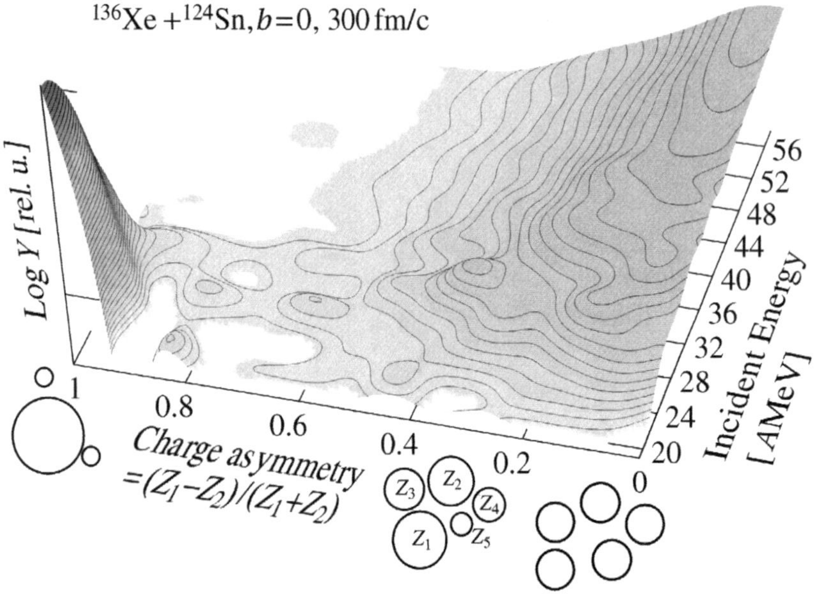

Figure 5. Distribution of exit channels, represented by the charge asymmetry $(Z_1 - Z_2)(Z_1 + Z_2)$ as a function of the incident energy for the reaction ^{136}Xe+^{124}Sn for a central impact parameter.

a link to the thermodynamic picture of a first-order phase transition. We also mention that very close experimental observations have found bimodalities in fragment observables in heavy-ion collisions at Fermi energies from the analysis of peripheral impact parameters [12, 13]. The present theoretical study encourages to extend experimental investigations of bimodal behaviour in fragment observables also to central collisions.

6. Overview on a landscape of exit channels

An even more general survey extends the set of calculations also to the full range of impact parameters, and completes the dynamical calculation with a further decay process of the hot fragments through a statistical evaporation model [14]. We still keep the same reaction ^{136}Xe+^{124}Sn at 32AMeV but we additionally vary the impact parameter b.

The left panel of fig. 6 is still restricted to central collisions and shows the evolution of the multiplicity of fragments larger than Be as a function of the incident energy (coloured distribution with the average marked by a solid line). The distribution of the fragments resulting from the evaporation process reaches a maximum above about 35AMeV and a fall above 50AMeV due to the more and more reduced size of the fragments. The right panel of fig. 6 shows the average dependence of the fragment multiplicity as a function of the impact parameter. We observe that the maximum corresponds to central collisions for large incident energies because the multifragmentation channel dominates the reaction cross section. For smaller incident energies the maximum moves to more peripheral collisions and it is reduced to events with three or four fragments: this indicates that around 30AMeV and for semiperipheral impact parameters ($\sim$ 6fm) the system oscillates between binary mechanisms and ternary splits, where a third smaller fragment is formed in the neck region.

7. Conclusions

With the purpose of studying the link between a given entrance channel and a variety of possible exit channels in situations where a nuclear system experiences instabilities, we built a one-body

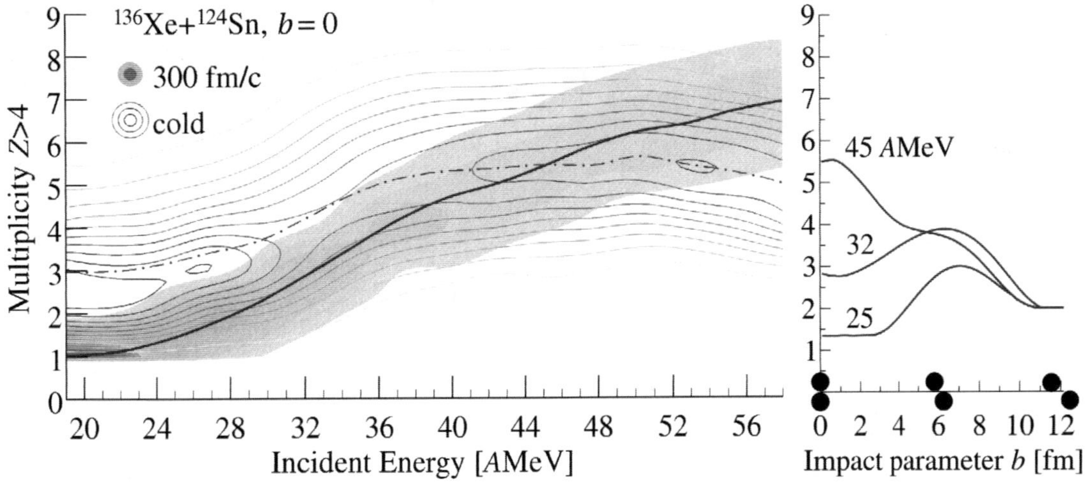

Figure 6. Multiplicity of fragments larger than Be in the reaction ^{136}Xe+^{124}Sn as a function of the incident energy for a central impact parameter (left panel) and as a function of the impact parameter for 25, 32 and 45 *A*MeV (right panel).

description based on the solution of the Boltzmann-Langevin equation in three dimensions. The method is able to introduce fluctuations of correct amplitude in terms of dispersion relation.

Thanks to this transport model we could draw a dynamical description of the origin of trajectory bifurcations and bimodality in head-on collisions down to the multifragmentation threshold. In particular, we observe that this phenomenology results from the competition between instability growth and the resilience of the mean field. Within the same picture we might also explain asymmetric binary channels. Analysing the variation of the exit channel with the impact parameter it is also possible to determine the most favourable conditions for producing fragments from the neck regions in peripheral collisions.

References

[1] Reinhard P-G and Suraud E 1992 *Ann. Phys.* **216** 98
[2] Baran V, Colonna M, Greco V, Di Toro M 2005 *Phys. Rep.* **410** 335
[3] Balian R, Alhassid Y and Reinhardt H 1986 *Phys. Rep.* **131** 1
[4] Randrup J, Burgio G F and Chomaz Ph 1992 *Nucl. Phys.* A **538** 393c, Ono A and Randrup J 2006 *EPJ* A **30** 109
[5] Napolitani P and Colonna M 2013 *Phys. Lett.* B **726** 382
[6] Rizzo J, Chomaz Ph and Colonna M 2008 *Nucl. Phys.* A **806** 40
[7] Napolitani P and Colonna M 2012 *EPJ Web of Conf.* **31**, 00027 (IWM2011)
[8] Chomaz Ph, Colonna M and Randrup J, 2004 *Phys. Rep.* **389** 263
[9] Colonna M and Chomaz Ph 1994 *Phys. Rev* C **49** 1908
[10] Gagnon-Moisan F et al. (INDRA coll.) 2012 *Phys. Rev* C **86** 044617
[11] Danielewicz P 2002 *Acta Phys. Polon.* **B33** 45, Coupland D D S, Lynch W G, Tsang M B, Danielewicz P and Zhang Y 2011 *Phys. Rev* C **84** 054603
[12] Bonnet E et al. (INDRA coll.) 2009 *Phys. Rev. Lett.* **103** 072701
[13] Pichon M et al. (INDRA coll.) 2006 *Nucl. Phys.* A **779** 267
[14] Durand D, 1992 *Nucl. Phys.* A **541** 266

Difference in evaporation residue yields in the cold and hot fusion reactions

Avazbek Nasirov[1,2], Giorgio Giardina[3,4,5], Giuseppe Mandaglio[3,4,5], Kyungil Kim[6] and Youngman Kim[6]

[1]BLTP, Joint Institute for Nuclear Research, Joliot-Curie 6, 141980 Dubna, Russia
[2]Institute of Nuclear Physics, Ulugbek, 100214, Tashkent, Uzbekistan
[3]Dipartimento di Fisica e di Scienze della Terra dell' Università di Messina, Salita Sperone 31, 98166 Messina, Italy
[4]Istituto Nazionale di Fisica Nucleare, Sezione di Catania, Italy
[5]Centro Siciliano di Fisica Nucleare e Struttura della Materia 95125 Catania, Italy
[6]Rare Isotope Science Project, Institute for Basic Science, Daejeon 305-811, Republic of Korea

E-mail: nasirov@jinr.ru

Abstract. An interest to study the mechanism of the complete fusion with massive nucleus reactions was caused by difficulties of synthesizing superheavy elements due to their small cross sections. Theoretical models used to obtain knowledge about fusion mechanism in heavy ion collisions at low energies are requested. We present the theoretical results of the compound nucleus (CN) formation and evaporation residue (ER) cross sections obtained for the ^{48}Ca+^{248}Cm and ^{58}Fe+^{232}Th reactions leading to the formation of CN with $A = 296$ and $A = 290$ of the superheavy element Lv ($Z = 116$), respectively. The ER cross sections, which can be measured directly, are determined by the complete fusion and survival probabilities of the heated and rotating CN. Those probabilities can not be measured unambiguously but the knowledge about them is important to study the formation mechanism of the observed products and to estimate the ER cross sections of the expected isotopes of elements. For this aim, the ^{48}Ca+^{249}Cf and ^{64}Ni+^{232}Th reactions are also considered. The survival probability of the heated and rotating CN is sensitive to the used mass and shell correction values of superheavy nuclei. The ER cross sections obtained by using the macroscopic-microscopic model with mass and fission barrier values calculated by Warsaw group lead to smaller ER values for all reactions in comparison with the case when the masses and shell correction calculated by P. Möller *et al* are used.

1. Introduction

In recent decades, interest in nuclear physics has increased due to observation of events in the synthesis of new superheavy elements (SHE) with the atomic numbers $Z = 114$—118 [1,2]. The achievements in hot fusion reactions led to the discovery of new elements Fl ($Z = 114$) and Lv ($Z = 116$) after those results have been confirmed in other similar experiments. Isotopes ^{286,287} Fl were obtained in the reaction ^{48}Ca with ^{242}Pu in an experiment at Berkeley gas-filled separator Lawrence National Laboratory (LBNL) [3] while isotopes $^{288-291}$Fl were obtained in the bombing of ^{244}Pu with ^{48}Ca at the recoil separator TASCA experiment of GSI [4] in Darmstadt. Events corresponding to the synthesis of isotopes ^{292,293}Lv were observed in the ^{48}Ca+^{248}Cm reaction by the velocity filter SHIP at GSI [5]. Two reasons have contributed to the success of the synthesis of superheavy elements with $Z = 113$—118: 1) large cross sections of the compound nucleus formation for hot fusion reactions and 2) increase of the fission barrier as they approach to the island of stability [6,7] (see Fig. 1). As Fig. 1 shows, the synthesis cross sections of SHE with $Z = 104$—111 in cold fusion are sufficiently larger than ones obtained in the hot fusion reactions. Starting from SHE $Z = 112$ the ER cross sections measured in the hot fusion become larger. The possibility of cold fusion have been exhausted for the synthesis of SHE with $Z \geq 114$. Experimental results showed that the ER cross section ($\sigma_{\rm ER}$) in cold fusion reactions fall sharply with increasing the electric charge of the formed compound nucleus [8] (see Fig. 1). The last SHE synthesized in cold fusion is Z=113, which was obtained at RIKEN (Japan) by using the ^{70}Zn + ^{209}Bi reaction. After repeated and prolonged experiments reached the lowest value $\sigma_{\rm ER}$ equal to 22 femtobarn (fb) [9], (1fb=10^{-15} barn).

The name cold fusion is associated with a low value of the minimal excitation energy ($E^*_{\rm CN} <$ 20 MeV) of the compound nucleus. The smallness of the excitation energy $E^*_{\rm CN} = E_{\rm c.m.} + Q_{\rm gg}$ follows from the large negative values of $Q_{\rm gg}$, which is the reaction energy balance is: $Q_{\rm gg} = B_1 + B_2 - B_{\rm CN}$. The binding energies of the initial projectile and target nuclei (B_1 and B_2) are obtained from the mass tables in Ref. [10], while the one of CN ($B_{\rm CN}$) is obtained from the theoretical mass tables [11, 12]. The use of nuclear binding energies including shell effects in calculations of the potential energy surface (PES) (see Fig. 2) and driving potential U$_{dr}$ of DNS lead to the appearance of hollows on the PES around the charge and mass values corresponding to the constituents of DNS with the magic proton or/and neutron numbers (see Figs. 3 and 4).

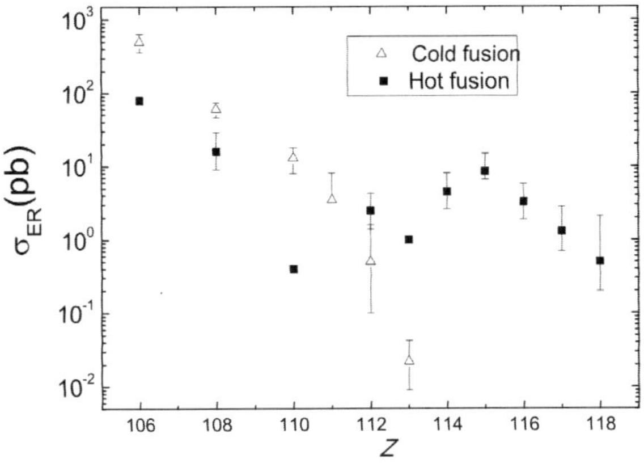

Figure 1. The maximum values of the excitation functions measured in synthesis of the superheavy elements (Z) in the cold (opening triangles) [8] and hot (full squares) fusion [1, 2]reactions.

2. Outline of the approach

The ER formation process is the last stage of the reaction mechanism in heavy ion collisions at $E_{\rm c.m.}$ energies near the Coulomb barrier:

$$\sigma_{\rm ER}(E^*_{\rm CN}) = \sigma_{\rm cap}(E_{\rm c.m.}) P_{\rm CN}(E^*_{\rm DNS}) W_{\rm sur}(E^*_{\rm CN}), \quad (1)$$

where $\sigma_{\rm cap}(E_{\rm c.m.})$ is the capture cross section, $P_{\rm CN}(E^*_{\rm DNS})$ is the complete fusion probability and $W_{\rm sur}(E^*_{\rm CN})$ is the survival probability of the heated and rotating CN against fission; $E^*_{\rm DNS} = E_{\rm c.m.} - V_{\rm min}$ is the excitation energy of DNS which is determined by an amount of the dissipated part of the relative kinetic energy; $V_{\rm min}$ is the minimum of the potential well where DNS has been trapped at capture.

Our experience of calculations shows that the capture probability is larger for the hot fusion reactions because the size of the potential well is larger due to the weaker Coulomb repulsion

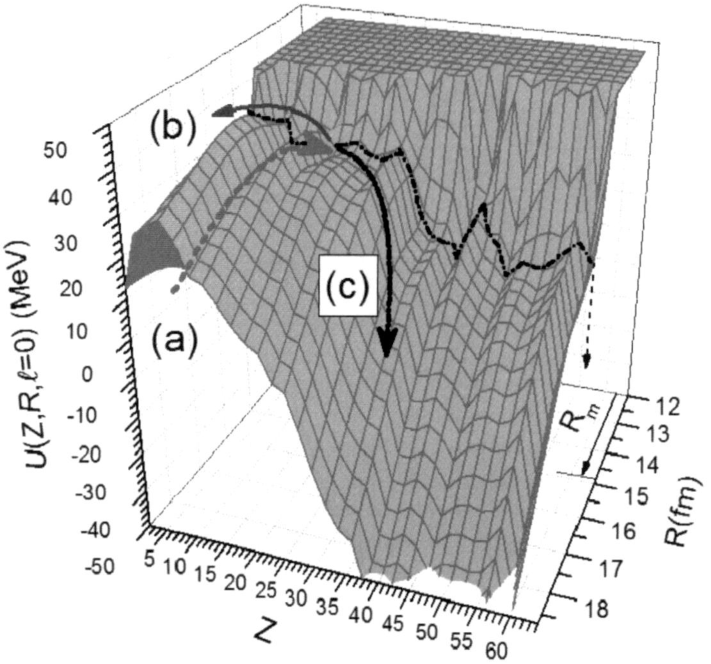

Figure 2. (Color online) Potential energy surface calculated for the DNS leading to formation of the ^{296}116 CN as a function of the relative distance between the centers of mass of interacting nuclei and charge number of a fragment. The capture stage path for the ^{48}Ca+^{248}Cm reaction is shown by arrow (a) and complete fusion by multinucleon transfer occurs (b) if DNS overcomes intrinsic fusion barrier. Arrow (c) shows one of possibilities of DNS decay into two fragments without reaching complete fusion stage (quasifission process). The broken dot-dashed line corresponds to the driving potential $U(Z, R_m)$ which is determined by the minimum values of the potential wells for each charge value Z. R_m is the position of the minimum value of interaction potential on the relative distance R.

in comparison with the one in cold fusion reaction: $(Z_1 \times Z_2)_{\rm hot} < (Z_1 \times Z_2)_{\rm cold}$ since the hot fusion reactants are more charge asymmetric than the ones of cold fusion.

A huge hindrance to transformation of the DNS, which is formed at capture, appears in the cold fusion reactions. The nature of this phenomenon is connected by the PES ($U(Z, R)$) landscape which is used to estimate the complete fusion probability in competition with quasifission. In Fig. 2, the capture stage path is shown by arrow (a) and complete fusion by multinucleon transfer occurs (b) if system overcomes intrinsic fusion barrier. Arrow (c) shows one of possibilities of the DNS quasifission from its more charge symmetric configurations.

PES is calculated as a sum of the reaction energy balance ($Q_{\rm gg}$) and the nucleus-nucleus potential ($V(R)$) between interacting nuclei:

$$U(Z, A, \ell, R) = Q_{\rm gg} + V(Z, A, \ell, R), \quad (2)$$

where $Z = Z_1$ and $A = A_1$ are charge and mass numbers of a nucleus of the DNS while the ones of another nucleus are $Z_2 = Z_{\rm tot} - Z$ and $A_2 = A_{\rm tot} - A$, where $Z_{\rm tot}$ and $A_{\rm tot}$ are the total charge and mass numbers of a reaction, respectively.

The driving potential $U_{\rm dr}(Z, R_m)$ is shown by the broken dot-dashed line in Fig. 2 and it is determined by the minimum values of the potential wells for each charge value Z. The position of the minimum value of interaction potential on the relative distance is denoted as R_m. The values of $U_{\rm dr}(Z, R_m)$ as a function of angular momentum ℓ are found from the data of PES calculated by formula

$$U_{\rm dr}(Z, A, \ell, R_m) = Q_{\rm gg} + V(Z, A, \ell, R_m). \quad (3)$$

If there is no potential well of $V(Z, A, \ell, R)$ at a given value of angular momentum or for symmetric massive nuclei, we use R_m corresponding to the smallest value of the derivation $|\partial V(Z, A, \ell, R_m)/\partial R|$ in the contact area of nuclei.

In Fig. 3 we compare the driving potentials calculated for the systems formed in the ^{48}Ca+^{248}Cm and ^{58}Fe+^{232}Th reactions which can lead to the CN formation with ^{296}Lv and ^{290}Lv, respectively, as a function of the fragment charge number. PES and driving potential U_{dr} are functions of the orientation angles α_1 and α_2 of axial symmetry axes of interacting nuclei. The results reported in figures 5-7, 14 and 15 have been obtained for the orientation angles $\alpha_1 = 45°$ and $\alpha_2 = 60°$.

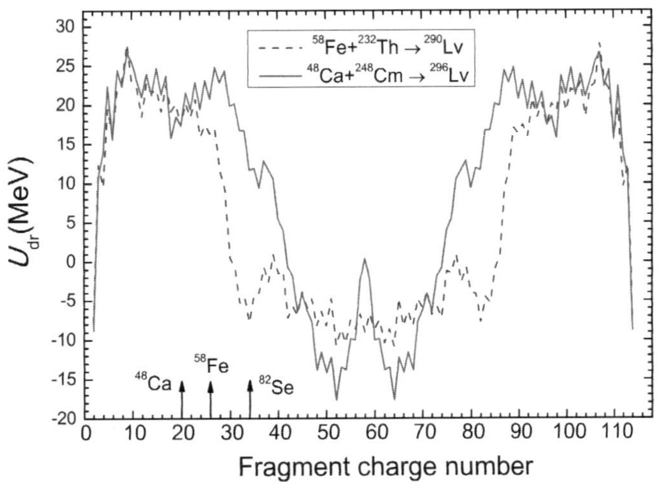

Figure 3. (Color online) Comparison of the driving potentials calculated for the systems formed in the ^{48}Ca+^{248}Cm and ^{58}Fe+^{232}Th reactions which can lead to formation of isotopes A=296 and 290 of the superheavy element Lv (Z=116), respectively, as a function of a fragment charge number.

It is seen from Fig. 3 that the driving potential decreases abruptly for the fragment with charge number larger than Z =30 for both the ^{48}Ca+^{248}Cm and ^{58}Fe + ^{232}Th reactions. If the value of the driving potential corresponding to the entrance channel is very low in respect to its maximum value in the fusion direction $Z \to 0$, the intrinsic fusion barrier $B^*_{\rm fus}$ becomes larger and the hindrance to complete fusion will be very strong. The method of calculating $B^*_{\rm fus}(Z, A, \ell)$ is shown in Fig. 4. It is determined as a difference between the maximum value of the driving potential between $Z = 0$ and $Z = Z_P$ and its value corresponding to the initial charge value:

$$B^*_{\rm fus} = U^{\rm max}_{\rm dr} - U_{\rm dr}(Z_P), \quad (4)$$

where is $U^{\rm max}_{\rm dr} = U_{\rm dr}(Z = 9)$; while are $Z_P = 20$ and $Z_P = 26$ for the reactions with ^{48}Ca and ^{58}Fe, respectively. The values of $B^*_{\rm fus}$ are equal to 8.5 MeV and 11.5 MeV for the ^{48}Ca+^{248}Cm and ^{58}Fe+^{232}Th reactions, respectively (see Fig. 3). If we consider for example the ^{82}Se+^{208}Pb

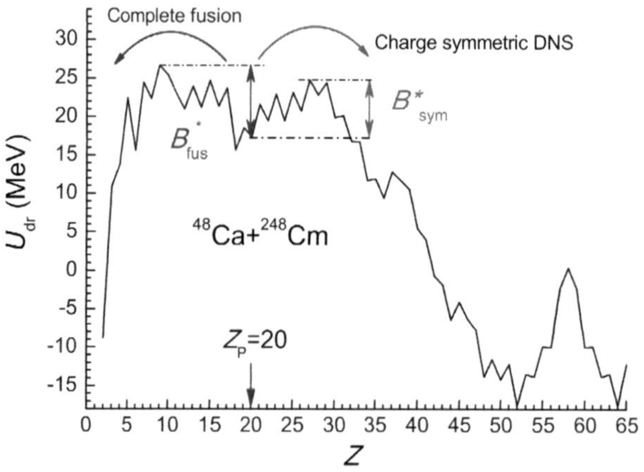

Figure 4. (Color online) Intrinsic fusion barrier $B^*_{\rm fus}$ and intrinsic barrier $B^*_{\rm sym}$ against to symmetry configurations from the initial charge asymmetry Z=20 on the driving potential (solid line) for the ^{48}Ca+^{248}Cm reaction.

$\rightarrow^{290}$Lv cold fusion reaction where Z_p is equal to 34, such a reaction is strongly hopeless of reaching complete fusion since $B^*_{\rm fus}$=34 MeV (please check if this value is correct).

The hindrance to evolution of DNS in direction of the symmetric charge distributions is determined by the barrier $B^*_{\rm sym}$ which is determined by the similar way as in the case of $B^*_{\rm fus}$ but the maximum value of the driving potential from symmetric charge region is used (see Fig. 4).

The study of dynamics of the heavy ion collisions near the Coulomb barrier energies showed that complete fusion does not occur immediately in the case of massive nucleus collisions [13–17]. In Ref. [18], authors estimated the excitation functions of the ER formation in the ^{50}Ti+^{249}Bk and ^{50}Ti+^{249,252}Cf reactions by using a newly developed DNS model with a dynamical potential energy surface. The authors concluded that taking into account dynamical deformation of DNS fragments, the calculation by this model leads to a decrease of the fusion probability and, consequently, to a decrease of the ER formation cross section.

The evolution of DNS may end with the CN formation with excitation energy $E^*_{\rm CN}$ in the statistically equilibrated state or it may lead to formation of two fragments without reaching the CN state. The latter process is quasifission. It takes place when DNS prefers to break down into two fragments instead of being transformed into fully equilibrated CN. The quasifission products may have characteristics similar to ones of fusion-fission products. For example, the total kinetic energy of the quasifission fragments is close to that of fission fragments. The mass and angular distributions of the fragments depend on the entrance channel properties and may overlap, causing difficulties in identification of a mechanism which produces the corresponding reaction products [19, 20]. We should stress the dominant role of the quasifission channel in the cold fusion reactions that causes strong hindrance to the formation of CN during the evolution of DNS.

The mononucleus survived to quasifission can reach the equilibrium shape, and it may be transformed into CN. In this case, it is important to know how large is the angular momentum $L_{\rm MN}$ of mononucleus for its transformation to the equilibrated state of CN since the fission barrier B_f of a being formed CN depends on its angular momentum. The macroscopic fission barrier $B_f(L)$ of the rotating nucleus disappears at large values of $L_{\rm MN} > L_f$ [21] and the mononucleus undergoes to fast fission leading to formation of two fragments similar to that from fusion-fission or quasifission. So, fast fission is splitting of mononucleus before reaching equilibrated CN if $L_{\rm CN} > L_f$. We remain that quasifission process can occur at all values of $L_{\rm DNS}$ of DNS. This is one of main differences between fast fission and quasifission; the other difference is that the fast fission process is the decay of the non-equilibrated complete fusion system, while the quasifission process is the decay of the excited DNS without to reach the stage of complete fusion.

The heated and rotating CN must survive against fission to be registered by detector as the recoil nucleus by emission of mainly neutrons, protons, α-particles, and gamma-quanta. The fusion-fission channel can be enriched by the fast fission products and both processes are characterized by the properties of the fission barrier described by the liquid drop model for the intermediate mass and heavy nuclei. For the very massive nuclei the properties of fission barrier is connected by the shell corrections which is determined by the microscopic models. This last contribution is decisive for the stability of very heavy atomic nuclei with charge number $Z > 106$ against fission because the contribution to the fission barrier due to the liquid drop model is disappeared. So, the possible existence of some value for the fission barrier may be due to the contribution of shell effects in the binding energy [22], which is sensitive to the L and $E^*_{\rm CN}$ values. Therefore, fission barrier of nucleus related with both nature (liquid drop and shell effect) disappears at large values of angular momentum L [21] and when shell effects have been completely damped [23] as a function of the excitation energy and angular momentum. Consequently, a mononucleus, which has survived against quasifission, can not reach CN equilibrium shape and suffers fast fission. Finally, the CN, which survived against fission by emission of particles during its cooling, forms the ER which can be registered as a superheavy element.

The main scope of this work is to reproduce the measured data for the superheavy elements with $Z = 116$ and $Z = 118$ formed in the ^{48}Ca+^{248}Cm and ^{48}Ca+^{249}Cf reactions, respectively, and to make predictions for $\sigma_{\rm ER}$ in the ^{58}Fe+^{232}Th and ^{64}Ni+^{232}Th reactions which can be used in future experiments.

2.1. Capture and fusion cross section in collisions of deformed nuclei

The final results of the partial capture and complete fusion cross sections are obtained by averaging the contributions from the different orientation angles α_1 and α_2 (relatively to the beam direction) of the projectile and target nuclei, respectively [17]:

$$< \sigma_{\rm fus}(E_{c.m.}, \ell) >= \int_0^{\pi/2} \sin\alpha_1 \int_0^{\pi/2} \sigma_{\rm fus}(E_{\rm c.m.}, \ell; \alpha_1, \alpha_2) \sin\alpha_2 d\alpha_1 d\alpha_2. \qquad (5)$$

The partial fusion cross section $\sigma_{\rm fus}(E_{\rm c.m.}, \ell; \alpha_1, \alpha_2)$ is determined by the product of the partial capture cross section $\sigma_{\rm cap}(E_{\rm c.m.}, \ell; \alpha_1, \alpha_2)$ and probability $P_{\rm CN}$ of the transformation of DNS into CN:

$$\sigma_{fus}(E, \ell; \alpha_1, \alpha_2) = \sigma_{cap}(E, \ell; \alpha_1, \alpha_2) P_{\rm CN}(E, \ell; \alpha_1, \alpha_2). \qquad (6)$$

The calculations of capture and fusion cross sections were performed in the framework of the DNS model. The capture probability and the largest value of orbital angular momentum (ℓ_d), leading to capture at the given values of the orientation angles α_1 and α_2, are calculated by solving equations of the relative motion of nuclei and orbital angular momentum of collision [16,17,19,24]. As a result we have partial capture cross sections (angular momentum distribution of DNS) for a given value of the collision energy $E_{\rm c.m.}$. Certainly, partial capture cross section is equivalent to cross section of formation of DNS with given angular momentum $L_{\rm DNS}$ and excitation energy E^*_Z.

The excitation energy E^*_Z of interacting nuclei of DNS at given collision energy $E_{\rm c.m.}$ and orbital angular momentum ℓ is calculated by taking into account the changes of the mass and charge numbers from the initial values A_P and Z_P for projectile (A_T and Z_T for target-nucleus) to the current values A and Z:

$$\begin{aligned} E^*_Z(A,\ell;\alpha_1,\alpha_2) &= E_{\rm c.m.} - V(Z,A,\ell,R_m;\alpha_1,\alpha_2) \\ &+ \Delta Q_{\rm gg}(Z,A). \end{aligned} \quad (7)$$

Here $\Delta Q_{\rm gg}(Z,A)$ is the change of binding energies of the DNS nuclei due to nucleon transfer between nuclei:

$$\Delta Q_{\rm gg}(Z,A) = B_Z(A) + B_{Z_{\rm tot}-Z}(A_{\rm tot}-A) - (B_{Z_P}(A_P) + B_{Z_T}(A_T)), \quad (8)$$

where $Z_{\rm tot} = Z_P + Z_T$ and $A_{\rm tot} = A_P + A_T$;
$V(Z,A,\ell,R_m;\alpha_1,\alpha_2)$ is the minimum value of potential well in the nucleus-nucleus interaction.

During evolution of DNS an intense mass transfer takes place and, in dependence on the landscape of potential energy surface and entrance channel, the mass asymmetry degree of freedom may be fully or partially equilibrated [25]. The duration of the quasifission is one order of magnitude larger than the time which is needed to reach the thermal equilibrium. It is more than $5 \cdot 10^{-21}$ s which was estimated by the analysis of experimental data on quasifission reactions [26, 28, 29]. The validity of using of the statistical method is righteous due to the fact that a full relaxation of the relative kinetic energy and mass (charge) asymmetry between the two fragments takes place at quasifission process [30].

Therefore, while DNS exists, we have an ensemble $\{Z\}$ of the DNS configurations which contributes to the competition between complete fusion and quasifission with probabilities $\{Y_Z\}$. The dependence of barrier $B^*_{\rm fus}$ and excitation energy of DNS E^*_Z for given charge Z and mass asymmetry A on angular momentum and orientation angles α_i (i =1,2) of the symmetry axis of interacting nuclei is connected with the method of calculation of the interacting potential between nuclei of DNS which is sensitive to those variables. Consequently, the fusion factor $P_{\rm CN}$ for the given reaction depends on the same variables through the charge distribution $Y_Z(E^*_Z)$ and fusion factor $P^{(Z)}_{\rm CN}$ from charge asymmetry configuration Z:

$$P_{\rm CN}(E^*_Z,\ell;\alpha_1,\alpha_2) = \sum_{Z_{sym}}^{Z_{max}} Y_Z(E^*_Z,\ell) P^{(Z)}_{\rm CN}(E^*_Z,\ell;\alpha_1,\alpha_2), \quad (9)$$

where $P^{(Z)}_{\rm CN}(E^*_Z,\ell;\alpha_1,\alpha_2)$ is the fusion probability for DNS having excitation energy E^*_Z at charge asymmetry Z and orientation angles of symmetry axis of its fragments are equal to α_1 and α_2. The evolution of Y_Z is calculated by solving the transport master equation. The details of calculations of $P^{(Z)}_{\rm CN}(E^*_Z,\ell;\alpha_1,\alpha_2)$ can be found in Refs. [16, 17, 19, 24].

2.2. Quasifission and fast fission of mononucleus evolving to equilibrium shape of compound nucleus

Another binary process which leads to the formation of two fragments similar to that from fusion-fission or quasifission is the fast fission. The fast fission occurs only at large values of angular momentum $\ell > \ell_f$ causing disappearance of the macroscopic fission barrier $B_f(\ell)$ of the rotating nucleus [21].

The partial cross section of fast fission is determined as cross section fusion with $\ell > \ell_f$ because mononucleus has survived against quasifission. The fast fission cross section is calculated by summing the contributions of the partial waves corresponding to the range $\ell_f \le \ell \le \ell_d$ leading to the formation of the mononucleus:

$$\sigma_{\rm ff}(E_{\rm c.m.};\beta_P,\alpha_2) = \sum_{\ell_f}^{\ell_d}(2\ell+1)\sigma_{\rm cap}(E_{\rm c.m.},\ell;\beta_P,\alpha_2)P_{\rm CN}(E_{\rm c.m.},\ell;\beta_P,\alpha_2). \quad (10)$$

The capture cross section in the framework of the DNS model is equal to the sum of the quasifission, fusion, and fast fission cross sections:

$$\sigma_{\rm cap}(E_{\rm c.m.};\beta_1,\alpha_2)=\sigma_{\rm qf}(E_{\rm c.m.};\beta_1,\alpha_2)+\sigma_{\rm fus}(E_{\rm c.m.};\beta_1,\alpha_2)+\sigma_{\rm ff}(E_{\rm c.m.};\beta_1,\alpha_2). \quad (11)$$

It is clear that the fusion cross section includes the cross sections of ERs and fusion-fission products. Obviously, the quasifission cross section is defined by

$$\sigma_{qf}(E_{\rm c.m.};\beta_P,\alpha_2)=\sum_{\ell=\ell_f}^{\ell_d}(2\ell+1)\sigma_{cap}(E_{\rm c.m.},\ell;\beta_P,\alpha_2)(1-P_{CN}(E_{\rm c.m.},\ell;\beta_P,\alpha_2)), \quad (12)$$

i.e the quasifission process can take place in the whole range of the orbital angular momentum values leading to capture, including central collisions ($\ell = 0$). This is important conclusion since the separation of the ranges of the angular momentum corresponding to the fusion-fission and quasifission products by some critical value $\ell_{\rm cr}$ in the analysis of the angular distribution of the fissionlike products is doubtful. The results of our calculations show that the ranges of the quasifission and fusion-fission overlap.

3. Comparison of the ^{48}Ca+^{248}Cm and ^{58}Fe+^{232}Th reactions leading to superheavy element Lv

The importance of the charge asymmetry in the CN formation is seen from the comparison of the ^{48}Ca+^{248}Cm and ^{58}Fe+^{232}Th reactions leading to the ^{296}Lv and ^{290}Lv CN, respectively. Fusion excitation functions, which have been obtained for these reactions, are presented by solid line in Figs. 5 and 6, respectively. We stress two main differences: 1) the cross section of the CN formation is larger in the reaction with ^{48}Ca in comparison with the ^{58}Fe+^{232}Th reaction; 2) CN formed in the former reaction has smaller excitation energy.

In Figs. 5 and 6, capture and quasifission cross sections are nearly equal to each other due to the dominant contribution of the quasifission cross section to capture in comparison with the sum of the fast fission and compound nucleus formation cross sections.

The number of events going to quasifission increases drastically with the increasing Coulomb interaction and rotational energy in the entrance channel [27, 31]. The advantage of the ^{48}Ca+^{248}Cm in synthesis of superheavy element is seen already in the CN formation stage.

One can see in Fig. 6 that the fusion excitation function decreases strongly at low energies $E_{\rm c.m.}$ <250 MeV. This effect is connected with the increase of hindrance to fusion since at these low energies the collisions with small orientation angles (α_1-projectile and α_2-target) can only contribute to the CN formation [17, 32]. Collisions with larger orientation angles α_1 and α_2 can not lead to capture since the collision energy in the center-of-mass system is not enough to overcome the Coulomb barrier of the entrance channel. For the DNS formed in collisions with small orientation angles α_1 and α_2, the intrinsic fusion barrier $B^*_{\rm fus}$ is large [17]. So, the hindrance to complete fusion depends on the orientation angles: more elongated shape of the DNS formed at collisions with small orientation angles (tip-to-tip configurations) promotes the quasifission rather than the formation of the CN [17,32].

Theoretical results of the ER cross sections for the synthesis of the element $Z = 116$ are compared with the experimental data in Fig. 7. In this figure, the full triangles and squares show experimental data of the ER cross sections measured in 3n- and 4n-channels, respectively, in the ^{48}Ca+^{248}Cm reaction [2]; the curves show theoretical results obtained in this work for the 2n-(dot-dashed line), 3n-(dashed line), 4n-(solid line), and 5n-channel (dotted line) by the DNS and advanced statistical models [33, 34] using the mass tables of Möller and Nix [11] (thin lines) and of Muntian *et al.* [12] (thick lines).

The mass values of the Warsaw group [12] are larger than ones of Möller and Nix [11] by 2-3 MeV for the isotopes of superheavy nuclei with $Z > 110$ and fission barriers [6, 7] are

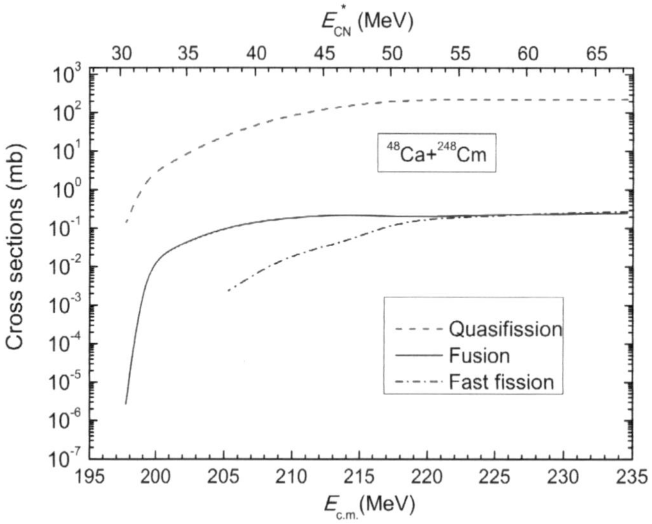

Figure 5. (Color online) (Color online) Quasifission (dashed line), complete fusion (solid line) and fast fission (dot-dashed line) excitation functions calculated by the DNS model [17, 24, 31] for the ^{48}Ca+^{248}Cm reaction leading to formation of the ^{296}Lv CN. The excitation energy $E^*_{\rm CN}$ (top axis) is calculated by the use of the Möller and Nix mass table [11].

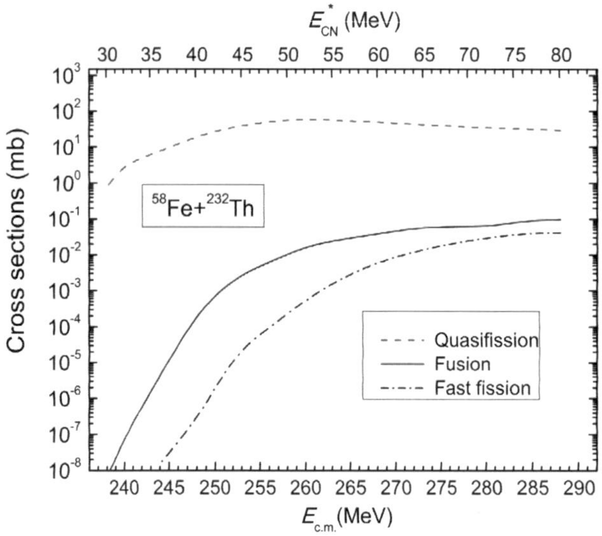

Figure 6. (Color online) As Fig. 5 but for the ^{58}Fe+^{232}Th reaction leading to formation of the ^{290}Lv.

smaller by 2-3 MeV in comparison with the similar values of Möller and Nix [11]. As a result, the Warsaw group results lead to two main consequences: 1) the excitation energy of the CN ($E^*_{\rm CN} = E_{\rm c.m.} + Q_{\rm gg} - V_{\rm rot}$) will be lower since the absolute value of $Q_{\rm gg} = B_{\rm proj} + B_{\rm targ} - B_{\rm CN}$ (negative) is larger; 2) the fission probability will be large in comparison with the case of using fission barrier of the Möller and Nix [11] model. When the binding energies and fission barriers of the Warsaw group [12] are used, the total score is that the survival probability $W_{\rm sur}$ becomes

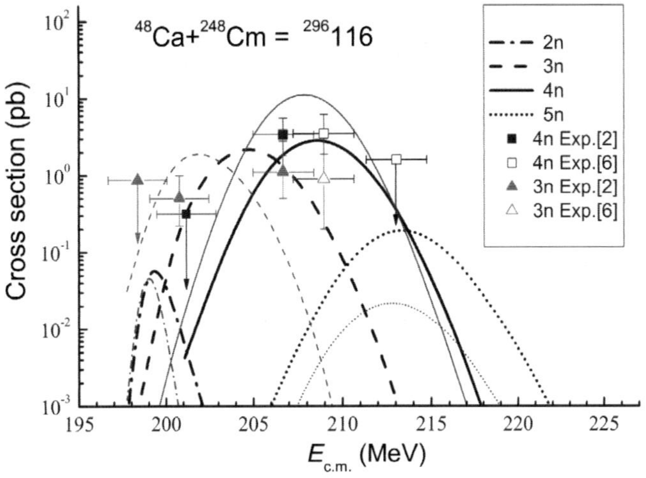

Figure 7. (Color online) Comparison between the ER excitation functions for the ^{48}Ca+^{248}Cm reaction calculated by using mass tables of Möller and Nix [11] (thin lines) and of the Warsaw group [12] (thick lines) for the 2n (dot-dashed lines), 3n (dashed lines), 4n (solid lines), and 5n (dotted lines) channels calculated by the advanced statistical model [33, 34]. The experimental data corresponding to the 3n (triangles) and 4n (squares) channels are obtained from Refs. [1, 2] (filled symbols) and [5] (open symbols).

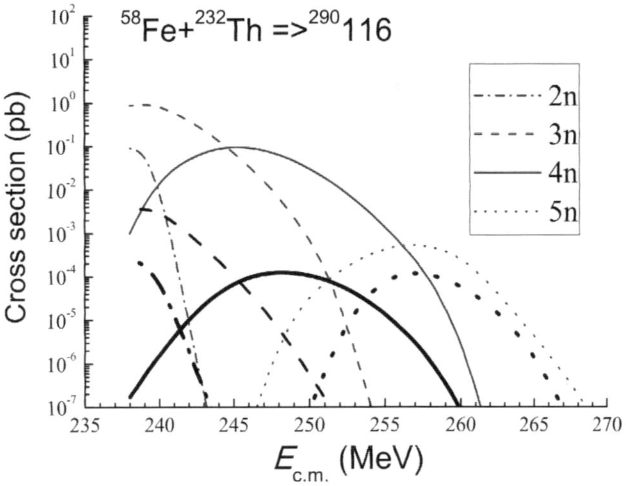

Figure 8. (Color online) As Fig. 7 but for the ^{58}Fe+^{232}Th reaction.

smaller in comparison with the case of using fission barrier of the Möller and Nix [11] model.

The results of σ_{ER} for the ^{48}Ca+^{248}Cm reaction, calculated by the use of the mass tables of the Warsaw group, better describe the experimental data than the ones obtained by using Möller *et al.* The largest cross section of the yield of superheavy element corresponds to the 4n-channel is about 10 pb when the collision energy is in the range $E_{c.m.}$ =205–212 MeV (thin solid line in Fig. 7).

Concerning the ER formation in the more symmetric ^{58}Fe+^{232}Th reaction, the results indicate that this reaction is less favorable in comparison with the ^{48}Ca+^{248}Cm reaction to be used in the synthesis of superheavy element $Z = 116$. The largest cross section for the 3n-channel is about 1 pb (see Fig. 8) if we use the mass tables of Möller and Nix [11] (thin dashed lines), while the use of the mass table and fission barriers of the Warsaw group leads to the two orders of magnitude lower cross section (thick dashed line) in comparison with the one in case of the using the mass tables of Möller and Nix.

4. Comparison of ^{48}Ca+^{249}Cf and ^{64}Ni+^{232}Th reactions leading to superheavy element Z=118

The capture, quasifission, fusion, and fast fission cross sections calculated in this work for the ^{48}Ca+^{249}Cf and ^{64}Ni+^{232}Th reactions are presented in Figs. 9 and 10, respectively.

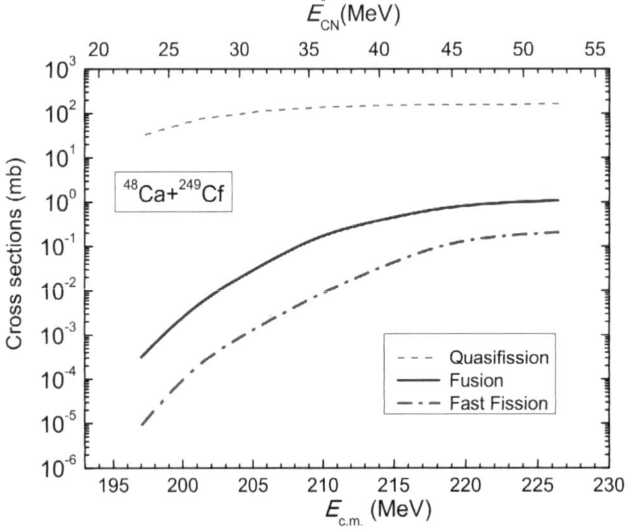

Figure 9. (Color online) As Fig. 5 but for the ^{48}Ca+^{249}Cf reaction leading to formation of the ^{297}118 CN.

The comparison between these figures shows that, at low energies, the capture cross section in the ^{64}Ni+^{232}Th reaction is much smaller than that in the ^{48}Ca+^{249}Cf reaction. The curve of capture cross section for the ^{64}Ni+^{232}Th reaction goes down at larger energies $E_{\mathrm{c.m.}} > 263$ MeV. But fusion cross section is very small due to drastic hindrance to complete fusion since at low energies only collisions with small values of orientation angles α_2 lead to the capture, *i.e* to formation of the long living DNS. The latter formed in collisions with small values of α_2 has large intrinsic barrier B^*_{fus} (see Fig. 11) and small quasifission barrier B_{qf} for this reaction in comparison with ^{48}Ca+^{249}Cf (see Fig. 12). This means that the potential well in the nucleus-nucleus interaction is shallow since the Coulomb interaction is stronger for more symmetric reactions (z= 248.41 for ^{64}Ni+^{232}Th reaction). These two reasons, which are unfavorable for complete fusion [17], lead small fusion probability for the ^{64}Ni+^{232}Th reaction at low energies.

The advantage of the charge asymmetric system (hot fusion reaction) in complete fusion appears at the second stage of the reaction mechanism leading to the CN formation: the intrinsic fusion barrier B^*_{fus} decreases and quasifission barrier B_{qf} increases with increasing the DNS charge asymmetry (see Fig. 4).

The theoretical ER excitation functions of different neutron-emission channels for these two systems are presented in Figs. 13 and 14. In each of the figures the ER cross sections are obtained

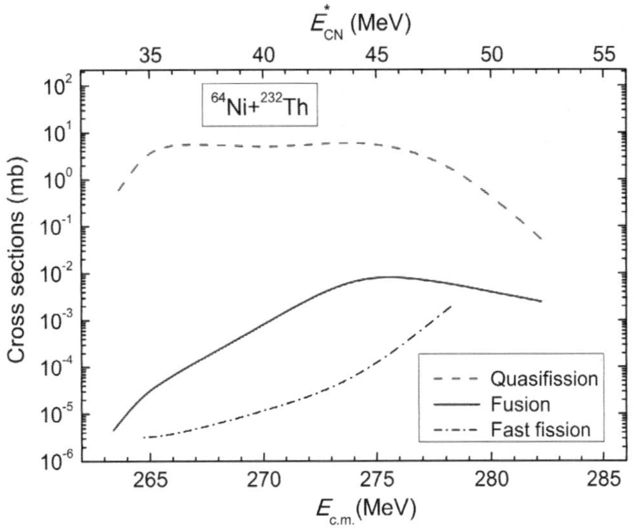

Figure 10. (Color online) Same as Fig. 5 but for the but for the ^{64}Ni+^{232}Th reaction which could lead to formation of the ^{296}118 CN.

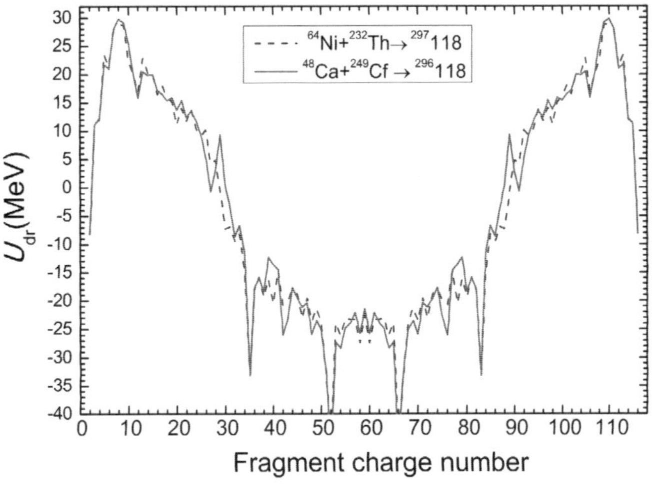

Figure 11. (Color online) Comparison of the driving potentials calculated for the DNS formed in the ^{48}Ca+^{248}Cf and ^{64}Ni+^{232}Th reactions which can lead to formation of isotopes A=297 and 296 of the superheavy element Z=118 as a function of the fragment charge number.

by using binding energies and fission barriers calculated in the microscopic-macroscopic models of Möller and Nix [11] and of the Warsaw group [12] are compared.

In Figs. 13 and 14, the results calculated for the ^{48}Ca+^{249}Cf reaction [23] leading to ^{297}118 CN and for the ^{64}Ni+^{232}Th reaction leading to the same superheavy element ^{296}118 with smaller mass number, respectively, are presented. In Fig. 13, the experimental data [2] corresponding to the synthesis of SHE ^{294}118 after emission of three neutrons from the ^{297}118 CN are shown by squares.

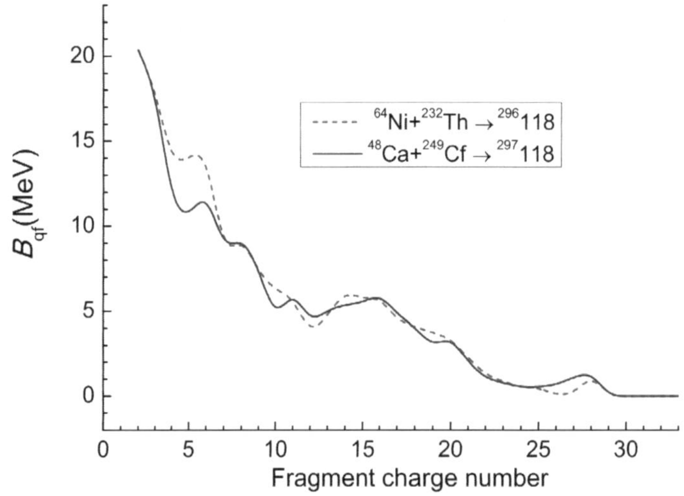

Figure 12. (Color online) The quasifission barriers of the DNS fragments as a function of their charge numbers for the ^{48}Ca+^{249}Cf(solid curve) and ^{64}Ni+^{232}Th (dashed curve) reactions.

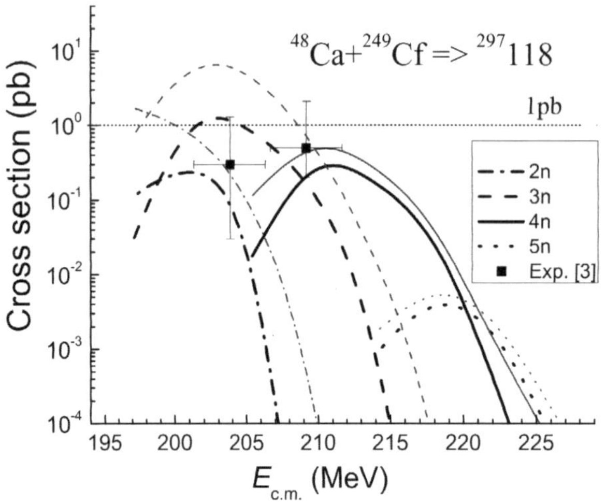

Figure 13. (Color online) Comparison between the ER excitation functions for the ^{48}Ca+^{249}Cf reaction calculated by using mass tables of Möller and Nix [11] (thin lines) and of the Warsaw group [12] (thick lines) for the 2n (dot-dashed lines), 3n (dashed lines), 4n (solid lines), and 5n (dotted lines) channels calculated by the advanced statistical model [33, 34]. The experimental data of the 3n channel from Ref. [2] are presented by squares.

One can see from Fig. 14 that the results obtained for all xn-channels of ER formation in the ^{64}Ni+^{232}Th reaction are very small. The largest point of the 4n-channel curve is about 10 fb (thin solid line) in the case when we used masses of the Möller and Nix table [11] in our calculations. The use of the masses obtained by the Warsaw group [12] leads to the ER cross sections (thick lines) which are more lower than ones obtained in the former case.

The ER cross sections of the 4n and 5n channels for the ^{64}Ni+^{232}Th reaction obtained by

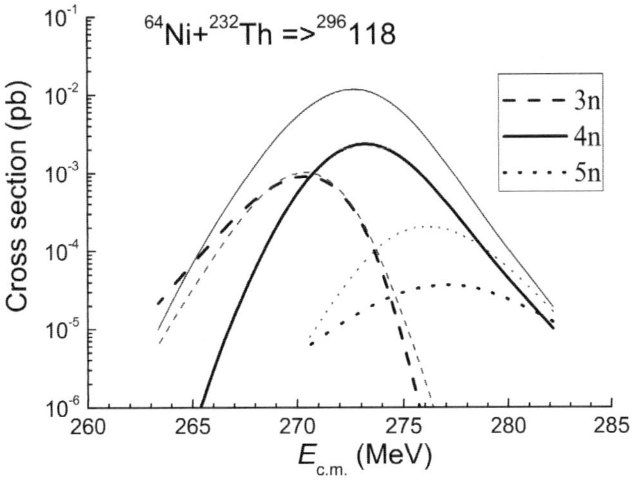

Figure 14. (Color online) As Fig. 13 but for the ^{64}Ni+^{232}Th reaction.

the using masses of the Möller and Nix table [11] are larger than those calculated by the using mass tables of Warsaw group [12], while the same behaviour does not appear by an evident way for the 3n channel. This is due to the effect of the different $E^*_{\rm CN}$ threshold values for the 3n channel when the masses of table [11] and Warsaw group [12] are considered. In the first case, the threshold energy for the 3n channel is at the $E^*_{\rm CN}$ =35 MeV, while for the second case it is at $E^*_{\rm CN}$ ≡30.5 MeV. Due to the above-mentioned conditions for the ^{64}Ni+^{232}Th reaction, the excitation function of the 3n channel is strongly different in comparison with the ones shown in Fig. 14 for the 4n and 5n channels.

5. Conclusions

In the framework of the combined DNS and advanced statistical models, the ER excitation functions for the ^{48}Ca+^{248}Cm and ^{58}Fe+^{232}Th reactions leading to the superheavy element Lv (Z=116) and the ^{48}Ca+^{249}Cf and ^{64}Ni+^{232}Th reactions leading to the superheavy element Z=118 have been calculated. The ER cross sections of the 3n- and 4n-channels for the ^{48}Ca+^{248}Cm reaction presented Refs. [1, 2] are well described when the Warsaw group mass tables [12] are used, while in both cases the use of the Möller and Nix [11] mass tables leads to overestimation of the experimental data.

The theoretical values of ER cross sections for the ^{48}Ca+^{249}Cf reactions (Fig. 13) are in good agreement with the experimental data presented in Ref. [1, 2] when we use the mass table and fission barriers of Warsaw group.

The results of the ER excitation functions for the ^{58}Fe+ ^{232}Th reaction indicate that this reaction is less favorable to be used in the synthesis of superheavy element Z = 116. The strong hindrance at low energies is caused by the large intrinsic fusion barrier $B^*_{\rm fus}$ for the small orientation angles of the symmetry axis of the deformed projectile ^{58}Fe (β_2 = 0.2) and target ^{232}Th (β_2 = 0.26) nuclei. Therefore, the curve of the fusion excitation function increases slightly with the increasing beam energy. The largest cross section for the 3n-channel is about 1 pb (see Fig. 8), if we use the mass tables of Möller and Nix [11] (thin dashed line), while the use of the mass table and fission barriers of Warsaw group leads to much lower cross sections (thick dashed line), no more than 40 fb.

The results obtained for all xn-channels of ER formation in the ^{64}Ni+^{232}Th reaction are very small (see Fig. 14). The largest point of the 4n-channel curve is about 10 fb (thin solid line) when masses from the Möller and Nix table [11] were used in our calculations. The use of the Warsaw group mass tables [12] leads to values of the ER cross sections, which are sufficiently lower than ones of the former case. That means this reaction is not favorable to observe an event corresponding to the synthesis of any isotopes of the superheavy element 118.

5.1. Acknowledgments
A. K. Nasirov is grateful to the Rare Isotope Science Project of the Institute for Basic Science of the Republic of Korea for the support the collaboration between the Dubna and Daejeon groups, and he thanks the Russian Foundation for Basic Research and Fondazione Bonino-Pulejo for the partial financial support in the performance of this work. The work of Y. Kim and K. Kim was supported by the Rare Isotope Science Project funded by the Ministry of Science, ICT and Future Planning (MSIP) and National Research Foundation (NRF) of Korea.

References

[1] Yu.Ts. Oganessian *et al.*, Phys. Rev. C **70**, 064609 (2004).
[2] Yu.Ts. Oganessian *et al.*, Phys. Rev. C **74**, 044602 (2006).
[3] L. Stavsetra *et al.*, Phys. Rev. Lett. **103**, 132502 (2009).
[4] J.M. Gates *et al.*, Phys. Rev. C **83**, 054618 (2011).
[5] S. Hofmann *et al.*, Eur. Phys. J. A **48**, 62 (2012).
[6] M. Kowal, P. Jachimowicz, A. Sobiczewski, Phys. Rev. C **82**, 014303 (2010).
[7] M. Kowal and A. Sobiczewski, Int. J. Mod. Phys. E **18**, 914 (2009).
[8] S. Hofmann, Rep. Prog. Phys. **61**, 639 (1998).
[9] K. Morita *et al.*, Jour. Phys. Soc. Japan, **81**, 103201 (2012).
[10] G. Audi, A.H. Wapstra, *Nucl. Phys.* **A595**, 509 (1995).
[11] P. Möller and J.R. Nix, J. Phys. G: Nucl. Part. Phys. **20**, 1681 (1994).
[12] I. Muntian, Z. Patyk, A. Sobiczewski, Phys. At. Nuclei **66**, 1015 (2003).
[13] B.B. Back *et al.*, Phys. Rev. C **32**, 195 (1985).
[14] N.V. Antonenko *et al.*, Phys. Rev. C **51**, 2635 (1995).
[15] G.G. Adamian, N.V. Antonenko, W. Scheid, Phys. Rev. C **68**, 034601 (2003).
[16] G. Fazio *et al.*, Eur. Phys. J. A **19**, 89 (2004).
[17] A.K. Nasirov *et al.*, Nucl. Phys. A **759**, 342 (2005).
[18] N. Wang, E.G. Zhao, W. Scheid, Sh.G. Zhou, Phys. Rev. C **85**, 041601(R) (2012).
[19] A.K. Nasirov *et al.*, Phys.Rev. C **79**, 024606 (2009).
[20] A.K. Nasirov *et al.*, Int. Jour. of Mod. Phys. E, **18**, 841 (2009).
[21] A.J. Sierk, Phys. Rev. C **33**, 2039 (1986).
[22] A. Sobiczewski, K. Pomorski, Prog. Part. Nucl. Phys. **58**, 292 (2007).
[23] G. Mandaglio, G. Giardina, A.K. Nasirov, A. Sobiczewski, Phys. Rev. C **86**, 064607 (2012).
[24] G. Fazio *et al.*, Mod. Phys. Lett. A **20**, 391 (2005).
[25] R. Bock *et al.*, Nucl. Phys. A **388**, 334 (1982).
[26] J. Tõke *et al.*, Nucl. Phys. A**440**, 327 (1985).
[27] G. Giardina, S. Hofmann, A.I. Muminov, A.K. Nasirov, Eur. Phys. J. A **8**, 205 (2000).
[28] D.J. Hinde *et al.*, Phys. Rev. C**45**, 1229 (1992).
[29] K. Siwek-Wilczynska, J. Wilczynski, R. H. Siemssen, H. W. Wilschut, Phys. Rev. C **51**, 2054 (1995).
[30] B.B. Back *et al.*, Phys. Rev. C **53**, 1734 (1996).
[31] G. Fazio *et al.*, Phys.Rev. C **72**, 064614 (2005).
[32] D.J. Hinde*et al.*, Phys. Rev. Lett. **74**, 1295 (1995).
[33] A. D'Arrigo, G. Giardina, M. Herman, A. Taccone, Phys. Rev. C **46**, 1437 (1992).
[34] R.N. Sagaidak *et al.*, J. Phys. G **24**, 611 (1998).

Production and investigation of neutron-rich Osmium isotopes with and around N=126 using gas flow transport method

K Novikov[1], E M Kozulin[1], I M Harca[1,2], S Dmitriev[1], A Flueras[1,3], P Greenlees[4], F Hanappe[5], S V Khlebnikov[6], T Loktev[1], J Maurer[7], A Di Nitto[7], J Pakarinen[4], P Ruotsalainen[4], M Sandzelius[4], J Sorri[4], M Sin[2], W H Trzaska[4], E Vardaci[8] and V I Zagrebaev[1].

[1]Flerov Laboratory of Nuclear Reactions, Joint Institute for Nuclear Research,(FLNR JINR) Dubna, Russia
[2]Faculty of Physics, University of Bucharest - P.O. Box MG 11, RO 77125, Bucharest-Magurele, Romania, EU
[3]University of Manchester, Oxford Road, Manchester M13 9PL, UK
[4]Accelerator Laboratory of University of Jyväskylä (JYFL), Jyväskylä, Finland.
[5]Universite Libre de Bruxelles (ULB), Bruxelles, Belgium
[6]Khlopin Radium Institute(KRI), St. Petersburg, Russia
[7]GSI Helmholtzzentrum für Schwerionenforschung, 64291 Darmstad, Germany
[8]Dipartamento di Scienze Fisiche and INFN (INFN-Na), Napoli, Italy

E-mail: kiryuha-ya@mail.ru

Abstract. Neutron-rich isotopes of heavy nuclei are until now poorly studied. In this work we investigate neutron-rich osmium isotopes produced in multi-nucleon transfer reactions. The reaction $^{136}Xe+^{208}Pb$ at energy near Coulomb barrier is used for production of osmium isotopes. The CORSAR-V setup is used to record the characteristics of osmium isotopes. The separation of the reaction products is based on their respective volatility. Experimental results are presented and discussed.

1. Introduction

One of the main problems of modern nuclear physics is the determination of the extreme conditions for the existence of nuclei. In recent years, great attention has been paid to the production and the study of the properties of nuclei located far away from the line of stability. At the present time, only the "north-east" region of the nuclide chart with nuclei having a neutron number N≈126 is not properly studied. In the framework of this project we propose to fill this region using multi-nucleon transfer and quasi-fission reactions [1].

The reaction $^{136}Xe+^{208}Pb$ is selected. The main idea of this experiment is to use the stabilizing effect of closed neutron shells in both nuclei. Transfer of protons from lead to xenon may be predominant as the light product of the reaction is a strongly bound nucleus and, therefore, Q value of reaction is ≈0.

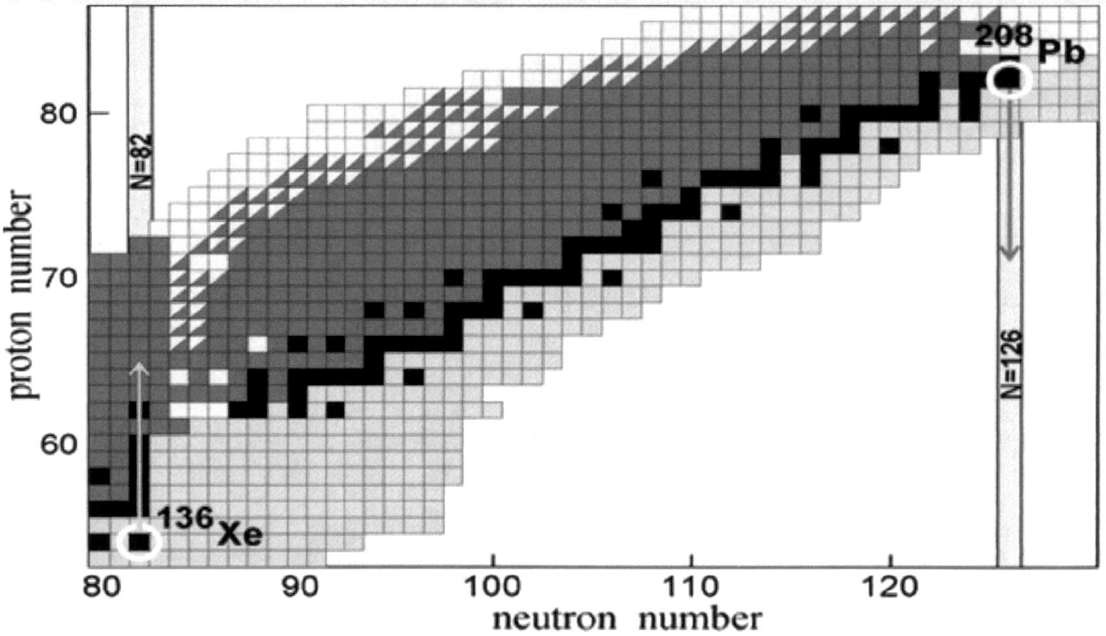

Figure 1. Chart of nuclide demonstrates possibility of proton transfer in reaction of $^{136}Xe+^{208}Pb$ at energy near Coulomb barrier (black squares – stable nuclei).[1]

1. Setup

The experimental set up devoted to investigations of multi-nucleon transfer reactions is described. It uses the radiochemical method of gas transport.
Measurements were made using the CORSAR-V setup (correlation setup for the reaction volatile products) specially designed for identification and investigation of properties of volatile heavy nuclei.

Measurements were made with the CORSAR-V setup (correlation setup for the reaction products registration (volatile products)) which was designed for identification and investigation of properties of neutron-rich heavy nuclei in the region of nuclei near N = 126. The setup has a closed gas system with recycling. A gas mixture (80%He + 20%O) at the pressure of 1.5 atm is fed to the gas catcher (figure 3) which is located in the reaction chamber. A rotating target of ^{208}Pb (450 microg/cm2) is located in front of the gas catcher (figure 4). Gas catcher system is a stainless steel cylinder with a 36mm diameter hole in its centre (allowing the ^{136}Xe beam to pass through). The catcher's face which is in front of the rotating target has windows covered with Mylar foils of 25 micron thick. The gas catcher was designed so that the products of the reaction going at~35° (with $74 \leq Z \leq 76$, $N \geq 125$ and kinetic energy $400<E<600$ MeV) can pass through the Mylar foil to be captured by the gas mixture flow. Elastic scattered beam of ^{136}Xe (with kinetic energy $E > 540$ MeV) will not be stopped by the gas flow because it has energy enough to reach the back wall of the gas catcher container. Reaction products with kinetic energy $E<400$ MeV are stopped in the Mylar foil. Catcher's efficiency for the capture of the studied reaction products and their transport to a cooled surface is roughly 60-70%. The products of the reaction ^{136}Xe (820MeV) + ^{208}Pb are gathered inside the gas mixture and are transported through an electric furnace which provides oxidation of Os isotopes OsO_4. Quartz filter is installed for more efficient oxidation of reaction products in the output tube of furnace.

Then gas mixture with OsO_4 flows through heated Teflon tubes to the detection system and is deposited on copper surfaces kept at the temperature of -200°C. Special cap foils with nano-structures are used. These foils allow to increase the surface area of deposition of the reaction products by a factor 1000. Two gamma ray detectors are used with an angle of 135 degrees between them to reduce the effect of Compton scattering. A nitrogen trap is installed in the gas system to clean the system of water vapors. β-radiation is measured by semiconductor detectors. The registration of beta rays is triggered by γ rays germanium detectors. After purification, the gas mixture returns back to the gas catcher.

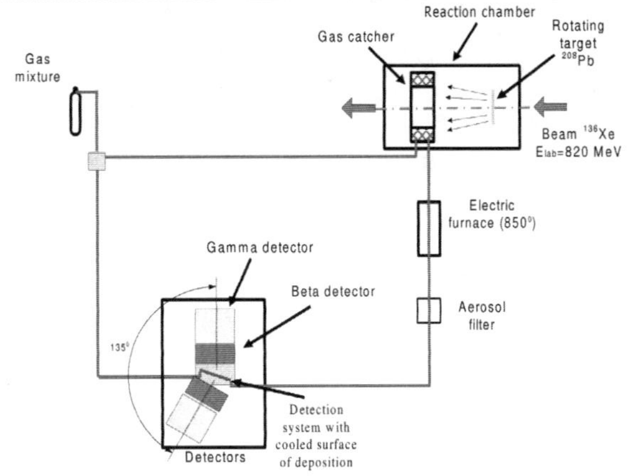

Figure 2. Scheme of the experiment

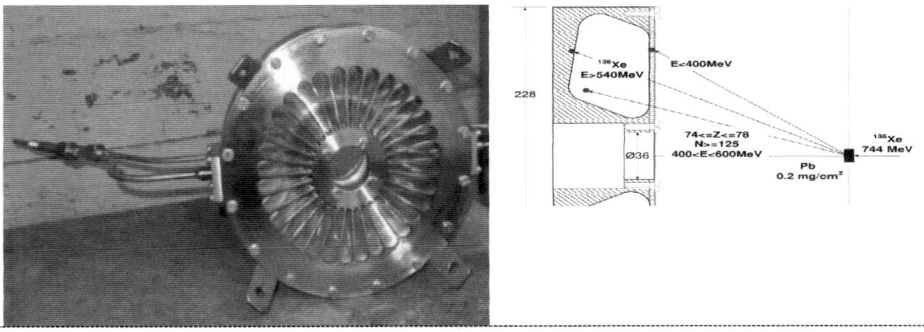

Figure 3. Gas catcher

Figure 4. Reaction chamber with rotating target and gas catcher.

2. Data analyses

As only osmium oxides are volatile products, only the chains of decay of osmium isotopes can be registered. (figure 5).

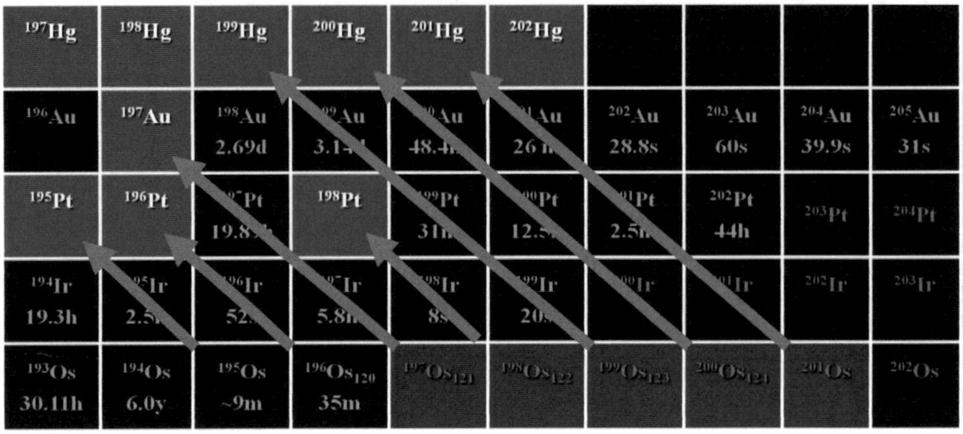

Figure 5. Chart of studied nuclides

Theoretical estimations and calculations for cross section of products of reaction $^{136}Xe+^{208}Pb$ were carried out at FLNR JINR by V.I. Zagrebaev and W. Greiner [2, 3, 4].

figure 6 (right panel) shows cross-sections of production for heavy neutron-rich elements and their survival probability in reaction $^{136}Xe+^{208}Pb$ at energy E_{cm}=450MeV, which is near to Coulomb barrier (Bass – barrier is ≈434MeV). The lower panel shows cross-sections for production of the nuclei, which are located near the closed neutron shell (N=126).

The calculations demonstrate that production of unknown neutron-rich heavy nuclei is quite possible in multi-nucleon transfer reactions at low energies close to the Coulomb barrier).

Large numbers of new nuclides may be produced in the region Z=74-80, as we can see in figure 6 (left panel) [2].

Experiment was performed on the reaction $^{136}Xe+^{208}Pb$ at the energy 820MeV at the JYFL accelerator K-130 during July 2012. This work is within collaboration between the FLNR, the Accelerator Laboratory of the University of Jyväskylä and the Department of Physics of the University of Napoli (Italy).

It should be noted that during the experiment two systems of data acquisition were used. The first system is a system of CAMAC standard with Kmax software and the second system is a system of cPCI standard with LyrTech software. The first system of data acquisition has a resolution of about 40 keV for beta- and 4 keV for gamma-detectors. System does not have a timestamp on events. It allows you to make quickly rough calculations for optimization of experimental conditions. The second system of data acquisition has a much higher resolution than the first system, namely 25 keV for beta- and 1.5-2 keV for gamma-detectors. This system has a timestamp for each registered event. It allows to estimate the lifetime of isotopes and accurately determine the time of decay of isotopes. Thereby, estimated results can be obtained from the data collected by the first system and more accurate results can be obtained from data from the second one.

A source of ^{207}Bi is used for energy calibration of detectors. Using the data collected by the second system, matrices of $\gamma - \gamma$ and $\beta - \gamma - \gamma$ energy - time were created in order to investigate coincident rays. By gating energy transitions, taking into consideration the detector's resolution, one-dimensional gamma spectra were created, which could be further processed and analyzed. In these spectra events which occurred due to background radiation contribution are subtracted using a method based on Sensitive Nonlinear Iterative Peak (SNIP) clipping algorithm [5]. Final gamma spectra were smoothed in order to eliminate the noise by a function based on Markov chain method [6], [7].

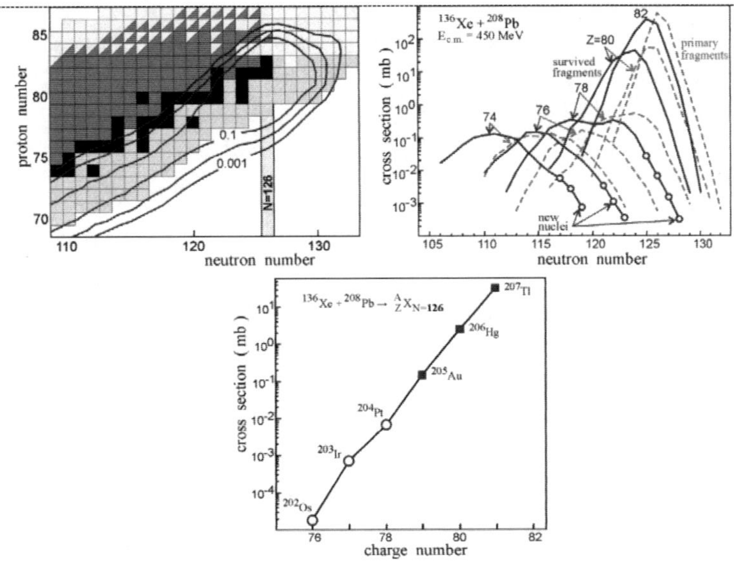

Figure 6. (The Left panel) - full section of formation of heavy fragments $d^2\sigma/dZdN$ (mb, numbers on a planimetric line) for reaction $^{136}Xe+^{208}Pb$ at energy E_{cm}=450MeV. (The Right panel) - section of formation of heavy neutron-rich nuclei in reaction Xe+Pb at energy E_{cm}=450 MeV, blue lines - correspond to the survived nuclei; open points - unknown isotopes. (The bottom panel) - exit of the nuclei near closed neutron shall N=126 in reaction $^{136}Xe+^{208}Pb$ at energy E_{cm}=450MeV; (open points) - unknown isotopes.[2].

Investigation was carried out by the $\beta-\gamma-\gamma$ coincidence method. Using this method, registered events analyzed taking into consideration the resolution and the efficiency of detectors. The results are shown for the isotopes produced by the decay of isotopes of ^{196}Os. Decay scheme of ^{196}Os and spectra resulting from processing by $\beta-\gamma-\gamma$ coincidence of experimental data are presented in figure 7

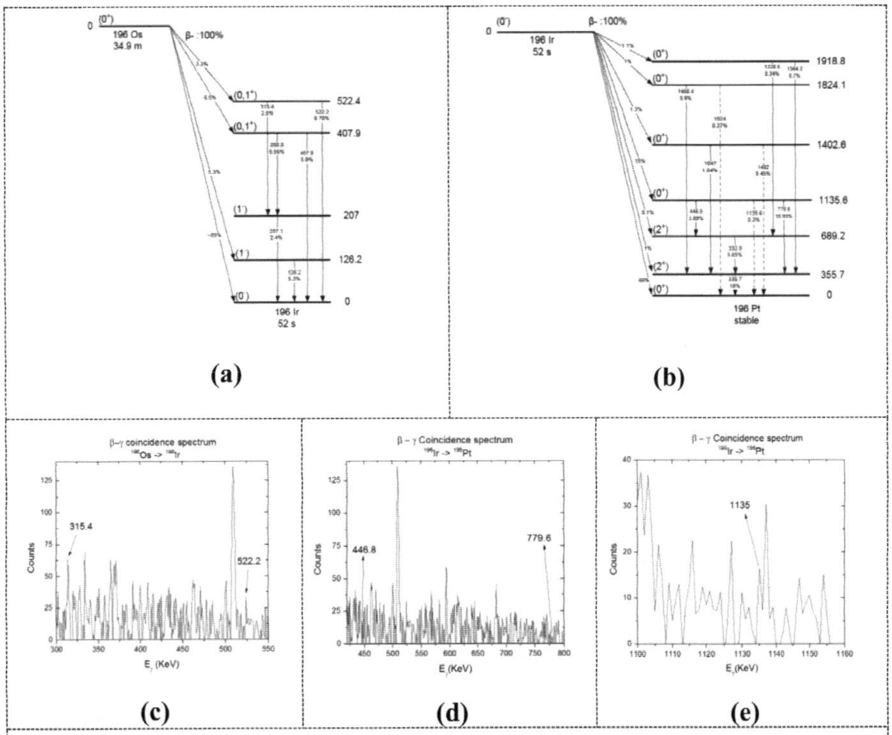

Figure 7. Level schemes of decay of ^{196}Os (a) and its daughter isotope ^{196}Ir (b) and gamma spectra in coincidence with beta (c),(d),(e).

Similar processing of data was performed until the ^{199}Os isotopes. The results of the identification of the daughter isotopes of ^{199}Os are shown in figure 8.

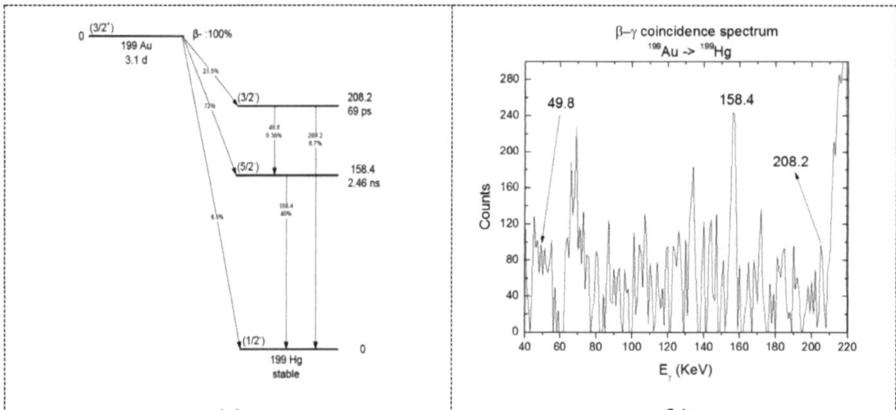

(a) **(b)**

Figure 8. Level scheme of decay of daughter isotopes of ^{199}Os, namely ^{199}Au (a) and gamma spectra in coincidence with beta (b).

Comparison of the relative yield of osmium isotopes with theoretical estimates is made. figure 9 shows that the relative yield of isotopes produced in the decay of isotopes of Osmium is consistent with the theory.

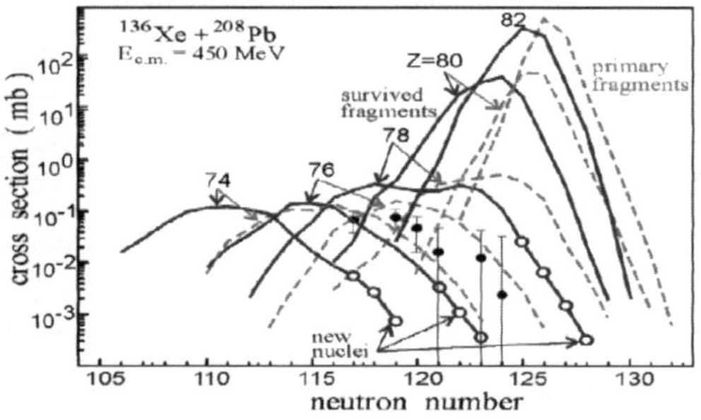

Figure 9. Cross-Section of formation of heavy neutron-rich nuclei in reaction Xe+Pb, blue lines - correspond to the survived nuclei (theory); open points - unknown isotopes; black points - experimental data of the relative yield of osmium isotopes based on decay chain of platinum isotopes.

3. Conclusion

1. The ability of separating volatile reaction products from non-volatile ones by CORSAR-V is demonstrated.

2. Low-energy multi-nucleon transfer reactions represent a good way to produce heavy and extra-heavy neutron-rich nuclei located in the unexplored region of nuclei near the closed shell N=126.

3. Different Osmium isotopes were produced and analyzed.

4. The relative yields of isotopes reproduce the trends of the theoretical predictions.

Acknowledgements

The authors would like to express their gratitude to the staffs of the JYFL Accelerator Laboratory. This work was supported by the Russian Foundation for basic Research (Grant No. 11-02-01507-a and Grant No. 13-02-01282-a), by the Department of Physics of the University of Jyväskylä and by the Flerov Laboratory of Nuclear Reactions, JINR, Dubna.

References

[1] Karpov A V et al 2007 Symp. on Exotic Nuclei (Khanty-Mansiysk, Russia, 2006) edited by Penionzhkevich Yu E and Cherepanov E A (Melville, New-York: *AIP*, 2007) p. 286-298
[2] Zagrebaev V et al. 2008 *Physical Review Letters* **101** 122701
[3] Zagrebaev V and Greiner W 2007 *J. Phys*. G **34** 1
[4] Zagrebaev V and Greiner W 2007 *J. Phys*. G **34** 2265
[5] Morhac M et al. 1997 *Nucl. Instrum. and Methods in Physics Research* A **401** 113–132
[6] Morhac M et al. 1997 *Nucl. Instrum. and Methods in Physics Research* A **401** 385–408
[7] Morhac M et al. 2000 *Nucl. Instrum. and Methods in Physics Research* A **443** 108–125

ECHIC2013 IOP Publishing
Journal of Physics: Conference Series **515** (2014) 012017 doi:10.1088/1742-6596/515/1/012017

Iso-Vectorial interaction and many-body correlations in Nuclear ynamics

[1] INFN - Sezione di Catania, Via S. Sofia, Catania, Italy

M Papa

E-mail: massimo.papa@ct.infn.it

Abstract. Comparisons involving Nuclear Matter calculations based on the Constraint Molecular model CoMD and semi-classical Mean-Field approximation using an effective interactions of the Skyrme type are presented. The performed study shows that specific correlations induced by the iso-vectorial interaction investigated in the framework of the molecular dynamics approach strongly affect parameter values of the effective interaction to be used to reproduce the standard saturation properties of Nuclear Matter. Moreover an example showing consequences in the balance between reaction mechanisms in the $^{48}Ca +^{48} Ca$ at 25 MeV/nucleon system is also discussed.

1. Introduction

The description of many-body systems is one of the most difficult problems in nuclear physics due to the complexity of this kind of systems which are quantum objects described by a large number of degrees of freedom. A large variety of theoretical models have been developed using mean-field based and beyond-mean field approaches. In these approaches which use the independent particle approximation as a starting point, phenomenological effective interaction like Skyrme and Gogny forces are widely used taking advantage of their simple form[1, 2, 3, 4].

At the Fermi energy and beyond semiclassical methods become necessary to describe the produced processes in which practically all the degrees of freedom are involved. Many-body correlations are responsible for the reorganization of the hot Nuclear Matter(NM) in to clusters. Two main classes of approaches have been developed to handle this complex scenario. The first one is based on the Boltzman transport equation including, at the higher energy, three body collisions term[5, 6, 7]and stochastic forces (typical of Langevin processes) [8, 9]. In this case the theoretical approach is able to describe the time evolution of the one-body distribution function in phase-space.

In these models the main ingredient concerning the interaction is the energy density which for a Skyrme type phenomenological interaction has a rather manageable form. The second line of development of theoretical semiclassical approaches to deal the nuclear many-body problem is represented by the so called Molecular Dynamics models. The point of view of these methods is some sense antithetic compared to the ones based on the mean-field concept. However also in these cases, due to their simplicity the same kind of phenomenological effective interactions Skyrme and Gogny are widely used. The starting point of all these models is a basic assumption on the wave function describing the single nucleon which is represented through a wave packet well localized in phase-space with uncertainty satisfying the uncertainty principle. The many-body wave fucntion can be expressed through a direct product, leading to the so called Quantum Molecular Dynamics (QMD)-like models [10, 11, 12, 13] or an anti-symmetrized product like in FMD and AMD [14, 15].

Within their own approximation schemes these models treat the many-body problem without the usage of the mean-field approximation. Many-body correlations are spontaneously produced and in general are able to describe the main features concerning the multi-breakup processes [16] observed in heavy ion collisions at intermediate energy and beyond. Due to these deep conceptual differences between the mean-field based models and the molecular dynamics ones it becomes interesting, and necessary in our opinion, to investigate the differences between the energy-density functionals which are produced with these two classes of approaches when one uses, in both the cases, the same kind of effective elementary interaction between nucleons. Clearly strongly correlated to this aspect are the differences produced on the behavior of simple observables obtained in the study of heavy ion collisions.

The study presented in this work is performed by using the Constraint Molecular Dynamical Model CoMD-II [11, 12].

The quantitative results obviously will be related to some specificity of the model and to the used effective interaction, but as due to the common features which links most of the molecular dynamics models (semi classical wave packet dynamics), the obtained results could assume a more wide meaning at a qualitative level. The work is organized as follows: in Sec. II we illustrate the choice of the effective interaction and the related total energy functional are introduced. In Sec.III we describe the results obtained from NM simulations. In Sec.IV we describe how the discussed many-body correlation can affect the balance between different reaction mechanisms in the $^{48}Ca+^{48}Ca$ collision at 25 Mev/nucelon. Sec.V contains the summary and the concluding remarks.

2. The effective interaction and the total energy

In the present work the elementary interaction between two nucleons with spacial coordinates $\mathbf{r}$,$\mathbf{r}'$ and third isospin component τ,τ' is of the Skyrme type and has the following form, around the saturation density:

$$V(\mathbf{r},\mathbf{r}') = V^{(2)}\delta(\mathbf{r}-\mathbf{r}') = \frac{T_0}{\rho_0}\delta(\mathbf{r}-\mathbf{r}') + \frac{2T_3\rho^{\sigma-1}}{(\sigma+1)\rho_0^{\sigma}}\delta(\mathbf{r}-\mathbf{r}') + \frac{T_4}{\rho_0}(\frac{\rho}{\rho_0})^{\gamma-1}(2\delta_{\tau,\tau'}-1)\delta(\mathbf{r}-\mathbf{r}') \quad (1)$$

The first term of eq.(1) represents the iso-scalar contribution to the two-body interaction, the second one is the usual 3-body effective interaction. For this term the ρ density dependence is modeled through the parameter σ.

The third term represents instead the Iso-Vectorial contribution. Also in this case the density dependence is introduced to reproduce the results of more complex microscopic calculations based on realistic interaction concerning the symmetry energy[18, 19, 20, 21, 22] and reflects effects beyond the two-body interaction.

2.1. *The Semiclassical Mean Field Approximation*

At a fixed time starting from eq.(1) (we are considering here a stationary problem) , in a rather general way, we can obtain the expression for the total energy W related to the potential interaction by folding the elementary interaction with the two-body density distribution in phase-space $D(\mathbf{r},\mathbf{r}',\mathbf{p},\mathbf{p}')$. By taking into account the usual truncation condition, D can be expressed trough the product of the one-body distributions and we obtain for the potential interaction

E_{pot}:

$$E_{pot} = U_{twb} + U_{trb} + U_{asy} = \frac{1}{2}V^{(2)}\rho = \frac{1}{2}\frac{T_0\rho}{\rho_0} + \frac{T_3\rho^\sigma}{(\sigma+1)\rho_0^\sigma} + \frac{1}{2}T_4F_k(\rho)\beta^2 \tag{2}$$

$\beta = \frac{\rho_n - \rho_p}{\rho}$ represents the charge/mass asymmetry parameter evaluated from the neutron and proton density ρ_n, ρ_p respectively. The total binding energy E is obtained by adding the kinetic contribution coming from the Fermi motion

$$E = E_{pot} + E^F_{kin} \tag{3}$$

$$E^F_{kin} = \frac{3}{5}\frac{\hbar^2}{2m_0}(\frac{3\pi^2\rho}{2})^{2/3}[1 + \frac{5}{9}\beta^2] \tag{4}$$

In particular we see how the iso-vectorial vectorial forces with strength proportional to T_4 contribute to the symmetry energy E_{sym} depending on the asymmetry parameter β. For quadratic form in β we get:

$$E_{sym}(\rho) = e_{sym}\beta^2 \tag{5}$$

$$e_{sym} = \frac{1}{2}(\frac{\partial^2 E}{\partial\beta^2}) = \frac{1}{6}\frac{\hbar^2}{m_0}(\frac{3\pi^2\rho}{2})^{2/3} + \frac{1}{2}T_4F(\rho) \tag{6}$$

The common accepted bulk value of e_{sym} at the saturation density is about30 MeV even if relativistic Hartree-Fock models can predict higher values up to about 40 MeV [24, 25]. However, still under study is the density dependence of this quantity which is able to affect neutron-skin thickness in nuclei and the Giant Monopole Resonances [24, 25, 26]. Finally we note that the structure of the obtained energy-density functional in Se-MFA makes independent the choice of the parameter describing the main properties of the symmetry energy from the other ones which instead are fixed from the saturation properties of the symmetric nuclear matter. We will show that this is no longer true in the CoMD approach.

In molecular dynamics approaches a strong assumption is done on the wave functions describing the nucleonic degree of freedom. It is commonly assumed that the wave function is a gaussian wave packet with width σ_r in coordinate space.

The widths in momentum and space satisfy the minimum uncertainty principle condition $\sigma_r\sigma_p = \frac{1}{2}\hbar$. Another assumption concerns the N-body Wigner distribution that in CoMD model [11](like in QMD approach[10]) is a direct product of the single particle distributions. In this case therefore for a system formed by A nucleons the one-body and 2-body distributions above introduced have a multi-component structure that is:

$$D_1(\mathbf{r},\mathbf{p}) = \sum_1^A f_i(\mathbf{r},\mathbf{p}) \tag{7}$$

$$D(\mathbf{r},\mathbf{p},\mathbf{r}',\mathbf{p}') = \sum_{i\neq j=1}^A f_i(\mathbf{r},\mathbf{p})f_j(\mathbf{r}',\mathbf{p}') \tag{8}$$

f_i represents the Gaussian Wigner transform of the wave packet associated to the generic nucleon i. From the above relations we can see that in general $D \neq D_1(\mathbf{r},\mathbf{p})D_1(\mathbf{r}',\mathbf{p}')$. In particular, for the case in which these distributions are expressed as a sum of different localized components inside a volume V_g, we can indicate with a_g the number of components which give non negligible contributions in V_g. The relative difference $1 - \frac{D_1(\mathbf{r},\mathbf{p})D_1(\mathbf{r}',\mathbf{p}')}{D}$ can be estimated to be of the order of $1/a_g$ which corresponds to the ratio between diagonal ($i = j$) and off-diagonal elements ($i \neq j$) within the ensemble of $a_g(a_g - 1)$ terms. In semiclassical mean-field models a_g can be enough high to make the difference negligible, in fact the single particle distribution usually spreads over the whole system (test particles methods) and therefore the truncation condition can be retained valid [5, 6, 7]. On the contrary this is not surely the case for the molecular dynamics approaches for which the typical spreading volume is of the order of 2-10 fm^3. Localization and therefore coherence of the wave-packed used to describe the single-particle wave-functions allows to keep memory of the two-body nature of the inter-particles interaction, and at same time, allows for the spontaneous appearance of the clustering phenomena in simulations concerning low-density and excited portion of nuclear matter [27].

With these assumptions on the 2-body phase-space distribution, taking in to account the properties of the δ function, we can obtain the explicity expression for the different terms concerning the total energy; for the two-body isoscalar contribution W_{twb} we get:

$$W_{twb} = \frac{T_0}{2\rho_0}\sum_{i=1}^{A} S_v^i \quad (9)$$

$$S_v^i = \sum_{j\neq i=1}^{A} \frac{1}{(4\pi\sigma_r^2)^{3/2}} exp[-\frac{(\mathbf{r}_i - \mathbf{r}_j)^2}{4\sigma_r^2}] \quad (10)$$

In the above expression S_v^i is the normalized sum of the Gaussian terms and it represents just a measure of the overlap between the nucleonic wave-packets. Its two body character is quite explicit. In the calculations concerning the NM simulation that we will illustrate in the next sections , the large number of particles A involved in the system allows us to write the above quantity in a simpler way by introducing the average overlap per nucleon $\overline{S_v^i} = \overline{S_v}$ whose dependence on the particle index in the ideal case can be omitted:

$$E^C_{twb} = \frac{W_{twb}}{A} = \frac{T_0}{2\rho_0}\overline{S_v} \quad (11)$$

Concerning the three-body term, taking in to account the previous obervations we get:

$$E^C_{trb} = \frac{T_3}{(\sigma+1)\rho_0^\sigma}\overline{S_v}^\sigma \quad (12)$$

For the term related to the iso-vectorial interaction in the limit $A, N, Z >> 1$ (N and Z represent the number of neutron and protons) we obtain:

$$E^C_{isv} = \frac{T_4}{2\rho_0}\tilde{\rho}_A(\frac{\overline{S_v}}{\overline{S_{v0}}})^{\gamma-1}[(1+\frac{\alpha}{2})\beta^2 - \frac{\alpha}{2} \quad (13)$$

$$E_{bias} = -\frac{T_4}{4\rho_0}F'_k(\overline{S_v})\tilde{\rho}_A\alpha \quad (14)$$

where $\tilde{\rho}_A \equiv A\tilde{\rho}$ is related ot the average overlap for couples of homonym nucleons. From eq.(13) we obtain that for $\alpha \neq 0$ ($\alpha \sim 0.15$ as CoMD calculations predicts) the iso-vectorial force produces a term which we name E_{bias}. It has no explicit dependence on β (likewise iso-scalara term) but depends on the same parameter and density form factor which characterize the symmetry energy (first term in eq.(13)) The related effect is not-negligible if compared to the balance of the different terms appearing in the expression of the total energy. The correlation coefficient α responsible for the appearance of the E_{bias} term represents the difference in percentage of the overlap between the neutron-proton couples (n-p) from the one related to the couples formed by homonym nucleons. It is positive because n-p pairs experience a larger overlap with respect the other ones due to the more attractive interaction. α also increases with the strength of the iso-vectorial forces (T_4 parameter).

Finally, the kinetic contribution E_{kin} is obtained in a self-consistent way for ground state configurations by applying the cooling-warming procedure coupled with the constraint on the occupation numbers (see Ref.[12]) given by the Pauli principle. The total energy per nucleon is therefore obtained by adding all the contributions discussed in this section and it will be indicated as $E^C = E^C_{pot}(\overline{S_v}, \alpha, \tilde{\rho}_A) + E_{kin}(\rho)$. In particular we note that, at this level, it depends on the new defined primary quantity $\overline{S_v}$, α, $\tilde{\rho}_A$. It can be shown that these quantity are strictly related to the spatial correlation function between nucleon pairs and to the density. However a complete and explicit expression for the density dependence of these quantities can not be obtained in an analytical way due to the problem complexity. Therefore in the following we will briefly illustrate the obtained results which are based on the dynamical simulation of very large slabs of nuclear matter.

3. The nuclear matter simulation

As mentioned in the introduction the main goal of the present work is to understand also at a quantitative level and for a given simple form of the effective interaction, what are the consequences produced by the correlations above discussed by taking as a reference point the results obtainable in the framework of the Se-MFA approach. The study will be performed in a narrow region of density around the saturation density ρ_0. To this aim in the following we will try to find the set of parameters for the effective interaction which reproduce in both the cases some of the commonly accepted saturation properties of symmetric nuclear matter. In particular we will refer to ρ_0=0.165 fm^{-3}, binding energy $E(\rho_0) = -16$ MeV, compressibility modulus for symmetric nuclear mater $K_{NM}(\rho_0) = 220$ MeV. The strength of the iso-vectorial interaction T_4 will be chosen in such way to produce a symmetry energy value $e_{sym}(\rho_0) \simeq 28.6$ MeV in the case of the Se-MFA, i.e. $T_4 = 32$ MeV. From the functional E reported in eqs.(3-4) we can find easily the parameter values satisfying the above conditions by solving the system formed by the following set of equations for symmetric NM:

$$\rho_0 = 0.165; \quad E(\rho_0) = -16MeV; \quad \frac{dE}{d\rho}/_{\rho_0} = 0; \quad 9\rho_0^2\frac{d^2E}{d^2\rho}/_{\rho_0} = K_{NM}(\rho_0) = 220MeV \tag{15}$$

We get the following values for the parameters: $T_0 = -263 \pm 1.3$ MeV, $T_3 = 208 \pm 1$ MeV and $\sigma = 1.25 \pm 0.02$. The evaluation of the total energy per nucleon E^C related to the CoMD calculations requires the solution of the many-body problem using the equation of motion regulating the wave-packet dynamics in large portion of nuclear matter and also correction for surface effects. Details on the self-consistent and the iterative procedure used can be found in ref.[27]. In particular these calculation shows that as expected $\overline{S_v(\rho)}$ increases with the density, while $\alpha(\rho)$ decreases. The density dependence of this quantities is not linear. In particular, quadratic corrections with the density are not negligible around the saturation density and are enough to well represent the global behavior.

In the left-panel of Fig.1 we show the final values of the interaction parameters which are able within 10-20% to reproduce the above mentioned EOS properties. To satisfy these condition the γ parameter describing the degree of stiffness of the iso-vectorial interaction has to change in a rather restricted range of values, approximately 0.85-1.5

We also note that as due to the coupling effects generated by the many-body correlations (including also the one derived from the Pauli principle) the new set of parameters which describes the iso-scalar interaction show a dependence on γ. Furthermore, the magnitude of the new parameter values is about a factor two smaller than the one obtained in the limit of the semiclassical approximation. Calculations have been performed also for asymmetric nuclear matter. The obtained value of the symmetry energy at the saturation density is about 1-1.6 MeV lower than the value obtained in the case of the mean-field approximation. Larger differences are instead observed for the density dependence of the symmetry energy. In particular it can

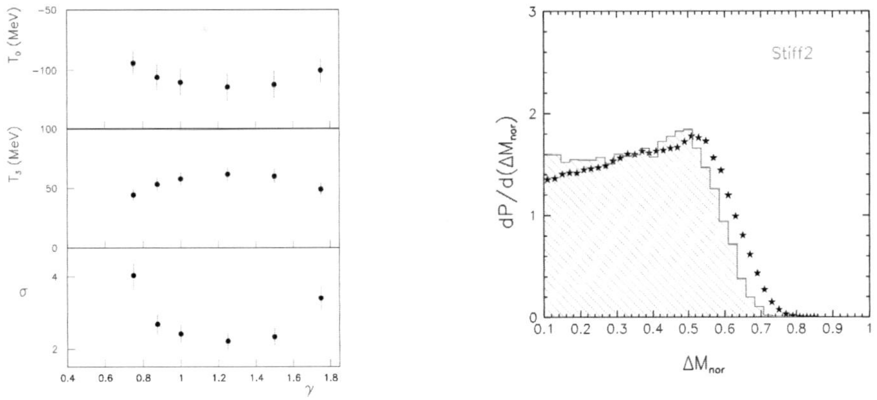

Figure 1. (Left-panel) The values of the parameters T_0 T_3 and σ (panel a,b,c respectively) obtained through the iterative procedure applied to CoMD calculations are shown as a function of the stiffness parameter γ. The bar errors represent the global uncertainty on the parameter values related to the numerical procedures-(Right-panel) Experimental Δ_{nor} yield distribution for the ^{48}Ca+^{48}Ca collision at 25 MeV/nucleon is shown with marked points. The red line represents CoMD-II calculations with $\gamma = 1$ (Stiff2 option)

be shown that the larger values(about a factor 3) [27] of K_{sym}(proportional to the second derivative of the symmetry energy at ρ_0) are strongly related to the Gaussian shape of the wave function used in molecular dynamics approaches. This result surely will affect the predicted the incompressibility modulus for asymmetric NM and an evaluation of this effect is in progress.

4. Effects on the nuclear dynamics

Up to now we have discussed the EOS limit, it is however very interesting to illustrate the effect of the correlation discussed also at a more qualitative level on the nuclear dynamics.

As an example in the right panel of Fig.1 we show experimental results on the $^{48}Ca +^{48} Ca$ at 25 Mev/A. Data have been collected with the Multi-detector CHIMERA [28]. In the figure we report the relative yields of well reconstructed events ordered trough the parameter $\Delta M_{nor} = \frac{m_1 - m_2}{m_{tot}}$, where m_1 and m_2 represent the mass of the biggest and the second biggest fragments produced. The yield of essentially binary processes including IMF production therefore can be located at small values of ΔM_{nor} while complete and incomplete fusion events are located at the larger ΔM_{nor} values. The continues line represents CoMD-II calculations which well fit the data using a γ value about 1. Consistently, same results are obtained when studying the similar systems $^{40}Ca +^{40} Ca$, $^{40}Ca +^{48} Ca$ and $^{48}Ca +^{46} Ti$ [29]. In particular the collected data for all these systems show that incomplete fusion processes are more suppressed for systems with the highest N/Z asymmetry. In the left panel of Fig.2 we show with different lines CoMD-II calculations performed by using different γ values and same set of parameter values concerning the iso-scalar interaction. The different curves indicate that the increasing of the stiffness associated to the iso-vectorial interaction (increasing values of the γ parameter) produce an increasing attractive effect with a correlated enhance of the incomplete fusion process. In the right upper panel of Fig.2 as an example we show a corresponding estimation of the coupling term E_{bias} in the EOS limit (see eq. 14) as function of the density for two extremal values of the

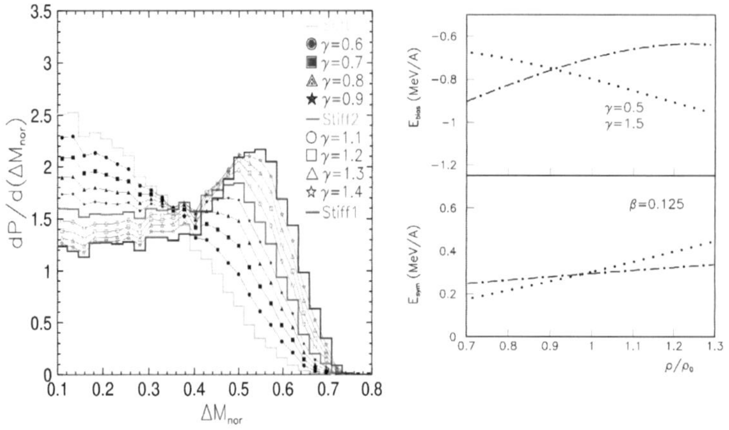

Figure 2. (Left panel) ΔM_{nor}(see the text) distribution evaluated through CoMD-II calculations for the ^{48}Ca+^{48}Ca collision at 25 Mev/nucleon and corresponding to different γ values. Stiff1 label indicates $\gamma \simeq 1.5$ while Soft $\gamma \simeq 0.5$ The Stiff2 option which fit well the experimental data (see Left panel of Fig.1) corresponds to $\gamma = 1$. (Right-upper panel) Estimated value of the E_{bias} term in NM limit corresponding to the above mentioned colliding system for two differnt extremal values of γ. (Right-bottom) Symmetry potential evaluated in the NM limit in the framework of the Se-MF approximation. We note that in this approximation E_{sym} is the only term related to the Iso-vectorial interaction

γ parameter. In the right-bottom panel, the symmetry potential E_{sym} (see last term in eq.2) is shown in the semiclassical Mean-Field limit. As can be seen from a comparison between the panels the repulsive/actractive behavior of the iso-vectorial interaction, (magnitude of the slope as function of the density) and the one related to the Δ_{nor} in the nuclear dynamics obtained by decreasing/increasing the values of the γ parameter is strongly correlated with the behavior of the E_{bias} correlation term rather than with the symmetry energy itself which shows instead an opposite trend. This comparison therefore shows that the many-body correlations discussed in this work concerning the iso-vectorial interaction strongly influence the behavior of observable related to the nuclear dynamics and surely they can strongly affect the results obtained from the fit procedure of experimental data with model based on the semiclassical wave packets dynamics.

5. Summary and Concluding Remarks

Many-body correlations produced in Molecular Dynamics approach based on the CoMD model have been discussed and their connection with the used effective interaction have been analyzed in the case of nuclear matter simulations. This study has been performed by comparing the results obtained for the total energy functional and the nuclear matter (NM) main saturation properties with the ones obtainable in the case of a semiclassical mean field approximations (Se-MFA). This comparison has been performed by using the same kind of simple effective Skyrme interaction. While we can expect the two approaches to produce large differences at low density due to the cluster formation process, the following study shows that noticeable differences are obtained also in a narrow range of densities around the saturation one where no cluster production has been observed. For asymmetric NM, in the range of the investigated asymmetry parameters ($\beta = 0. - 0.2$), the values of the obtained symmetry energies around the saturation density are only slightly different, but rather large differences are instead obtained for the slope L and especially for the K_{sym} curvature parameter. In CoMD calculations K_{sym} assume in fact rather higher and positive values. This result can be attributed to the shape of the single particle wave-packets which determines the behavior of the average overlap per nucleon as a function of the density. An example showing the effects of these correlations on a simple observable related to the balance between different reaction mechanisms in the $^{48}Ca +^{48} Ca$ at 25 MeV/nuleon collision has been also discussed. Finally we conclude by observing that even if from a numerical point of view the obtained results are strictly valid for the CoMD model, the performed study shows that the observed differences in the parameter values describing the effective interaction can have a more wide meaning. In fact, they are strictly linked to some rather general properties of the semiclassical wave packets dynamics.

References

[1] D.Vautherin and D.M.Brink 1972, *Phys.Rev.C* **5** 626.
[2] J.Meyer 2003, *Ann.Phys.Fr.* **28** n.3.
[3] M.Bender, P.-H. Heener. and P.-G. Reinhard 2003, *Rev.Mod.Phys* **75** 121.
[4] P.-G. Reinhard, M.Bender, W.Nazarewicz, and T.Vertse 2006, *Phys.Rev.C* **73** 014309.
[5] A.Bonasera, F.Gulminelli, and J.Molitoris 1994, *Phys.Rep.* **243** 1.
[6] G.Bertsh and S.Das Gupta 1991, *Phys.Rep.***202** 233.
[7] B.A.Li, W.Bauer and G.F.Bertsch 2095, *Phys.Rev C* **44** 2095.
[8] J.Randrup, G.F.Burgio, Ph.Chomaz 1992, *Nucl.Phys.* **A538** 393.
[9] M. Colonna, et al. 1998, *Nucl.Phys.* **A642** 449.
[10] J.Aichelin 1991, *Phys.Rep.* **201** 233.
[11] M. Papa, T. Maruyama and A. Bonasera 201, *Phys.Rev.C* **64** 024612.
[12] M.Papa, G.Giuliani A.Bonasera 2005, *Jou.of Comp.Physics* **208** 403.
[13] Ning Wang, Zhuxia Li, and Xizhen Wu 2002,*Phys.Rev.C* **65** 064608.
[14] H.Feldmeier and J.Schnack 2000, *Rev.Mod.Phys.* **72** 655.
[15] A.Ono 1999, *Phys.Rev.C* **59** 853(1999).
[16] M.Papa et al 2007, *Phys.Rev.C* **75** 054616 and references therein.
[17] M. Papa and G. Giuliani 2009, *Eur.Phys.Jour.* **A 39** 117.
[18] M. Prakash, T.L. Ainsworth, J.L. Lattimer 1988, *Phys.Rev.Lett.* **61** 2518.
[19] R.B. Wiringa, V. Fiks, A. Fabrocini 1988, *Phys.Rev.C* **38** 1010.
[20] A. Akmal, V.R. Pandharipande, D.G. Ravenhall 1998, *Phys.Rev.C* **58** 1804.
[21] W.Zuo, I.Bombaci, and U.Lombardo 1999, *Phys.Rev.C* **60** 024605.
[22] S. Yoshida, H. Sagawa, N. Takigawa 1998, *Phys.Rev.C* **58** 2796.
[23] J.Rikovska Stone, J.C.Miller, R.Koncewicz, P.D.Stevenson, M.R.Strayer 2003, *Phys.Rev.C***68** 034324.
[24] V.Baran, M.Colonna, V.Greco, M.Di Toro 2005, *Phys.Rep.* **410** 335.
[25] Bao-An Li, Lie-Wen Chen, Che Ming Ko 2008, *Phys.Rep.* **464** 113.
[26] M.B.Tsang et al 2009, *Phys.Rev.Lett.* **102** 122701.
[27] M.Papa 2013, *Phys.Rev.C* **87** 014001.
[28] A.Pagano et al 2004 *Nucl.Phys.* **A734** 504.
[29] F.Amorini et al 2009, *Phys.Rev.Lett.* **102** 112701.
[30] G.Cardella et al 2012, *Phys.Rev.C* **85** 064609.

Decay competition in IMF production in the collisions ^{78}Kr+^{40}Ca and ^{86}Kr+^{48}Ca at 10 AMeV

S Pirrone[1], G Politi[1,2], M La Commara[3,4], J P Wieleczko[5], E De Filippo[1], B Gnoffo[1,2], G Ademard[6], M Vigilante[3,4], F Amorini[7], L Auditore[8,9], C Beck[10], I Berceanu[11], E Bonnet[5], B Borderie[6], G Cardella[1], A Chbihi[5], M Colonna[7], A D'Onofrio[3,12], J D Frankland[5], E Geraci[1,2], E Henry[13], E Laguidara[1], G Lanzalone[7,14], P Lautesse[56], D.Lebhertz[5], N Le Neindre[17], I Lombardo[3,4], K Mazurek[5], A Pagano[1], M Papa[1], E Piasecki[17], M Quinlann[13], F Rizzo[2,7], E Rosato[3,4], P Russotto[1], W U Schroeder[13], G Spadaccini[3,4], A Trifirò[8,9], J Toke[13], M Trimarchi[7,8] and G Verde[1]

[1]INFN Sezione di Catania, Italy
[2]Dipartimento di Fisica e Astronomia, Università di Catania, Italy
[3]INFN Sezione di Napoli, Italy
[4]Dipartimento di Fisica, Università Federico II Napoli, Italy
[5]GANIL Caen,France
[6]IN2P3 - IPN Orsay, France
[7]INFN Laboratori Nazionali del Sud, Italy
[8]Dipartimento di Fisica, Università di Messina, Italy
[9]INFN Gruppo Collegato di Messina, Italy
[10]IN2P3 - IPHC Strasbourg, France
[11]IPNE, Bucharest, Romania
[12]Dipartimento di Matematica e Fisica - Seconda Università di Napoli,Caserta, Italy
[13]University of Rochester, USA
[14]Università Kore, Enna, Italy
[15]IN2P3 - IPN Lyon, France
[16]IN2P3 - LPC Caen, France
[17]University of Warsaw, Poland

E-mail: sara.pirrone@ct.infn.it

Abstract. De-excitation modes of compound systems ^{118}Ba and ^{134}Ba, produced respectively in the ^{78}Kr+^{40}Ca and ^{86}Kr+^{48}Ca collisions at 10 A.MeV, are investigated. In particular, the competition between the various disintegration decay paths of medium mass compound nuclei, formed by fusion processes and the isospin (related to N/Z) of the entrance channel influence on the decay process, are studied. Data were taken at the INFN-Laboratori Nazionali del Sud (LNS) in ISODEC experiment, by using the CHIMERA array. The experiment complements and improves the previous experiment performed at GANIL where the same mechanisms were studied at lower excitation energies. The results show the presence of a relaxed component in the reaction mechanism, evident staggering effects in the Z distributions, as well as different isotopic composition and neutron enrichment for the reaction products in the two systems.

Journal of Physics: Conference Series **515** (2014) 012018 doi:10.1088/1742-6596/515/1/012018

1. Introduction

In the recent years, one of the principal aim in the field of the nuclear reactions is the study of the influence of the entrance channel N/Z ratio, on the reaction mechanism and on the formation and production of the fragments in the exit channel.

In this context, particular interest is focused on the heavy-ion reactions at low energy (E/A< 10-15 AMeV), where the fusion process are predominant. Indeed, one expects that the N/Z ratio, directly connected to the isospin degree of freedom, plays an important role on the compound nucleus decay and on the emission process, providing crucial information on fundamental nuclear parameters as the level density, the fission barrier and the viscosity. Thus the chemical composition influences the fission dynamics and a program of systematic measurements of fission cross-section for a large isotopic chain of compound nuclei, from neutron rich to neutron-poor, provides careful information on this mechanism.

In this work, we present the up to date results of the ISODEC experiment, realized to study the competition between the various disintegration modes of ^{118,134}Ba compound nuclei produced in the reactions ^{78}Kr+^{40}Ca (neutron poor system) and ^{86}Kr+^{48}Ca (neutron rich system) at 10 AMeV. The experiment was performed at the INFN Laboratori Nazionali del Sud (LNS) in Catania by using the CHIMERA detector. It complements and improves the data obtained at 5.5 MeV/A for ^{78,82}Kr+^{40}Ca reactions [1], previously realized with beams delivered by GANIL facility in Caen (France) and by using the INDRA detector.

The investigated systems allow to produce compound nuclei in a large domain of N/Z (from 1.11 to 1.39) considering the use of stable beams, and similar both spin distribution and excitation energy. Such a set of data also will provide new constraint on sophisticated models attempting to describe statistical and/or dynamical properties [2] of excited nuclei.

2. Experimental Setup

The presented scientific program needs the measurements of several key observables such as the cross sections, multiplicities, angular and kinetic energy distributions of the reaction products (Intermediate Mass Fragments - IMFs, Light Charge Particles – LCPs and Fission Fragments - FF). The measurement of these observables requires a good isotopic resolution and a low energy thresholds for LCPs and IMFs, a high granularity and a broad angular acceptance.

The presence at LNS of a second generation 4π multi detector, CHIMERA [3], and the availability of beams with very good intensity and timing characteristics, strongly pushed to perform the experiment in Catania.

Figure 1. The CHIMERA detector at LNS

Journal of Physics: Conference Series **515** (2014) 012018 doi:10.1088/1742-6596/515/1/012018

Briefly, CHIMERA consists of 1192 detector telescopes, arranged on 9 rings in the forward part and 17 rings, in spherical configuration, in the backward part; the covered polar angle goes from 1° to 176° and the efficiency is the 94% of the total solid angle. In figure 1, a picture of the detector in its experimental scattering chamber at LNS is presented.
The single detection cell consists of a silicon detector (Si, thickness about 300 μm) followed by a Caesium Iodine–Thallium doped crystal, (CsI(T1), thickness from 3 cm to 12 cm), coupled to a photodiode.
The identification methods employed are:
ΔE-E for charge identification of particles punching through the Si detector and stopped in the CsI(Tl) with also mass identification for particles with Z<10;
E-TOF (Time of Flight) for mass identification, velocity and energy measurement of the particles stopped in the Si detector;
PSD (Pulse Shape Discrimination) in CsI(Tl), for isotopic identification of more energetic light charge particles;
PSD (Pulse Shape Discrimination) in Si detector, for charge identification of the particles stopped in the Silicon detector.
This latter technique [4,5] was the last implementation made on the CHIMERA array and allows the use of this device to study reaction mechanism also in the low energy domain, extending the investigation dynamical range from fusion to multifragmentation reactions.
In figure 2 is reported as an example the Energy (E_{Si}) vs Rise Time (RT_{Si}) plot obtained by the PSD methods in silicon detector, for the n-poor system ^{78}Kr+^{40}Ca at 10 AMeV at ϑ =34°.

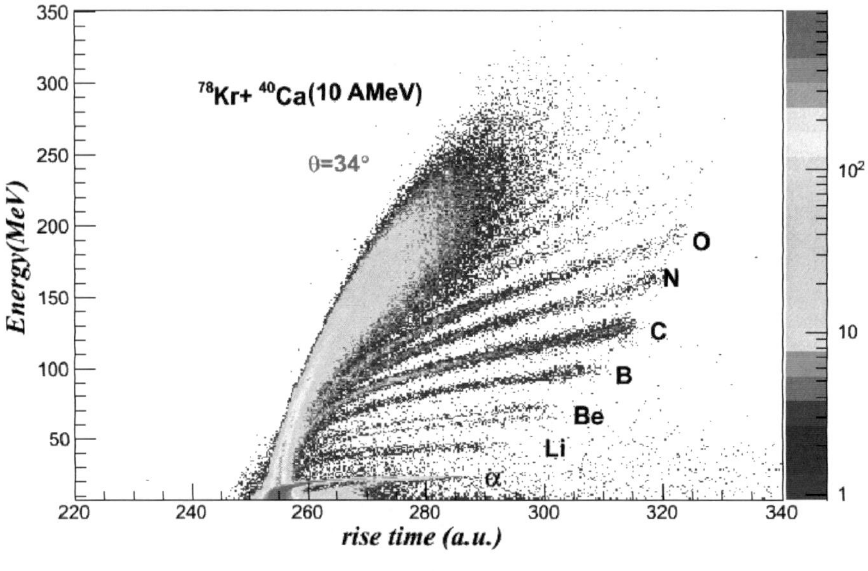

Figure 2. Energy-RiseTime plot for ^{78}Kr+^{40}Ca

Besides, the CHIMERA multidetector is characterized by a low energetic detection threshold, that is less of 0,5 MeV/A for heavy ions and 1 MeV/A for light particles. These characteristics allowed the complete identification of LCP in a wide energy range, the complete identification in charge and mass of the IMF (3<Z<8) products, the charge identification up to Z =14-16 for products stopped in the Silicon, and up to about Z = 30 for the most energetic particles stopped in CsI.

In figure 3 is shown an example of the typical matrix ΔE-E to show the good charge resolution discrimination of the array.

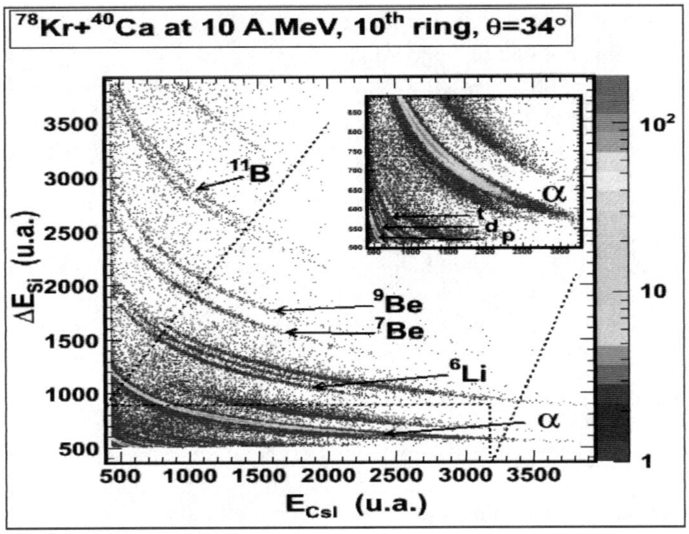

Figure 3. ΔE_{Si}-$E_{CsI(Tl)}$ plot for ^{78}Kr+^{40}Ca

Figure 4 shows an example of isotopic identification for particles punching-through the silicon detector. The mass distribution for Carbon isotopes is obtained at polar angle of 15° for the two studied reactions [6].

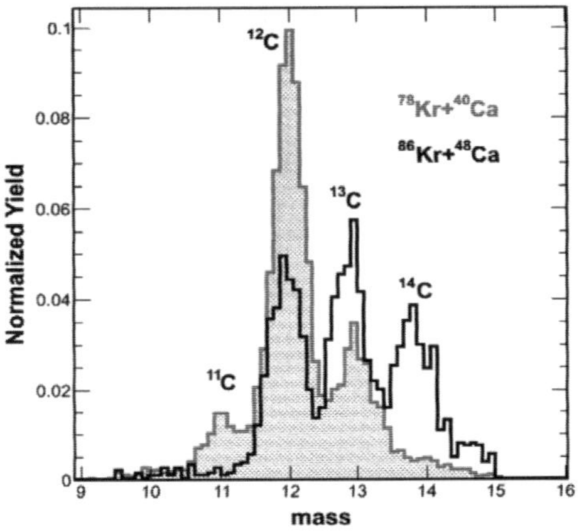

Figure 4. Carbon isotopes distributions at 15° for the two studied reactions

In the experiment, self-supporting 1 mg/cm^2 thick ^{40}Ca and ^{48}Ca targets, prepared in collaboration between INFN-LNL and INFN-LNS Target Laboratories, were bombarded by ^{78,86}Kr beams at 10 AMeV, delivered by the LNS Superconductive Cyclotron, with typical beam intensity of 800-1000 pA. The pulsed beam was delivered with a timing resolution about 800 ps - 1 ns. Inclusive and coincidence measurements were realized.

Journal of Physics: Conference Series **515** (2014) 012018 doi:10.1088/1742-6596/515/1/012018

3. Experimental Results

3.1. Mass distribution

In order to study the influence of the isospin degree of freedom it is essential to get the isotopic identification at least of part of the detected fragments. The capability of the CHIMERA detector allows us to get the mass distributions for the IMF $2 \leq Z \leq 8$.

In figure 5 the mass distribution for Z = 3, 4, 5 and for the two studied systems are reported[7,8].

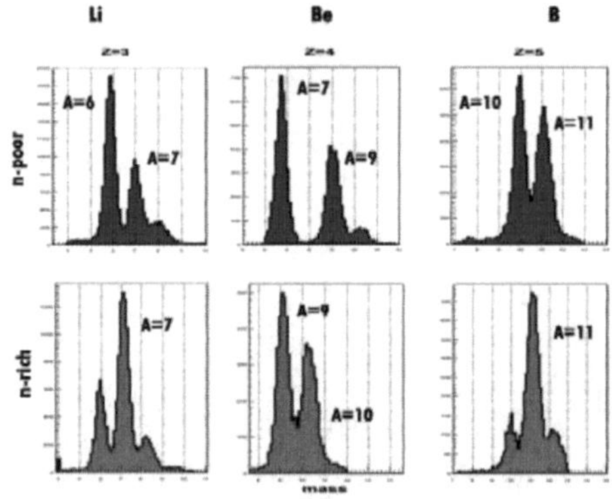

Figure 5. Mass distributions for Z=3,4,5, for the n-poor system and n-rich system

One can see the different isotopic composition and relative enrichment in correspondence of the same Z, for the compared systems. In particular this effect is evident for the Be element, in which the isotopic composition goes from A=7, 9, in the n-poor system, to A= 9,10, in the n-rich one. These differences in the two systems show that a memory of the entrance channel is still present in this class of reaction products, and this could be to an effects of the isospin influence on reaction mechanism or on the structure of the emitted fragments. The research on these effects is in progress.

3.2. Kinetimatical features

Relevant information on the reaction mechanism can be obtained from the kinetic energy spectra of the detected fragments. In figure 6, the Z=6 centre of mass energy spectra are reported for the neutron rich system, measured at the laboratory angles angle ϑ = 10.75°, 12.25°, 13.75°, 15.25°. For each angle, the spectra for the identified isotopes are shown.

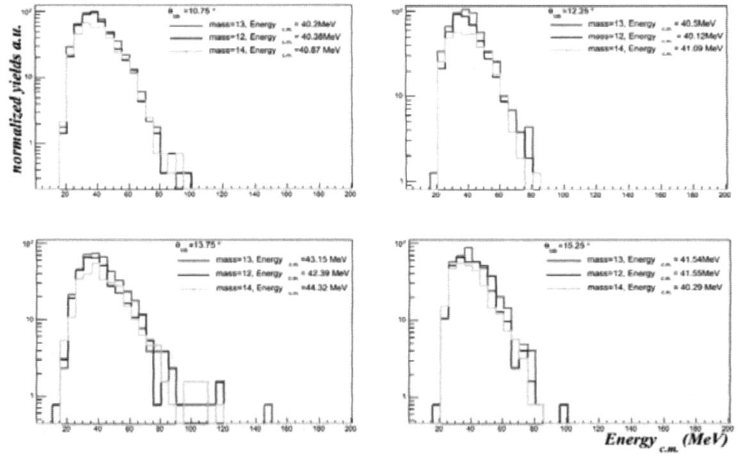

Figure 6. Center of mass Energy spectra for C isotopes for neutron rich system at different angles.

In figure 7 the centre of mass energy spectra for Z=12, at laboratory angle ϑ = 10.75°, 12.25°, 13.75°, 15.25° and for the neutron poor system is shown.
One notes the Gaussian-like shape of the spectra and that they are centred at the same energy value, independently from the detection angle. The same result is obtained for the energy spectra of the $1\leq Z\leq 17$ fragments and for both the studied systems.

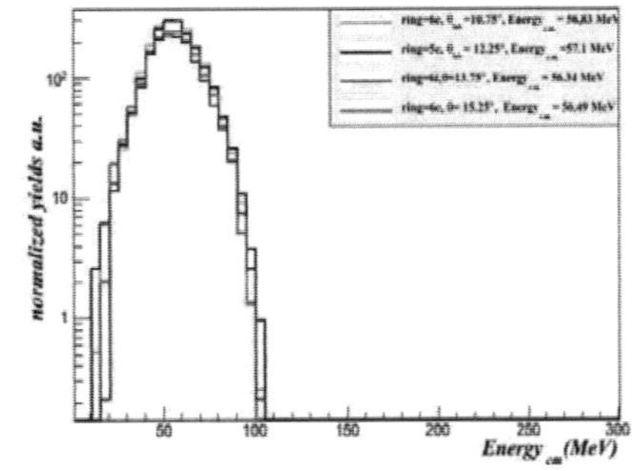

Figure 7. Centre of mass Energy spectra for Z=12 at different angles, for the neutron poor system

From the energy spectra the average velocity in the centre of mass $<v_{cm}>$ can be calculated for $3\leq Z\leq 17$. The results are presented in figure 8 for various laboratory angles and for the neutron poor system. One notes that for a given Z the average velocity in the centre of mass is essentially constant regardless of the emission angle.
The general behaviour observed for the emitted fragments, both in the energy spectra and in the average velocity, indicates an high degree of relaxation in the production mechanism.

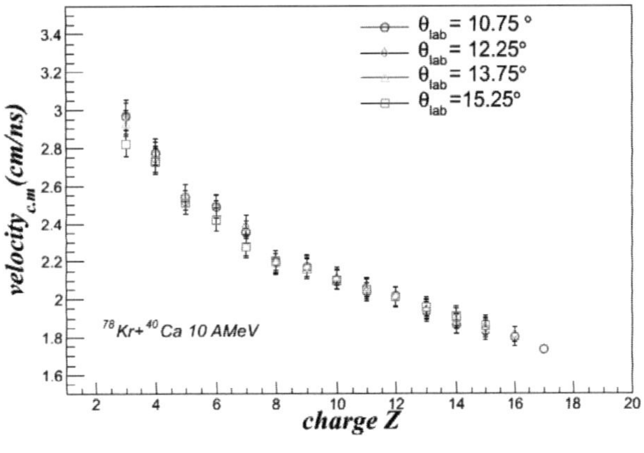

Figure 8. Average velocity distribution (see the text)

3.3. Angular distributions
Fundamental information can also be deduced from the angular distribution of the emitted fragments, allowing for example to distinguish totally equilibrated processes, as the compound nucleus formation followed by evaporation or fission, from partially equilibrated process as the fast-fission or the DIC (deep inelastic collisions) processes.
To get this kind of information the angular distribution for $8\leq Z\leq 14$ are compa red with the *1/sen* ϑ distribution behaviour, that is a clear indication of the isotropic emission typical of the compound nucleus emission. In figure 8, this comparison is shown for Z=8,10,12, 14 for the neutron poor

system; the *1/sen ϑ* distribution behaviour is reported with dashed line. This comparison gives a clear indication of the isotropic emission typical of the compound nucleus emission.

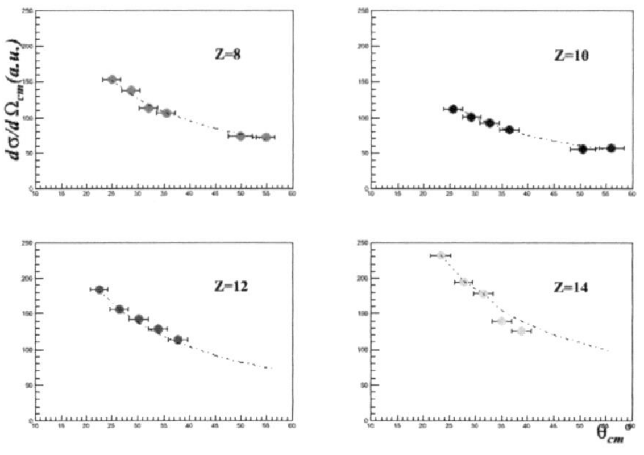

Figure 8. Angular distributions of Z=8,10,12,14 for the neutron poor system, compared with *1/sen ϑ* distribution behaviour (dashed line)

3.4. Charge distributions

As said before, one of the principal aim of the experiment is the study of the influence of the isospin or of the connected observables, on the reaction mechanism and on the fragments productions. This should appear also in the charge distribution, as already observed in the Ganil experiment.

A preliminary comparison between the fragments yields of the studied systems is presented in figure 9. The results are obtained by integrating the yields of particles with $3<Z<15$, identified with ΔE-E and PSD method on Si detector, in the angular range $10° \leq \theta \leq 16°$.

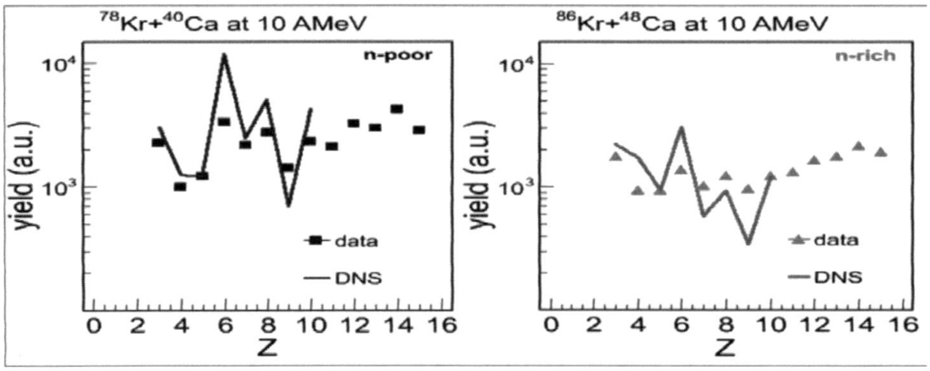

Figure 9. Charge distribution for the two studied systems and comparison with preliminary results from DNS-model.

In the figure, one can see a strong odd-even staggering of the Z yield for $Z<10$ detected fragments, and this effect persists for higher Z with a smaller amplitude.

Besides, comparing the two systems with different isospin, the yields of the IMF exhibit an even-odd staggering that is more pronounced for the neutron-poor system. This effect in the comparison appears also in other examples in the literature [9]

It seems that the neutron excess in the entrance channel affects the yields of the light fragments; the extraction of the absolute cross section of the reaction products, which is in progress, will confirm this

result. An influence can be originated also by structure effects [10] linked to the pairing forces and that is important for the studies on Symmetry Energy.
In figure 9 the lines are the preliminary results obtained by calculations with the dynamical DNS (Di Nuclear System) model, for Z<10. Following this model a dinuclear system is formed in the beginning of the reaction, that can go towards a not fully equilibrated fast-fission or towards an equilibrated fissioning compound nucleus. A detailed description of the model can be found in [11]. The calculation seems to be in a better agreement in the case of the neutron poor system, where the oscillation of the staggering of the yields is decreasing as the atomic number increases in agreement with the experimental findings.

4. Conclusions

Experimental methods and some results of the ISODEC experiment were illustrated. The experiment aims the study of the ^{78,86}Kr+^{40,48}Ca reactions at 10 AMeV incident energy and complements and improves the ^{78,82}Kr+^{40}Ca reactions at 5.5 MeV/A, realized at GANIL facility by using the INDRA detector.
Some of the results on the experimental features of the produced fragments were shown; in particular charge and mass distributions, as well as kinematical features and angular distributions were reported and compared for the two systems. These results indicate an high degree of relaxation of the system formed in the collision, before of its breakup, that is typical in compound nucleus and fission-like reaction mechanisms.
Staggering effects are evident in the Z distributions, as well as different isotopic composition and neutron enrichment for the reaction products in the two systems.
The observed effects could be due to the role of the N/Z degree of freedom on the decay channels. Data analysis and more refined comparisons with theoretical models are in progress.

5. References

[1] Wieleczko J.P. et al. *FUSION08 AIP Conf. Proc.* **1098** 64; Ademard G. et al. 2011 *Phys. Rev. C* **83** 054619
[2] Nadtochy P.N. et al. 2007 *Phys. Rev. C* **75** 064614
[3] Pagano A. et al. 2004 *Nucl. Phys.* **A734** 504
[4] Politi G. et al. 2006 *IEEE Nuclear Science Symposium Conference Record* **N28** 1140 ; Alderighi M. et al. 2005 *IEEE Trans. Nucl. Sci.* **52** 1624
[5] Bassini R. et al. 2006 *IEEE Nuclear Science Symposium Conference Record* **N14** 173
[6] Pirrone S. et al. *2011 EPJ Web of Conferences* **17** 16010
[7] Politi G. et al. 2012 *EPJ Web of Conferences* **21** 02003
[8] Pirrone S. et al. 2013 *AIP Conf. Proc.* **1524** 7, La Commara M. et al., 2013 *Proc. of INPC2013, Florence, Italy, (in press)*;
[9] Lombardo I. et al. 2011 *Phys. Rev. C* **84** 024613; Casini G. 2012 *Phys.Rev. C* **86** 011602(R)
[10] D'Agostino M. et al. 2011 *Nucl. Phys.* A **861** 47
[11] Kalandarov Sh.A., Adamian G.G., Antonenko N.V.and Scheid W. 2010 *Phys. Rev. C* **82** 044603; Zubov A.S., Adamian G.G., Antonenko N.V., Ivanova S.P. and Scheid W. 2003 *Phys. Rev. C* **68** 014616;

Journal of Physics: Conference Series **515** (2014) 012019 doi:10.1088/1742-6596/515/1/012019

Low Energy Reaction Mechanisms In a Transport Approach

C Rizzo[1], M Colonna[1], M Di Toro[1], V Baran[2]

[1] *Laboratori Nazionali del Sud INFN, I-95123 Catania, Italy.*
[2] *Physics Faculty, Univ. of Bucharest and NIPE-HH, Romania.*

E-mail: rizzoc@lns.infn.it

Abstract. Starting from Heavy Ion Collisions with exotic nuclear beams at low energy, the reaction path is investigated in a transport theory based on a stochastic mean field approach. We focus on the interplay between reaction mechanisms, fusion vs. break-up in HIC with exotic nuclear beams at low energies. Fusion probabilities for reactions induced by 132Sn on 64,58Ni targets at 10AMeV are evaluated looking at the evolution of the phase-space quadrupole collective modes. The break-up events appear sensitive to the stiffness of the symmetry energy. 197Au + 197Au collisions at 15 and 23 AMeV are simulated to investigate the main modes of re-separation, ternary-quaternary partition, of a heavy nuclear system.

1. Introduction

We focus the attention on the fusion vs. deep-inelastic mechanism interplay for dissipative Heavy Ion Collisions (HIC) with exotic nuclear beams at low energies, just above the Coulomb Barrier (between 5 and 20 AMeV). The exotic nuclei production, in laboratory conditions, can open the exploration of new nuclear structure and dynamics aspects: unstable ion beams with large asymmetry, i.e. up to extreme ratios of neutrons (N) to protons numbers (Z), and with good intensities in this energy range will be soon available in Radioactive Ion Beam facilities as SPES [1] or SPIRAL2 [2]. The goal of this analysis is to pin down specific observables which are sensitive to the symmetry energy to learn about its poorly known density behaviour.

We show that the reaction dynamics with HIC undergoes a fusion/break-up path bifurcation at early times. This bifurcation due to Coulomb and angular momentum effects, is essentially ruled by dynamical fluctuations and it could be sensitive to differences in the repulsive symmetry term, because the neutron-rich neck connecting the two colliding partners can survive a long time producing in the case of break-up very deformed final fragments, eventually leading to ternary/quaternary fragmentation events.

Usually the observed non fusion processes were found to be binary deep-inelastic reactions at energies below 20MeV/nucleon and at peripheral impact parameters. Investigating heavy ions collisions and symmetric systems, it is not possible to describe fusion exit channel, even in central collisions, due to strong Coulomb repulsion: a ternary-quaternary partitioning of the system can be observed. Starting from LNS data [3] we can evaluate the fast ternary and quaternary breakup mechanism of a very heavy collision, ^{197}Au + ^{197}Au, at 15MeV/nucleon: the colliding system will break into three or four fragments of comparable mass in inelastic collisions, which are nearly aligned along a common reseparation axis.

The observed ternary and quaternary breakup reactions should not be mixed up with the very asymmetric projectile partitioning observed at the Fermi-energy domain, interpreted as neck fragmentation processes [4] and described by the stochastic Boltzmann-Nordheim-Vlasov (BNV) [5] and the Constrained Molecular Dynamics (CoMD-II) [6] models: the observed fragments seem to originate from the nearly symmetric non equilibrium breakup of the primary products of deep-inelastic reactions.

2. Reaction dynamics

The reaction path followed by HIC with neutron-rich (or exotic) nuclear beams at low energies is investigated within a transport theory based on a Stochastic Mean Field (SMF) approach, extension of the microscopic Boltzmann-Nordheim-Vlasov transport equation [7, 8].

$$\frac{\partial f}{\partial t} + \frac{\mathbf{p}}{m}\frac{\partial f}{\partial \mathbf{r}} + \frac{\partial U}{\partial \mathbf{r}}\frac{\partial f}{\partial \mathbf{p}} = I_{coll}[f] + \delta I[f] \quad (1)$$

The stochastic fluctuating term δI[f] is implemented by fluctuations of the spatial density [9] in the SMF model; it is essential to allow the growth of dynamical instabilities. The mean field is built from Skyrme forces:

$$U_{n,p} = A\frac{\rho}{\rho_0} + B(\frac{\rho}{\rho_0})^{\alpha+1} + C(\rho)\frac{\rho_n - \rho_p}{\rho_0}\tau_q + \frac{1}{2}\frac{\partial C}{\partial \rho}\frac{(\rho_n - \rho_p)^2}{\rho_0} \quad (2)$$

where $q = n, p$ and $\tau_n = 1$, $\tau_p = -1$. The A, B and α coefficients, characterizing the isoscalar part of the mean-field, are fixed to reproduce the saturation properties of symmetric nuclear matter (ρ_0 = 0.145fm^{-3}, E/A = −16MeV), with a compressibility modulus around 200 MeV. The time evolution of the semi-classical one-body distribution function $f(r,p,t)$ is following a Boltzmann-Langevin evolution dynamics where two parameterizations are implemented for the density dependence of symmetry energy (Asy-stiff EOS, $C(\rho)/\rho_0 = ({}^{32}/_{\rho 0})({}^{2\rho}/_{\rho+\rho 0})$, and Asysoft choice, $C(\rho)/\rho_0 = 482 - 1638\rho$, where C(ρ) function give the potential part of the symmetry energy) [7].

3. Fusion dynamics for ^{132}Sn induced reactions

A systematic and detailed fusion dynamics study, focusing on the symmetry energy dependence, will be possible with the availability of intense low energy radioactive beams with a different N/Z asymmetry. We perform the ^{132}Sn + ^{64,58}Ni reactions at 10 AMeV for semi-peripheral impact parameters (from b =4.5 fm to b = 8.0 fm, with Δb= 0.5 fm) to study isospin and symmetry energy effects on the competition between fusion and break-up (deep-inelastic), exploiting the SMF approach where an in-medium dependent nucleon-nucleon cross section is employed in the collision integral, I_{coll}, via the local density [10]. The cross section is set equal to zero for nucleon-nucleon collisions below 50 MeV of relative energy. In this way we avoid an increasing numerical noise due to spurious effects that may dominate in this energy range when the calculation time becomes too large. We used 100 test particles per nucleon for a better description of the phase space, and Gaussian phase space wave packets of test particles are considered to map the particle occupation at each time step.

We show that the reaction dynamics undergoes a fusion/break-up path bifurcation at very early times due to Coulomb and angular momentum effects. In such critical transition stage the final outcome is essentially ruled by dynamical fluctuations and it will be rather sensitive to differences in the neutron-repulsive symmetry term.

Fusion probabilities for reactions induced by ^{132}Sn beam at 10AMeV are evaluated by the evolution of the phase-space quadrupole collective modes, to extract the fusion cross section at relatively early times: $Q(t) = \langle 2z^2(t) - x^2(t) - y^2(t) \rangle$ the quadrupole moment in coordinate space averaged over the space distribution in the composite system and $QK(t) = \langle 2p^2_z(t) - p^2_x(t) - p^2_y(t) \rangle$, the quadrupole moment in momentum space evaluated in a spatial region around the center of mass (z-axis coincides with the symmetry axis, the x-axis is on the reaction plane). The information on the final reaction path is deduced investigating the quadrupole fluctuations of the system at early times (200-300 fm/c), when the formation of composite elongated configurations is observed [11]. The method described can indicate when the reaction mechanism is changing into the SMF approach.

The Q(t) behaviour is different for more peripheral impact parameters (increasing trend, i.e. break-up) and for b=5-6 fm, where we have still a little oscillation in the time interval between 100 and 300

fm/c, indication of a fusion contribution. These observations can be interpreted assuming that starting from about b = 5 fm, we have a transition from fusion to a break-up mechanism, like deep-inelastic. The Q(t)-slope values should be associated with a quadrupole deformation velocity of the dinuclear system that is going to a break-up (Q(t)-slope > 0) or to a fusion (Q(t)-slope < 0) exit channel.
When increasing impact parameter the quadrupole deformation in momentum space QK(t), in a sphere of radius 3 fm around the center of mass, becomes more negative in the time interval between 100 and 300 fm/c. Since the angular momentum is large, the components perpendicular to the symmetry axis, which is rotating in reaction plane, are increasing leading to a separation of the deformed di-nuclear system. The break-up probability increases when the quadrupole moment in p-space is more negative [11].
The Fig. 6 of ref. [11] shows the Q slope-QK correlation plots for the systems $^{132}Sn + ^{64}Ni$ and the two asy-EOS. We can evaluate the normal curves of the displayed quantities and the relative areas for each impact parameter in order to select the events: break-up events will be located in the regions with both positive slope of Q and negative QK. In this way, we can evaluate the fusion events for each impact parameter by the difference between the total number of events and the number of break-up cases [11].
In Figure 1 we present the fusion distribution plots. For the $^{132}Sn + ^{64}Ni$ system (left panel), we note a difference between the σ-fusion corresponding to the two different asy-EOS in the centrality transition region, with larger values for Asysoft, a 4-5% effect in the neutron rich system: 1128 mb (Asysoft) vs. 1078 mb (Asystiff), and even smaller, 1020 mb vs. 1009 mb, for the ^{58}Ni target (Figure 1, right panel). However, through a selection in angular momentum, $130 \leq l \leq 180\hbar$, we find that the Asysoft cross section curve is significantly above the Asystiff one, 10-15% in the case of the more neutron-rich system. We like to note that for the neutron-rich case, $^{132}Sn+^{64}Ni$, our absolute value of the total fusion cross section presents a good agreement with recent data, at lower energy (around 5 AMeV), taken at the ORNL [12].

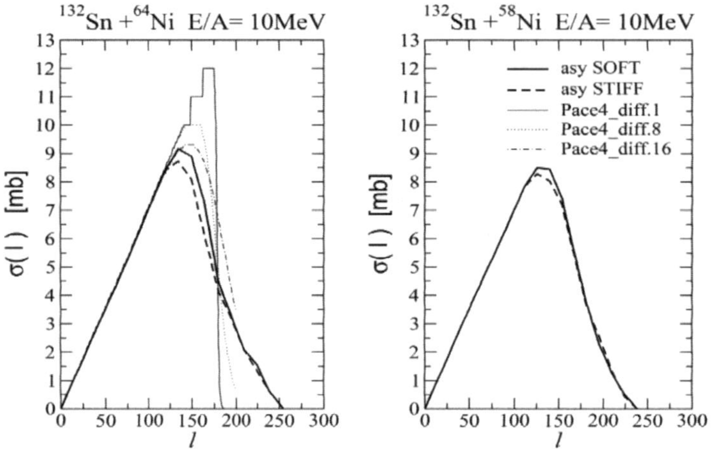

Figure 1. Angular momentum distributions of the fusion cross sections (mb) for the two reactions and the two choices of the symmetry term. For the $^{132}Sn + ^{64}Ni$ system (left panel), the results of *PACE4* [13, 14] calculations are also reported, for different l-diffuseness.

The larger fusion probability obtained with the Asysoft choice seems to indicate that the reaction mechanism is regulated by the symmetry term at suprasaturation density, where the Asysoft choice is less repulsive for the neutrons [7, 15].

A first transport approach analysis of symmetry energy effects on the structure of break-up events in peripheral collisions of ^{132}Sn +^{64} Ni at 10 AMeV has been reported in ref. [16]. The neck dynamics on the way to separation is found influenced by the symmetry energy below saturation. This can be observed in the deformation pattern of the Projectile-Like (PLF) and Target-Like Fragments (TLF) [16]: larger deformations are seen in the Asystiff case for peripheral impact parameters, corresponding to a smaller symmetry repulsion at the low densities probed in the separation region.

4. Analysis of ternary events on ^{197}Au + ^{197}Au reaction

A study of the ^{197}Au + ^{197}Au reaction at 15*A*MeV and 23*A*MeV energy was carried out [3, 17, 19] by the CHIMERA Collaboration at Laboratori Nazionali del Sud (LNS). For this system at an energy of 15AMeV and at relatively small impact parameters, corresponding to strongly damped collisions [18], the data contain predominantly binary events but a fast break-up into three or four massive fragments of comparable mass has been experimentally revealed.

The ternary events are dominated by processes in which the heaviest fragment, close to the ^{197}Au mass, is the remnant of the projectile fragment (PLF) or the target fragment (TLF), while the two fragments, F1 and F2, are induced from the complementary primary fragment breakup of the TLF or PLF: ^{197}Au + ^{197}Au → TLF + PLF → TLF + F1 + F2 or F1 + F2 + PLF. The mass spectra of F1 and F2 fragments differ significantly, that excludes the only pure statistical fission [3].

In the ternary partitioning case it is observed an alignment of the fragments close to the axis of the system re-separation into PLF and TLF: the PLF (or PLF) break-up is collinear with PLF-TLF separation axis. The smallest mass fragment, F1, has an energy greater than F2, as shown by the in-plane velocity distribution of fragments in ref. [3] (this excludes a purely statistical fission).

The partitioning results of a study of the same system at an energy of 23AMeV identified a variety of light, medium and heavy fragments.

An Intermediate-Mass Fragment (IMF) is observed with *A* less than 20 in addition to two main fragments, a TLF and a PLF. Emission of IMFs is a neck fragmentation process during the reseparation of the colliding system, that was observed in other experiments, but at higher energies. Ternary events are also observed going to a heavy fragment, TLF (or PLF) and two other fragments of comparable mass, originating from the reaction partner break-up or fission. The fig. 4 of ref. [19] of the two-dimensional transversal vs. longitudinal velocity distribution of three fragments, projected on the reaction plane, shows also the fragments F1 and F2 concentrated around a 'Coulomb ring' that indicated a sequential binary split mechanism of the primary PLF (i.e. after the TLF+PLF separation) in addition to a fragment alignment along the system re-separation axis. The experimental velocity distribution of three fragments could be decomposed in neck fragmentation, IMF, induced symmetric fission and F1-F2 breaking, like at 15AMeV.

The QMD model reproduces some evidences of the reaction; both TLF and the dynamical breakup of PLF into two fragments of comparable masses are correctly located in the velocity space, but it fails to reproduce the collinear emission experimentally observed [3].

In this work, we tried to reproduce the dynamical break up. The mechanism is investigated in a transport theory based on a SMF model, where the Asystiff parameterization for the density dependence of the symmetry energy is implemented. In order to map the particle occupation at each time step, triangular phase space wave packets of test particles are considered. The in-medium nucleon-nucleon cross section is fixed at 0.14fm^{-3} density in the collision integral. We run 100 events for each set of macroscopic initial conditions and we obtain the 70% of ternary events over this ensemble.

In Figure 2 (left) we present a theoretical simulation of the ^{197}Au + ^{197}Au system at 15 AMeV for b=4fm, left panel, and b=6fm, right panel, impact parameters. The reaction simulation is very sensitive to the test particles number and to the surface term used in the code, especially because of strong Coulomb repulsion. 100 test particles per nucleon have been employed in simulations for an accurate description of the mean field dynamics. Through a qualitative study of the time evolution of the space density distributions projected on the reaction plane, the reaction mechanism seems to be heavily dominated by the quadrupole and octupole modes. The neutron-rich neck connecting the two partners survives a rather long time, producing very deformed primary PLF/TLF especially for more central impact parameters, that lead to three or four massive fragments of comparable size.

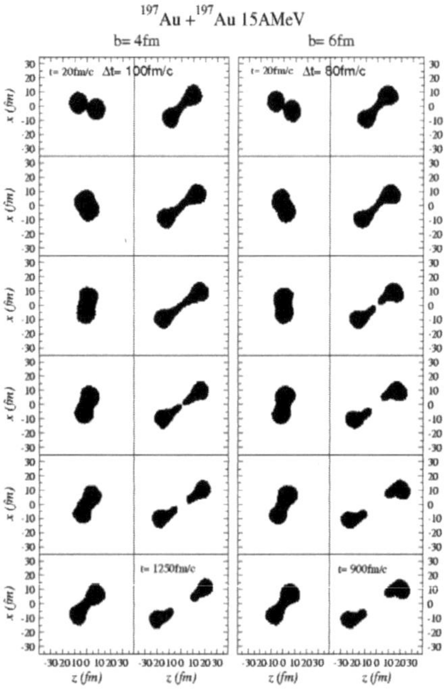

Figure 2. Time evolution of the space density distributions for the ^{197}Au +^{197}Au collisions at 15 AMeV beam energy, for b=4 and b=6fm impact parameter.

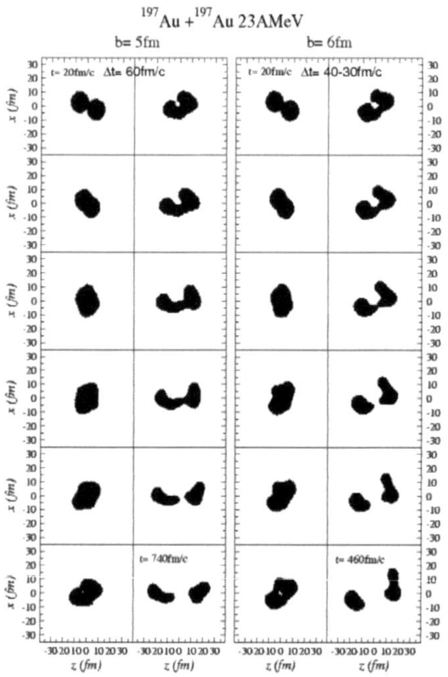

Figure 3. Like in Fig.2 but for ^{197}Au +^{197}Au system at 23 AMeV beam energy, b=5 and b=6fm centrality.

Assuming that the ternary events have a larger probability than quaternary events, we try in a simple way to evaluate the breaking fragment masses from the PLF or TLF residues with large deformation (with large quadrupole and octupole component): first we evaluate the center of the larger agglomerate inside the deformed PLF or TLF, then we obtain the nucleon number $N2_{part}$ integrating from the left extreme to the center (with fragment orientation as shown in figure 4).

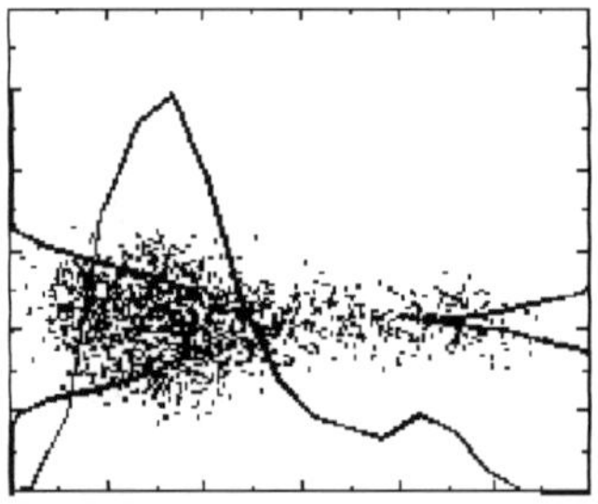

Figure 4. Scheme to evaluate the breaking fragment masses from the PLF or TLF (see text).

We can evaluate the F1 and F2 mass for each event by $F2= 2N2_{part}$ and $F1= N_{tot} - F2$. Figure 5 shows the reconstructed mass spectrum of lightest, called F1, and heaviest fragments, F2, obtained by this method, for two relatively small impact parameters where we observe almost all break up events.

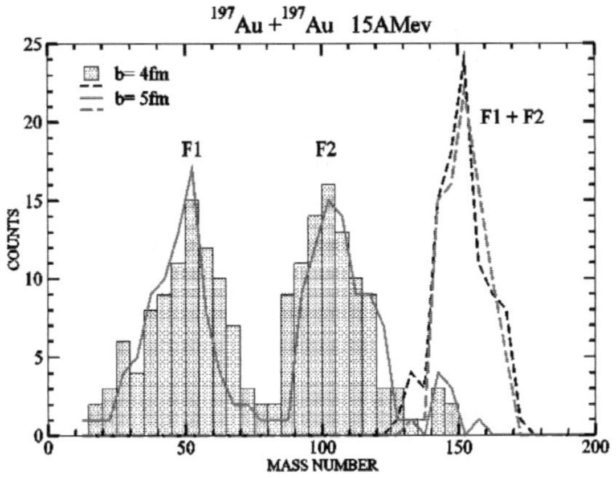

Figure 5. Reaction ^{197}Au+ ^{197}Au semi-peripheral at 15AMeV. The mass spectrum of mass numbers of fragments F1 and F2, for b=4fm and b=5fm impact parameters, reconstructed by the method describe above.

The mass distribution presents a good agreement with recent data at 15AMeV (shown in ref. [3], fig.4b), taken at LNS. In particular we can reproduce the distance between the F1 and F2 centroids and the ratio between the centroid and the variance of the distribution. This seems to be a nice evidence that the ternary partitioning in comparable masses comes from semi peripheral impact parameters. We underestimate the fragment mass, by about 20 units, with respect to the experimental data [3]. It is probably due to the overestimation of the evaporated spurious particles in the code because the calculation time for break up is very large at low energy especially for b=4fm and b=6fm. Indeed the evaluated time are 1250fm/c and 900fm/c, respectively. The final time choice is evaluated when the symmetry axis of the system is next to the deformed fragment axis, as shown by the experimental evidence, because the alignment is crucial to reproduce the velocities of the PLF and TLF fragments.
In this work we also report the partitioning study results of the same ^{197}Au + ^{197}Au system at the higher bombarding energy of 23AMeV. In addition to the partitions observed at low energies, an emission of intermediate mass fragments is also observed in the data at 23AMeV [19]. In Figure 3 (right) an increasing instability in the neck region is observed. The plot shows that a process of neck fragmentation during re-separation of the colliding system plays an important role, in which one of the two fragments is systematically enriched in mass at b=6fm. Increasing the beam energy and impact parameter, the statistical spontaneous fission is more favored while this effect could be not alive at 15AmeV energy because the system returns to a more spherical compound fragment. This can also be associated with enhanced quadrupole and reduced octupole modes that leads to a PLF (or TLF) statistical breaking. The observed strong instability could lead to lose the preferential collinear emission because we could have different fragment breaking times. This could also explain why the F1 and F2 fragment data are concentrated around a "Coulomb ring".
The velocity distribution of the primary PLF/TLF is reconstructed in figure 6 at 23 AMeV, b=6fm centrality, where the evaluated time is 750 fm/c. The simulated primary fragments are localized in a region of velocity space like in the data, that proves the good selection of the macroscopic initial condition set, in particular that the fragments originate from deep inelastic collisions in a region of semi-peripheral impact parameters.

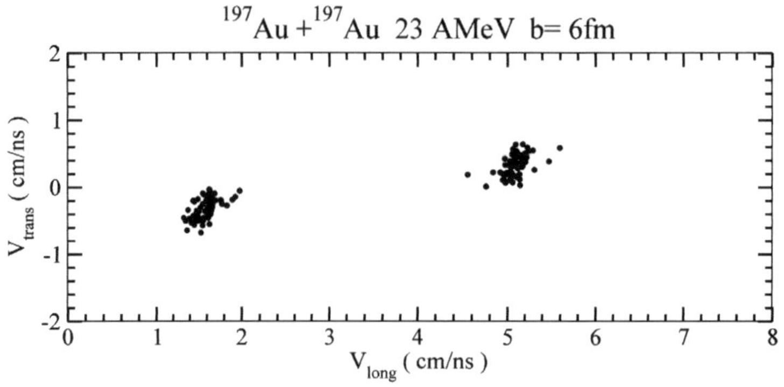

Figure 6. Velocity distribution of primary PLF and TLF fragments from partitioning of the ^{197}Au+^{197}Au system at 23AMeV energy beam and for b= 6fm centrality (in the laboratory reference frame as a function of the longitudinal and transversal components).

The mass spectrum of fragments F1 and F2, for b=5fm impact parameter, reconstructed by the method and approximations describe above is reported in left panel of figure 7. We notice an analogous behavior as observed at 15AMeV, especially for the centroid position, and a statistical event component between the F1 and F2 centroids, caused by the enhanced number of events with only quadrupole deformation. In the right panel, we report the spectra analysis excluding the fragments with only quadrupole component (we do not consider the statistical fission).

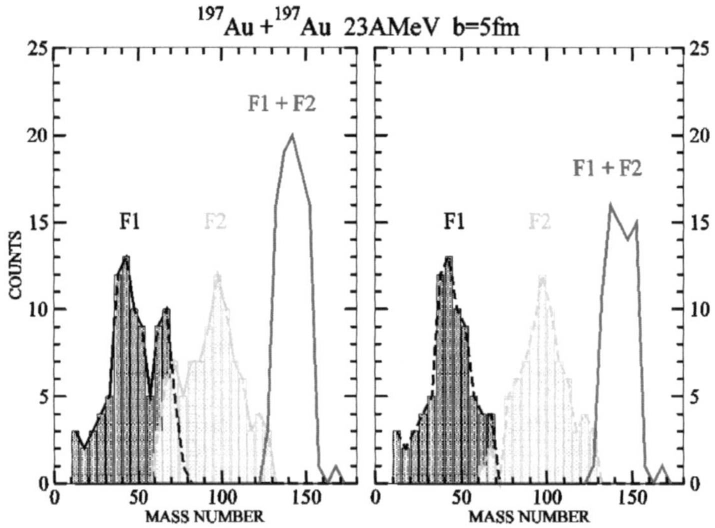

Figure 7. Reaction ^{197}Au+ ^{197}Au at 23AMeV for b=5fm impact parameter. The mass number spectrum of fragments F1 and F2 reconstructed by the method describe above. Left panel: events with quadrupole and quadrupole + octupole components. Right panel: quadrupole + octupole events (see text).

In addition, we repeated the analysis using the following approximations for b=5fm and b=6fm impact parameter: we identified the more energetic fragments with F1 and considered only the fragments with large quadrupole and octupole mode into the mass spectrum.

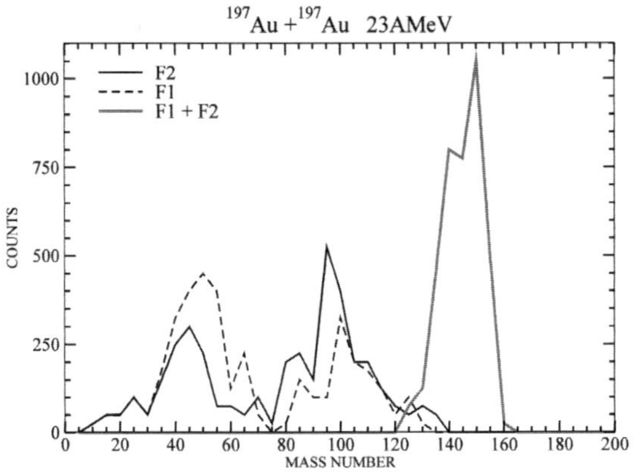

Figure 8. The weighted average mass number spectra of ^{197}Au+ ^{197}Au collisions at 23AMeV energy beam between b=5 and b=6fm impact parameters.

The figure 8 shows the weighted average spectra between b=5 and b=6fm. We notice that, with a simple impact parameter and energy selection, the plot reproduces the smooth behavior of the fragment masses, with a preferential emission direction as seen in the experimental data, especially at 15AMeV.

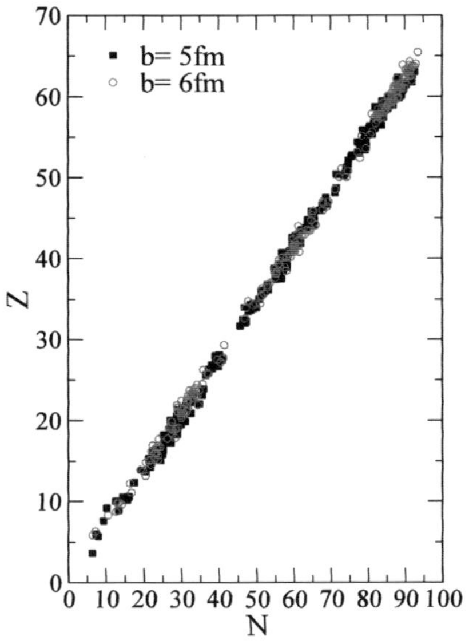

Figure 9. N - Z plot of PLF/TLF primary fragments and reconstructed fragments, by the method described in the text, at b = 5fm and b=6fm.

Let us move to discuss the fragment isotopic properties. The N - Z plot, figure 9, shows a linear behavior, with the same slope, for primary (region with $N \geq 70$ and $Z \geq 50$) and reconstructed fragments at b = 5fm. The lightest fragment exhibits a slightly larger slope at b=6fm. It seems that in this case, owing to the strong Coulomb repulsion in the composite system, protons are transferred to the low-density neck region, from which the lightest fragment emerges. On the other hand, the N/Z ratio seems to be more equilibrated at b=5fm, where the latter effect is not observed.

Conclusions and perspectives

The reaction study by stochastic BNV approach of collisions with exotic systems at 10AMeV beam energy, from fusion to dissipative binary events, showed that the fusion vs. break-up probability is influenced by the neutron repulsion because the observed densities are above the normal value during the approaching phase in the connecting region. Larger fusion cross sections are observed in asy-Soft case because we have a smaller symmetry energy value at supra-saturation densities. Our analysis is based on the fluctuations of quadrupole moments, in phase space, that develop during the di-nuclear phase and that essentially determine the final reaction channel. Isospin effects selecting the semi-peripheral impact parameters at low collision energy are revealed. The symmetry term study below and above saturation density can open new interesting perspectives for new experiments. We suggest the observable of the ternary/quaternary events probability too. We tested the ternary/quaternary reaction dynamics in the ^{197}Au + ^{197}Au collisions and the alignment along the axis of reseparation of PLF + TLF system in the first reaction phase. This study showed a larger ternary and quaternary partitioning probability at small impact parameters, where the characteristic collinear emission of the reseparation fragments is also observed, especially at 23AMeV, due to the increasing instability in the neck region.

References

[1] De Angelis G et al. 2011 *Journal of Physics Conference Series* **267** 012003
[2] http://pro.ganil-spiral2.eu/spiral2/
[3] Wilczynski J et al. 2010 *Phys. Rev.* C **81** 024605
[4] De Filippo E et al. 2005 *Phys. Rev.* C **71** 044602
[5] Baran V, Colonna M, Di Toro M 2004 *Nucl. Phys.* A **730** 329
[6] Papa M et al. 2007 *Phys. Rev.* C **75** 054616
[7] Baran V, Colonna M, Greco V, Di Toro M 2005 *Phys. Rep.* **410** 335
[8] Rizzo J, Chomaz P, Colonna M 2008 *Nucl. Phys.* A **806** 40
[9] Colonna M et al. 1998 *Nucl. Phys.* A **642** 449
[10] Li G Q, Machleidt R 1993 *Phys. Rev.* C **48** 1702; 1994 *Phys. Rev.* C **49** 566
[11] Rizzo C, Baran V, Colonna M, Corsi A, Di Toro M 2011 *Phys. Rev.* C **83** 014606
[12] Liang J F et al. 2007 *Phys. Rev.* C **75** 054607
[13] Gavron A 1980 *Phys. Rev.* C **21** 230
[14] Tarasov O B, Bazin D 2003 *Nucl. Inst. Methods* B **204** 174
[15] Colonna M et al. 1998 *Phys. Rev.* C **57** 1410
[16] Di Toro M et al. 2007 *Nucl. Phys.* A **787** 585c
[17] Skwira-Chalot I et al. 2008 *Phys. Rev. Lett.* **101** 262701
[18] Wilczynski J et al. 2010 *Phys. Rev.* C **81** 067604
[19] Cap T et al. 2013 *Phys. Scr.* T **154** 014007

Isospin Against Size Effects In Projectile Dynamical Fission For $^{112,124}Sn+^{58,64}Ni$ and $^{124}Xe+^{64}Zn$ Reactions At 35 A.MeV

P. Russotto[1], E. De Filippo[1], A. Pagano[1], E. Piasecki[2,3], L. Acosta[4], F. Amorini[4], A. Anzalone[4], L. Auditore[5], V. Baran[6], I. Berceanu[7], C. Boiano[8], B. Borderie[9], M. Bruno[10], T. Cap[11], G. Cardella[1], A. Castoldi[8,12], S. Cavallaro[4,13], M.B. Chatterjee[14], A. Chbihi[15], M. Colonna[4], M. D'Agostino[10], M. D'Andrea[1], M. Di Toro[4,13], F. Fichera[1], L. Francalanza[4,13], E. Geraci[1,13], R. Gianì[4,13], B. Gnoffo[13], A. Grimaldi[1], A. Grzeszczuk[16], C. Guazzoni[8,12], P. Guazzoni[8,17], N. Giudice[1], S. Kowalski[16], E. La Guidara[1], G. Lanzalone[4,18], G. Lanzanò[1,23], I. Lombardo[19,20], C. Maiolino[4], G.Marquínez-Durán[21], T. Minniti[13], M. Papa[1], E.V. Pagano[4,13], G. Passaro[4], S. Pirrone[1], R. Płaneta[22], G. Politi[1,13], F. Porto[4,13], L. Quattrocchi[5], M.F. Rivet[9], E. Rosato[19,20], F. Riccio[8,12], F. Rizzo[4,13], G. Saccà[1], K. Schmidt[16], K. Siwek-Wilczyńska[11], I. Skwira-Chalot[11], A. Trifirò[5], M. Trimarchi[5], G. Verde[1], M. Vigilante[19,20], J.P. Wieleczko[15], J. Wilczyński[23], P. Zambon[8,12], L. Zetta[8,17] and W. Zipper[16]

[1]INFN-Sezione di Catania, Catania, Italy
[2]Heavy Ion Laboratory, University of Warsaw, Warsaw, Poland
[3]A. Sołtan Institute for Nuclear Studies, Świerk/Warsaw, Poland
[4]INFN-Laboratori Nazionali del Sud, Catania, Italy
[5]INFN, Gr. Coll. di Messina and Dip. di Fisica e Scienze della Terra, Univ. di Messina, Italy
[6]Physics Faculty, University of Bucharest, Romania
[7]National Institute of Physics and Nuclear Engineering "Horia Hulubei", Bucharest, Romania
[8]INFN, Sezione di Milano, Milano, Italy
[9]Institut de Physique Nucléaire, CNRS/IN2P3, Université Paris-Sud 11, Orsay, France
[10]INFN, Sezione di Bologna and Dip. di Fisica, Univ. di Bologna, Italy
[11]Faculty of Physics, University of Warsaw, Warsaw, Poland
[12]Dip. Elettronica, Informazione e Bioingegneria, Politecnico di Milano, Milano, Italy
[13]Dip. di Fisica e Astronomia, Univ. di Catania, Catania, Italy
[14]Saha Institute of Nuclear Physics, Kolkata, India
[15]GANIL (DSM-CEA/CNRS/IN2P3), Caen, France
[16]Institute of Physics, University of Silesia, Katowice, Poland
[17]Dipartimento di Fisica, Univ. di Milano, Milano, Italy
[18]"Kore" Università, Enna, Italy
[19]Dip. di Fisica - Univ. di Napoli "Federico II", Compl. Un. Monte S.Angelo ed. 6, Via Cinthia, 80126 Napoli, Italy
[20]INFN-Sez. di Napoli, Compl. Un. Monte S.Angelo ed. 6, Via Cinthia, 80126 Napoli, Italy
[21]Departamento de Física Aplicada, Universidad de Huelva, E-21071 Huelva, Spain
[22]M. Smoluchowski Institute of Physics, Jagellonian Univ., Cracow, Poland

[23] deceased

[23]National Centre for Nuclear Research, Otwock-Świerk, Poland

E-mail: russotto@lns.infn.it

Abstract. In past experiments, mass asymmetric projectile-target combinations $^{124}Sn+^{64}Ni$ and $^{112}Sn+^{58}Ni$ were investigated at $E_{Lab}(^{112,124}Sn)$=35 A.MeV beam energy by using the 4π multi-detector CHIMERA. From a quantitative comparison of cross sections associated to Statistical and Dynamical Fission of the Projectile-Like Fragments, it resulted that Dynamical Fission process is about two times more probable in the neutron rich $^{124}Sn+^{64}Ni$ system than in the $^{112}Sn+^{58}Ni$ neutron poor one. In contrast, no sizable difference was found for Statistical Fission mechanism. The observed difference in the strength of the Dynamical effects could arise from the difference in entrance channel Isospin (N/Z) content. In order to disentangle Isospin effects from effects due to the different masses of the two systems, a new experiment $^{124}Xe+^{64}Zn$ at 35 A.MeV beam energy has been recently carried out.

1. Introduction

A characteristic experimental signature of Heavy-Ion reactions in the Fermi energy domain (20-100 A.MeV) is the abundant emission of relatively light fragments, the so called Intermediate Mass Fragments (IMF), having atomic number Z≥3. IMF show a broad velocity distribution ranging from Target-Like Fragments (TLF) velocity to Projectile-Like Fragments (PLF) one, indicating that their production comes from different reaction mechanisms with characteristic time scales, from dynamical and pre-equilibrium emission to statistical decay of excited systems at equilibrium. In particular, in semi-peripheral reactions, the production of IMF and Light Charged Particles (LCP) in the velocity region between the PLF and TLF (midrapidity) has been experimentally observed and the features of a dynamical origin (not related to a statistical emission from an equilibrated source) have been clearly evidenced in many works (see [1–6] and references therein). In addition, a clear signature of an emission chronology related to the IMF size was demonstrated [1–3, 7, 8]. It was found that lighter IMF (Z <∼ 9) in the mid-velocity region are likely to be emitted during the re-separation process of the primary interacting Projectile and Target Like Fragments (PLF* and TLF* respectively) in prompt neck-rupture mechanism (time <∼120 fm/c) [1, 2, 4, 7]. Many experimental results have shown a net neutron enrichment of fragments in the midvelocity region with respect to the ones coming from the sequential emission of PLF* or TLF* equilibrated nuclei [9–16]. In a recent paper [17] we have proved that the correlation between the isotopic content of the IMF, the alignment properties and the time-scale of their emission can be used to probe the symmetry term of the nuclear Equation of State (EOS) at densities below the saturation one [18].

Conversely, emission of heavier IMF (Z >∼ 9) was shown to happen at the late stage of the neck expansion process and was associated with the "Dynamical Fission" mechanism [3, 7, 8, 19, 20]; this is nearly a two-step (sequential) reaction: scattering of primary nuclei (PLF*-TLF*) followed by fast non-equilibrated fission-like process. Our previous analysis [2, 3, 8] suggested that Dynamical Fission occurs in a late stage of the re-separation of the PLF*-TLF* binary system (time >120 fm/c), but before the PLF* and TLF* have achieved full "equilibration", as it was demonstrated by the experimental anisotropy observed in the angular distribution of the IMF. A very fast (on a time scale of 70-100 fm/c) ternary and quaternary aligned breakup process, following deep inelastic binary reactions, has been also observed in Au+Au collisions at 15A.MeV [21, 22]. However, the reaction mechanism observed in these deep inelastic collisions at lower energies probably differs from that observed at higher energies [3, 6, 8, 20].

In addition to these fast-dynamical mechanisms, much slower equilibrated PLF* and TLF* de-excitation [23], characterized by clear isotropic angular distribution has been observed.

The observed experimental signature of IMF emission supports a time scale evolution of the reaction, in agreement with predictions of Stochastic Mean Field (SMF) [24, 25] and molecular dynamics (CoMD-II) [7] calculations.

In [8] cross sections associated to Dynamical and Statistical Fission mechanisms for neutron rich ^{124}Sn+^{64}Ni and neutron poor ^{112}Sn+^{58}Ni reactions at 35 A.MeV beam energy (REVERSE experiment) were evaluated. It was shown that Dynamical Fission process is about two times more probable in the neutron rich ^{124}Sn+^{64}Ni system than in the $^{112}Sn+^{58}Ni$ neutron poor one. This sizable difference in the cross section associated to Dynamical Fission indicates a strong influence of the entrance channel on the reaction mechanism. However, in order to disentangle Isospin effects from effects due to the different size of the two systems, a new experiment (InKiIsSy) with system ^{124}Xe+^{64}Zn at 35 A.MeV beam energy has been recently carried out.

In section 2 we briefly review the previous results of the REVERSE experiment, already published in refs. [3, 8], that constitute the motivations of the new InKiIsSy experiment, that will be briefly presented in section 3.

2. Previous experimental results and motivations of the InKiIsSy experiment

The REVERSE experiment was performed at the INFN-LNS Super-Conducting Cyclotron of Catania (Italy), bombarding thin ($\approx 300\mu g/cm^2$) self-supporting ^{64}Ni and ^{58}Ni targets with 35 A.MeV ^{124}Sn and ^{112}Sn beams, respectively. Reaction products were detected with the forward part of the 4π multi-detector CHIMERA [1] that is constituted by 688 Si($\approx$300 μm)-CsI(Tl) telescopes over a total of 1192, arranged in 18 rings and covering the angular range between 1° and 30°. Further information about CHIMERA multi-detector and REVERSE experiment can be found in [26]. Details about the experimental methods can be found in our previous works [1–3, 27] and refs. therein.

In order to select peripheral collisions, the method of Cavata [28] was used to estimate the impact parameter from the total charged-particle multiplicity. This method assumes a purely geometrical monotonic correlation between the total charged-particle multiplicity of the collision process and the impact parameter. Fig. 1 shows the correlation between the experimental total charged particles multiplicity and the estimated reduced impact parameter $b_{red} = b/b_{Max}$ for the two reactions ^{124}Sn+^{64}Ni and ^{112}Sn+^{58}Ni at 35 A.MeV, where b_{Max} corresponds to the total geometrical cross section. Note that at a fixed b_{red} the neutron poor system produces a higher (by $\approx$ one unit) multiplicity. In fact, as predicted by calculations [10, 25], in the neutron poor system the dynamical and pre-equilibrium phases lead to more light charged particles (mainly protons) emission than in neutron rich one where, in contrast, neutron emission is favored.

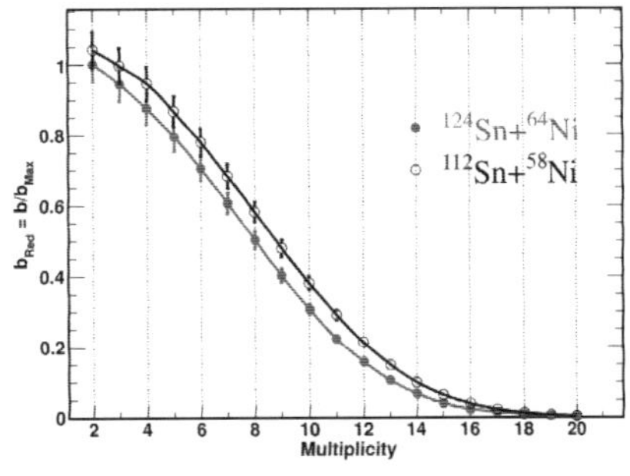

Figure 1. Estimated reduced impact parameter b_{red} plotted as function of the experimental total charged particles multiplicity for the ^{124}Sn+^{64}Ni and ^{112}Sn+^{58}Ni reactions at 35 A.MeV, as obtained by applying the method of Cavata [28].

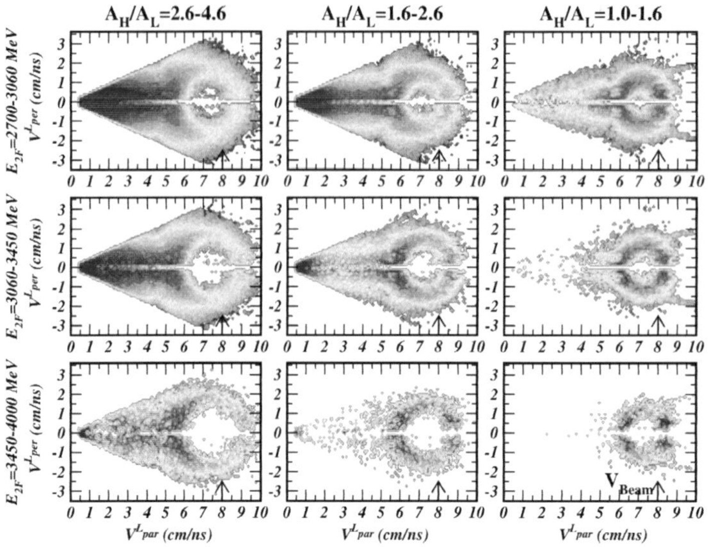

Figure 2. Invariant V_{par} versus V_{per} plots for the Light fragments in the ^{124}Sn+^{64}Ni reaction at 35 A.MeV. Different panels correspond to different values of mass asymmetry A_H/A_L and total kinetic energy $E_{2F} = E_H + E_L$. The distributions are shown in logarithmic scale. The red color corresponds to the largest cross sections and the arrows indicate the beam velocity.

After selecting, in both systems, peripheral collisions with condition $b_{red} > 0.7$, the two heaviest fragments of the chosen subset of events, named as Heavy (H) and Light (L) according to their atomic number Z, were analyzed in those events that satisfy the following conditions: (i) the combined charge of the two selected fragments $Z_{2F} = Z_H + Z_L$ is close to the charge of the projectile ($Z_{proj} = 50$), that is, $37 < Z_{2F} < 57$, and (ii) the Heavy-to-Light-fragment mass ratio is $A_H/A_L < 4.6$, so that the Light fragment has charge $Z_L >\sim 9$. Applying such conditions, the Heavy fragments have the component of the velocity parallel to the beam axis (V^H_{par}) very close to the value of ~ 7.5 cm/ns, that is, slightly below the beam velocity of ~ 8 cm/ns, while the Light fragments have a wider distribution of the parallel velocity, ranging from the velocity of the TLF (~ 1 cm/ns) up to velocities exceeding the PLF ones. This is shown in Fig. 2, where, using a logarithmic intensity scale, the V_{par} versus V_{per} Galilean-invariant plots for Light fragments produced in the ^{124}Sn+^{64}Ni reaction are presented, for three ranges of mass asymmetry A_H/A_L (columns), and for three ranges of the total kinetic energy of the two selected fragments $E_{2F} = E_H + E_L$ (rows). The quantity E_{2F} is a measure of the collision violence: larger values of E_{2F} are associated with more gentle (peripheral) collisions, while more violent collisions are associated with smaller E_{2F} values. In all panels of Fig. 2 it is possible to observe the characteristic Coulomb rings centered slightly below the beam velocity; the presence of such rings points to PLF* as a well-defined decay source and proves the scenario of two separate reaction steps: first the scattering of the PLF*, followed by its splitting into two fragments (H and L). In almost symmetric divisions after less dissipative collisions [E_{2F}=3450-4000 MeV and A_H/A_L =1.0-1.6], the Light fragments distribution is forward-backward symmetric, that is, the Light fragments have equal probability to be emitted forward or backward in the reference frame of the PLF* source. This result is characteristic of an equilibrated fission, where the nucleus is supposed to be completely equilibrated in all its degrees of freedom; there, the PLF* splitting is expected to occur a long time (1000 fm/c or more) after at least one complete rotation. In more

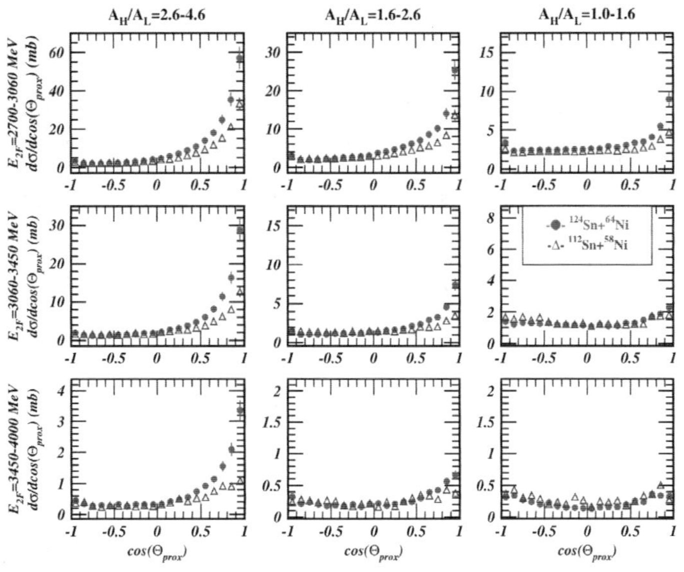

Figure 3. $cos(\theta_{prox})$ angular distributions of the PLF break-up fragments for the ^{124}Sn+^{64}Ni (circles) and ^{112}Sn+^{58}Ni (triangles) at 35 A.MeV, for three different bins of the total kinetic energy (E_{2F}) and mass asymmetry A_H/A_L.

dissipative collisions and/or more asymmetric splits, the population of the Coulomb ring is no longer forward-backward symmetric. In fact, Light fragments have the tendency to populate preferentially the low-velocity side of the Coulomb ring, which means that they are backward-emitted in the PLF* reference frame, that is, toward the TLF*. Nonetheless, a forward-backward symmetric component is still present. Therefore, the observed distributions can be interpreted as a superposition of a forward-backward symmetric component and an asymmetric one. The observed forward-backward asymmetry is the main signature of the Dynamical Fission; that indicates that PLF* fission-like splitting has to be a fast process.
After selecting Light fragments populating the Coulomb ring by gating on V^L_{par} (see ref. [3] for details), in order to disentangle and estimate isotropic and anisotropic fission-like splitting in the two Sn+Ni sytems, the differential cross sections $d\sigma/dcos(\theta_{prox})$ have been evaluated. That is shown in Fig. 3 for the three bins of the mass asymmetry A_H/A_L and three bins of the E_{2F}. The θ_{prox}, defined in left panel of Fig. 4, is the angle betwen the PLF* flight direction (in the CM reference frame) before fission-like splitting , and the breakup axis, defined by the relative velocity of the two fission-like fragments $\mathbf{V}^H - \mathbf{V}^L$. The value of $cos(\theta_{prox}) = 1$ corresponds to the Heavy fragment moving forward along the PLF* flight direction, while the Light fragment is emitted backward along the PLF*-TLF* direction. In the case of equlibrated splitting, a symmetric ditribution around the value of $cos(\theta_{prox}) = 0$ is expected, as observed in Fig. 3 for almost symmetric splitting in more peripheral collisions. Instead, by progressively increasing the mass asymmetry and inelasticity (lower E_{2F}), an increase of a forward peaked distribution, superimposed to a symmetric one around the value of $cos(\theta_{prox}) = 0$, is clearly seen; that is associated to the Dynamical Fission. In order to isolate the equilibrated component, a symmetrization around $cos(\theta_{prox}) = 0$ of the backward part, $cos(\theta_{prox}) < 0$, of the distribution has been done, by assuming that the latter is not influenced by Dynamical Fission. The Dynamical contribution is then obtained by subtracting the extrapolated Statistical Fission distribution from the total experimental one. A sketch of this procedure is shown in the right

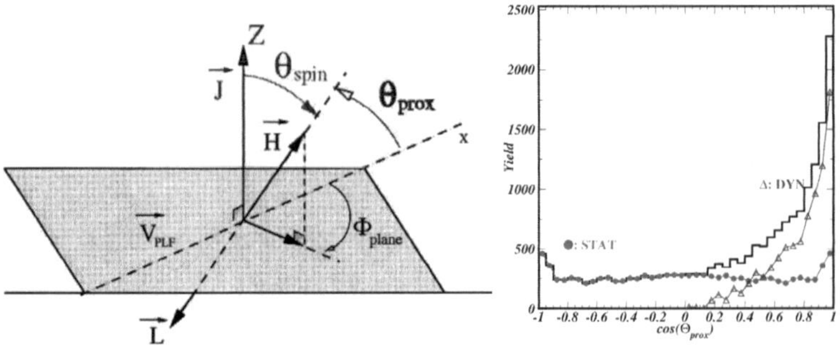

Figure 4. left panel: θ_{prox} definition: angle between the breakup axis, oriented from the light L to the heavy fragment H, and the recoil velocity in the center of mass of the projectile-like fragment (noted V_{PLF}) reconstructed with the two fission fragments, from [20]; right panel: example of extrapolation of Dynamical and Statistical Fission from $cos(\theta_{prox})$ distribution.

panel of Fig. 4. It results that the Statistical Fission cross section is approximately the same in the two systems. In contrast the Dynamical component is larger for the neutron rich system by a factor of about 2 as compared with the neutron poor one, as can be found in tables published in ref. [8]. This sizable difference in the cross section associated to Dynamical Fission indicates a strong influence of the entrance channel on the reaction mechanism. Since the two systems have a large difference in the Isospin content (see table 1) it appears quite natural to link this to an Isospin effect; however, a possible influence of the system size can not be a priori excluded.

3. The InKiIsSy experiment

In order to disentangle entrance channel Isospin effects from the possible dependence of these results upon the initial different mass of the two systems, in April 2013 we carried out a new experiment, named InKiIsSy (Inverse Kinematic Isobaric Systems) [29], using a projectile/target combination having the same mass as the neutron rich $^{124}Sn +^{64} Ni$ system and a N/Z near to value of the neutron poor ^{112}Sn+^{58}Ni one, that is ^{124}Xe+^{64}Zn at the same bombarding energy of 35 A.MeV; Isospins of investigated systems are given in table 1. The experiment has been

Table 1. Isospin (N/Z) of the systems investigated in the REVERSE and InKiIsSy experiments

System	*N/Z Projectile*	*N/Z Target*	*N/Z Compound*
^{124}Sn+^{64}Ni	1.48	1.29	1.41
^{112}Sn+^{58}Ni	1.24	1.07	1.18
^{124}Xe+^{64}Zn	1.30	1.13	1.24

performed at the INFN-LNS in Catania by using the 4π CHIMERA multi-detector. By using CHIMERA we will provide new data set to be compared to the old one. Besides, CHIMERA was coupled to 4 prototypes of the new correlator FARCOS (Femtoscope Array for COrrelation and Spectroscopy) [30]. The FARCOS detector has been designed as a compact high resolution array; it is composed of telescopes, each one made of two double-sided silicon-strip detectors (DSSSD) of 64x64 mm^2 area and thickness of 300 and 1500 μm, respectively, facing the target, followed by 4 CsI(Tl) crystals of 32x32x60 mm^3. The light produced by each crystal is read-out by a Photo-Diode with a thickness of 300 μm and an active area of 2.5x2.5 cm^2.

Concerning the silicon strip detectors, each DSSSD is segmented in 32 horizontal and 32 vertical strips, from which it is possible to define 1024 individual pixels each covering an area of 2x2 mm^2. A schematic drawing of the different stages of one FARCOS telescope is shown in upper panel of Fig. 5. In the InKiIsSy experiment, we used four telescopes placed at 25 cm from the target, covering the laboratory polar angles $\theta_{lab} \approx 15^o - 45^o$, and azimuthal interval $\Delta\phi_{lab} \approx 90^o$. In this angular region we will, however, detect LCP and IMF with a higher angular resolution. In addition, there we can measure with high relative momentum resolution p-p and fragment-fragment correlations, with an accurate event characterization (centrality, reaction mechanism,...) based on fragments detected in coincidence by CHIMERA. In fact one of the observable that has recently been suggested to be sensitive to the density dependence of the symmetry energy is represented by nucleon-nucleon correlation function [31, 32]. In particular it is expected that space-time properties of fast proton emitting sources in central collisions between N/Z-asymmetric nuclei depend on the stiffness of the symmetry energy, affecting proton and neutron emission times. The N/Z dependence of two-proton correlations has been recently investigated in $^{40}Ca+^{40}Ca$ vs $^{48}Ca+^{48}Ca$ collisions at beam energy of 80 A.MeV [33].

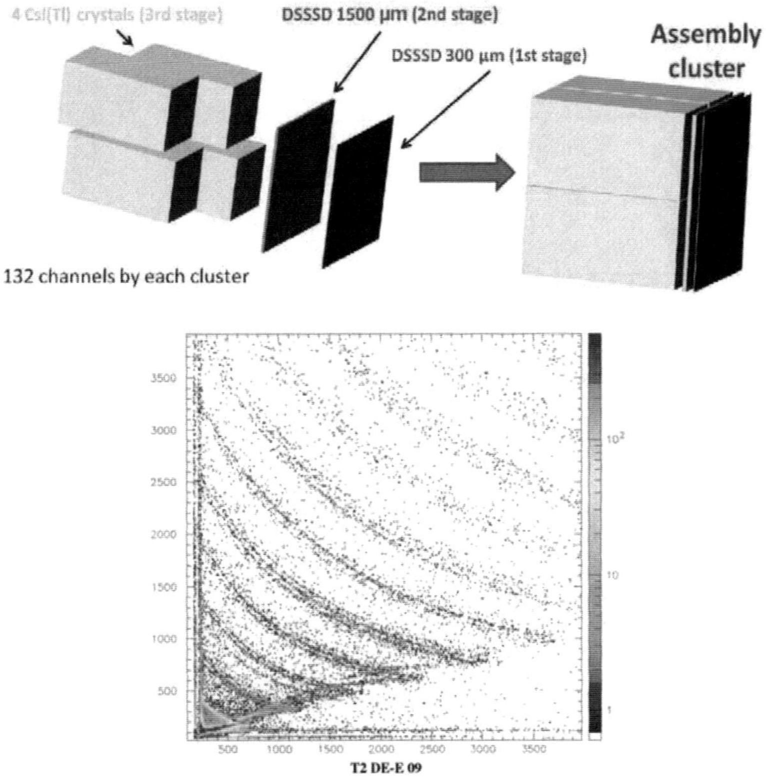

Figure 5. Upper panel: Structure of a FARCOS telescope. The two double-sided silicon strip detectors are followed by 4 CsI(Tl) crystals in a 2 rows x2 columns configuration. Lower panel: ΔE(Si 300 μm)-E(Si 1500 μm) scatter plot for a couple of strips as obtained by a FARCOS telescope prototype in the InKiIsSy experiment.

However, two-proton correlations are also sensitive to the size of the emitting system; thus, it is difficult to attribute the observed N/Z-effect to an Isospin phenomenon rather than to a system size effect. The InKiIsSy experiment may provide a first set of data to characterize FARCOS performance, study emission of LCP and IMF with high resolution and test feasibility of correlation measurements. Effective correlation measurements require, especially considering the demands in terms of statistics, dedicated experiments that we plan to perform in the future. As an example of FARCOS response, lower panel of Fig. 5 shows a ΔE(Si 300 μm)-E(Si 1500 μm) scatter plot for a couple of strips as obtained during the InKiIsSy experiment; a clear isotope identification is visible. Data analysis is in progress.

From theoretical side, the Dynamical Fission phenomenon has been studied by transport model simulations. The Stochastic Mean Field simulation (SMF) [24] is a good approach in order to describe the gross features of Dynamical Fission. However, the simulation can hardly follow the complete time evolution of the deformed PLF up to the scission point. Many features of the Dynamical Fission processes could be reproduced in quantum molecular dynamics models like the Constrained Molecular Dynamics (CoMD-II) [7]. The main feature of this last model is a self-consistent N-body approach that solves the equations of motion using procedures to satisfy event-by-event the Pauli principle and the total angular momentum conservation law. Then, we plan to perform future studies using both SMF and CoMD-II models in order to shed light on the observations here reported.

References

[1] A. Pagano *et al.*, Nucl. Phys. A **734**, 504 (2004).
[2] E. De Filippo *et al.*, Phys. Rev. C **71**, 044602 (2005).
[3] E. De Filippo *et al.*, Phys. Rev. C **71**, 064604 (2005).
[4] M. Di Toro *et al.*, Eur. Phys. J. A **30**, 65 (2006).
[5] S. Piantelli *et al.*, Phys. Rev. C **76**, 061601 (2007).
[6] A.B. McIntosh *et al.*, Phys. Rev. C **81**, 034603 (2010).
[7] M. Papa *et al.*, Phys. Rev. C **75**, 054616 (2007).
[8] P. Russotto *et al.*, Phys. Rev. C **81** 064605 (2010).
[9] Z. Kohley *et al.*, Phys. Rev. C **83**, 044601 (2011).
[10] R. Płaneta *et al.*, Phys. Rev. C **77**, 014610 (2008).
[11] S. Piantelli *et al.*, Phys. Rev. C **74**, 034609 (2006).
[12] J. Łukasik *et al.*, Phys. Rev. C **55**, 1906 (1997).
[13] D.V. Shetty *et al.*, Phys. Rev. C **68**, 054605 (2003).
[14] D. Theriault *et al.*, Phys. Rev. C **74**, 051602(R) (2006).
[15] S. Hudan *et al.*, Phys. Rev. C **86**, 021603(R) (2012).
[16] S. Barlini *et al.*, Phys. Rev. C **87**, 054607 (2013).
[17] E. De Filippo *et al.*, Phys. Rev. C **86**, 014610 (2012).
[18] M.B. Tsang *et al.*, Phys. Rev. C **86**, 015803 (2012).
[19] A.A. Stefanini *et al.*, Z. Phys. A **351**, 167 (1995).
[20] F. Bocage *et al.*, Nucl. Phys. A **676**, 391 (2000). 067604 (2010).
[21] I. Skwira-Chalot *et al.*, Phys. Rev. Lett. **101**, 262701 (2008).
[22] J. Wilczyński *et al.*, Phys. Rev. C **81**, 024605 (2010) and Phys. Rev. C **81**, 067604 (2010).
[23] R.J. Charity *et al.*, Phys. Rev. C **82**, 014610 (2010).
[24] V. Baran, M. Colonna and M. Di Toro, Nucl. Phys. A **730**, 329 (2004).
[25] V. Baran, M. Colonna, V. Greco, M. Di Toro, Phys. Rep. **410**, 335 (2005).
[26] A. Pagano *et al.*, Nucl. Phys. A **681**, 331 (2001).
[27] E. Geraci *et al.*, Nucl. Phys. A **732**, 173 (2004).
[28] C. Cavata *et al.*, Phys. Rev. C **42**, 1760 (1990).
[29] P. Russotto, E. De Filippo, A. Pagano and the EXOCHIM collaboration, InKiIsSy Experiment approved by the INFN-LNS PAC 2012.
[30] G. Verde *et al.*, Jour. of Phys. Conf. Series **420**, 012158 (2013).
[31] L.W. Chen *et al.*, Phys. Rev. C **68**, 014605 (2004).
[32] G. Verde, A. Chbihi, R. Ghetti and J. Helgesson, Eur. Phys. J. A **30**, 81 (2006).
[33] V. Henzl *et al.*, Phys. Rev. C **85**, 014606 (2012).

Journal of Physics: Conference Series **515** (2014) 012021 doi:10.1088/1742-6596/515/1/012021

Understanding the onset of incomplete fusion

Pushpendra P. Singh[1], Abhishek Yadav[2], Vijay R. Sharma[2], D. P. Singh[2], R. Kumar[3], R. P. Singh[3], S. Muralithar[3], B. P. Singh[2], R. K. Bhowmik[4], R. Prasad[2], and the AMU-IUAC collaboration

[1]Department of Physics, Indian Institute of Technology Ropar, Punjab - 140 001, India
[2]Accelerator Laboratory, Department of Physics, A. M. University, Aligarh-202 002, India
[3]NP-Group, Inter-University Accelerator Center, New Delhi-110 067, India
[4]Variable Energy Cyclotron Centre, 1/AF Bidhan Nagar, Kolkata 700 064, India

E-mail: pps@iitrpr.ac.in

Abstract. The entrance channel effect on the onset and strength of incomplete fusion (ICF) has been studied in the present work. Several inclusive experiments have been performed to measure the ICF strength function in ^{12}C,^{16}O+^{169}Tm systems at near and above barrier energies. Data obtained in these experiments suggest the existence of ICF even at slightly above barrier energies where complete fusion (CF) is supposed to be the sole contributor, and conclusively demonstrate strong projectile structure and energy dependence of ICF. The incomplete fusion strength functions for ^{16}O,^{12,13}C+^{159}Tb and ^{16}O,^{12,13}C+^{181}Ta systems are analyzed as a function of projectile α-Q-value at a constant ν_{rel} = 0.053c. It has been found that one neutron ($1n$) excess projectile ^{13}C (as compared to ^{12}C) results in less incomplete fusion contribution due to its relatively large negative α-Q-value. In order to understand the onset of ICF at such low energies, the driving input angular momenta (ℓ) involved in the production of different evaporation residues have been deduced from the analysis of experimentally measured spin-distributions for the same projectile-target combinations at the same incident energies. Higher ℓ-values, imparted into the system in non-central interactions, are found to be responsible for low energy ICF. The ICF-αxn/2αxn channels display involvement of higher ℓ-values than that observed in CF-xn/pxn/αxn/2αxn channels at the very same projectile energies. It has been observed that the mean value of ℓ increases with successively opened ICF channels and incident energy.

1. Introduction

Incomplete fusion (ICF) - where only a part of projectile fuses with target nucleus - in light heavy-ion (HI) induced reactions has recently been identified as a process of great significance even at slightly above barrier energies. According to the conventional picture of fusion, the interaction of two heavy nuclei at energies near and above the Coulomb barrier leads to a completely fused composite system. In this case, entire mass and charge of the interacting partners amalgamate with all nucleonic degrees of freedom to form a fully equilibrated compound nucleus which may eventually decay via light nuclear particles and/or characteristic γ-rays. This mode of reaction is termed as complete fusion (CF) [1, 2, 3, 4, 5, 6, 7, 8, 9, 10, 11, 12].

Recent studies, however, showed strong influence of ICF on CF at low incident energies, i.e., $E_{lab}\approx$3-7 MeV/nucleon [13, 14]. Unexpected show up of (ICF) at slightly above barrier energies has been justified as the natural extension of fusion due to excess input angular momenta (ℓ) imparted into the system in non-central interactions [4]. The fact that the CF and ICF events can be disentangled in terms of input angular momenta (ℓ-values) and on the basis of degree of linear momentum transfer (ρ_{LMT}) from projectile to target nucleus [11, 15, 16]. The interaction trajectories associated with ℓ-values $\rightarrow$ $\ell \leq \ell crit$ lead to CF events with predefined mass A, charge Z and recoil velocity v_{recoil}. However, for ℓ-values $\rightarrow$ $\ell \geq \ell_{crit}$, the interaction ℓ-balance inhabits fusion of entire projectile with traget nucleus [4]. As a consequence, the incident projectile breaks-up into its constituents to provide sustainable input angular momenta. A part of projectile fuses with target nucleus which leads to an incompletely fused composite system with relatively less A, Z and v_{recoil} as compared to CF events, and direct projectile-like-fragments mainly centered in forward cone.

The dynamics of ICF have been extensively investigated and several theoretical models have been proposed, see ref.[11]. Morgenstern *et al.* [17] correlated the fraction of ICF with μ_A, which suggests large ICF fraction for more mass-asymmetric system at a constant relative velocity (v_{rel}=$\sqrt{2(Ecm - Vb)/\mu}$/c). The peripheral nature of ICF has been emphasized by Inamura *et al.*[18]. However, the studies on ICF dynamics by Tserruya *et al.*[19] revealed the existence of both CF and ICF below and above the value of ℓ_{crit}. Despite a variety of existing studies (see ref.[12] for detail), the dynamics of ICF at low incident energies is still not fairly understood, thus continues to be an active area of investigations.

In the present work, the onset and strength of ICF have been studied in terms of (i) projectile energy (E_{lab}), (ii) mass-asymmetry ($\mu_A = A_T/A_{T+P}$) of interacting partners, (iii) projectile α-Q-value, and (iv) input ℓ-values to better understand the ICF dynamics. Experiments have been performed at the Inter-university Accelerator Center (IUAC), New Delhi. The percentage fraction of ICF, deduced from the experimental excitation functions of individual reaction residues in different projectile-target combinations, shows strong dependence on E_{lab}, α-Q-value and μ_A. Two particle-γ-coincidence experiments have been performed for the same projectile target combinations to understand the onset of ICF at slightly above barrier energies. The spin-distributions (SDs) of individual reaction residues have been measured. Experimental results from nearly systems are compared to display the entrance channel effect on ICF dynamics. The experimental and data reduction methodology are presented in section II. The finding of the present work and their interpretations in context of statistical model code PACE4 are given in section III, where the influence of ICF on CF and its dependencies on various entrance channel parameters are demonstrated. Section IV deals with the summary and conclusions of the present work.

2. Experimental and data reduction methodology

Experiments have been performed at IUAC New Delhi using general purpose scattering chamber (GPSC) and gamma detector array (GDA) coupled with charged particle detector array (CPDA) setups. The methodology and experimental setups have been discussed in refs.[10, 11, 12, 13, 14]. However, a brief account of experimental conditions is given here for the ready reference.

The EFs of individual reaction residues populated via CF and/or ICF in ^{12}C,^{16}O+^{169}Tm systems have been measured using stacked foil activation technique [10, 11, 12, 13, 14]. A stack of 3 target foils of natural ^{169}Tm (of abundance = 100%, and thickness $t_m \approx$ 1-1.8 mg-cm^{-2}) each backed by Al-foils of appropriate thicknesses to stop the most energetic recoiling products was bombarded in GPSC with ^{12}C and ^{16}O beams of energy range $\approx$ 4-7 MeV/nucleon with beam current $\approx$20-30 nA. The evaporation residues (ERs) produced during the irradiations were counted off-line with two pre-calibrated HPGe detectors. The ERs have been identified by their characteristic γ-lines and confirmed in their decay-curve analysis. A part of γ-ray spectra obtained at E_{lab} = 89.3 MeV in ^{12}C+^{169}Tm system is shown in Fig.1(a). Some of the peaks corresponding to different reaction residues are marked. Decay curve of ^{177}Re($t_{1/2}$ = 14 min) which is obtained by following 197 keV γ-line at increasing decay time is drawn in Fig.1(b). The production cross-sections ($\sigma_r(E)$) of ERs have been calculated using standard formulation [12].

The overall error in σ_r is estimated to be $\approx$ 14 %, excluding the uncertainty in branching ratios, decay constant, etc. Details of the error estimation and causing factors are given in ref.[12].

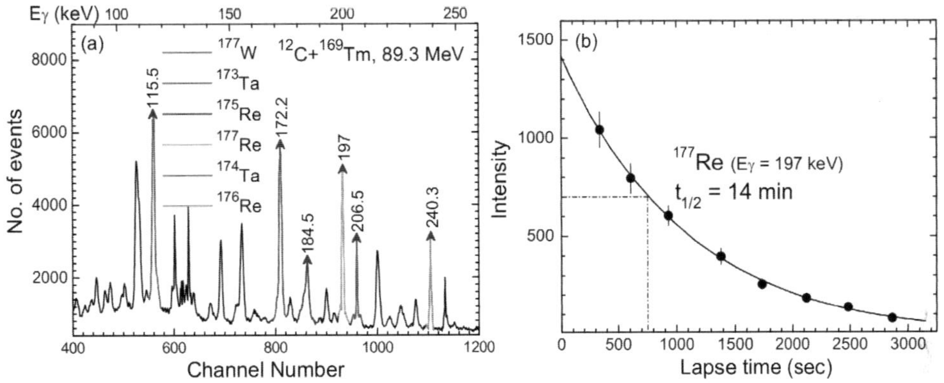

Figure 1. (*a*) A part of γ-ray spectra obtained at E_{lab} = 89.3 MeV in $^{12}C+^{169}Tm$ system. Some of the peaks corresponding to different reaction residues are marked. (*b*) Decay curve of $^{177}Re(t_{1/2}$ = 14 min) which is obtained by following 197 keV γ-line at increasing decay time.

Aiming to probe the role of high ℓ-values in the onset of ICF, the spin-distributions (SDs) of xn/pxn/αxn/2αxn-channels have been measured in $^{12}C,^{16}O+^{169}Tm$ systems at energies $\approx$ 4-7 MeV/nucleon using particle-γ coincidence technique [4]. Coincidences between particles(Z=1,2) and prompt-γs have been demanded using GDA and CPDA setups. Self-supporting ^{169}Tm (abundance = 100%) target of thickness $\approx$ 1mg/cm^2 has been bombarded with $^{12}C^{+5}$(E$\approx$54-90MeV) and $^{16}O^{+7}$(E$\approx$90MeV) beams. It may be pointed out that the forward cone (θ=10° – 60°) detectors are supposed to detect both; (*i*) CF α-component of energy $E_{\alpha-evaporation}$=18 MeV, and (*ii*) ICF direct-α-component belongs to the same velocity as that of incident projectile, e.g., $E_{\alpha-direct}$=22.5 MeV in case of 90MeV ^{16}O beam. In order to completely remove $\alpha - evaporation$ component detected in the forward cone, an Al-absorber of appropriate thickness has been kept on forward (10° – 60°) CPD's so that only $direct - \alpha$ component of energy E$\geq$18 MeV can be detected. The prompt γ-ray spectra have been recorded for different gating conditions, namely; α and 2α detected in the backward(B), forward(F) and 90°(S) angular zones. The SDs of xn-channels have been drawn by looking into the singles spectra. For the identification of pxn-channels, backward(B)-α-gated spectra has been subtracted from backward(B)-particles(Z=1,2)-gated spectra. However, αxn/2αxn(CN-α)-channels produced via CF have been identified from the backward(B)-α-gated spectra. Further, the direct-α-particles associated with ICF are expected to be emitted in forward cone. The αn/2αxn(Direct-α)-channels produced via ICF have been identified from forward(F)-α-gated spectra. The intensity and area under the photo-peak of the characteristic prompt γ-rays assigned to the particular reaction products were used to determine the relative production yield [4].

3. Entrance channel dependence of ICF: findings and interpretations

The effect of various entrance channel parameters, namely; (*i*) projectile energy, (*ii*) entrance channel mass-asymmetry of interacting p(μ_A), (*c*) α-Q-value, and (*d*) input ℓ-values, on the onset and strength of ICF have been systematically investigated in the present work. Finding of different experiments and their interpretations are presented in the following subsections.

3.1. Energy dependence of ICF

The experimental EFs of individual (ERs) have been analyzed in the framework of statistical model code PACE4 to probe the energy dependence of ICF. Details of code PACE4 can be found in ref.[12]. The value of level density parameter (a=A/K) is an important input parameter in this code which can be tuned to fit the experimental data. Different values of 'a' from A/10 - A/8 MeV^{-1} have been tested to reproduce experimental EFs. The ERs populated via xn/pxn-channels have been found to be in good agreement with that estimated via PACE4. However, the experimental EFs of α-emitting channels have been observed to be significantly enhanced as compared to the PACE4 predictions, indicating the existence of ICF. The experimental EFs of α-emitting channels are subtracted from the theoretical ones ($\sigma_{ICF} = \sigma_{exp}$ - σ_{pace4}) to get ICF contribution [11]. The percentage fraction of ICF (F_{ICF}) has been deduced and is plotted as a function of projectile relative velocity ($v_{rel}=\sqrt{2(Ecm - Vb)/\mu}/c$) for ^{12}C,^{16}O+^{169}Tm and ^{12}C+^{159}Tb systems in Fig.2(a), i.e., termed as ICF strength function.

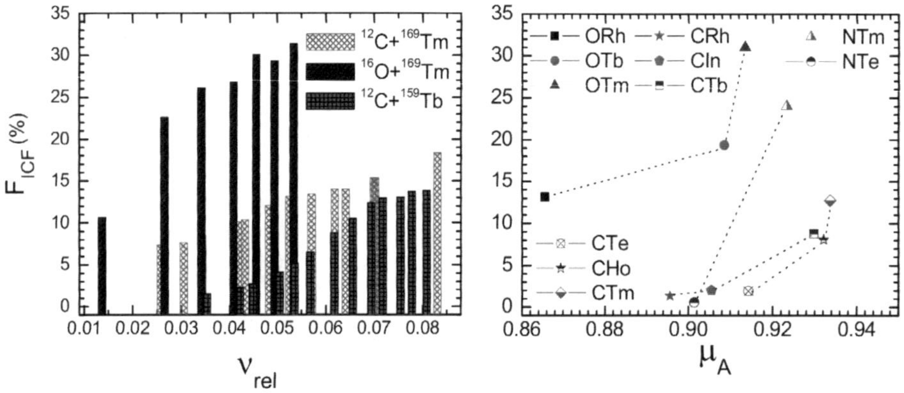

Figure 2. (*a*) The percentage fraction of ICF as a function of projectile relative velocity (v_{rel}) for ^{12}C,^{16}O+^{169}Tm and ^{12}C+^{159}Tb systems, (*b*) the percentage fraction of ICF as a function of μ_A for 11 projectile-target combinations at a constant value of v_{rel} ≈0.053

The F_{ICF} defines empirical probability of ICF at different values of v_{rel}. As shown in this figure, the F_{ICF} increases linearly with normalized projectile energies (i.e., v_{rel}). According to Morgenstern's systematics [17], ICF contributes significantly above v_{rel} ≈0.06 (6 % speed of light). As indicated in this figure, the values of v_{rel} are in the range from ≈0.027 (2.7 % of c) to ≈0.084 (8.4 % of c) for ^{12}C, and from ≈0.014 (1.4 % of c) to ≈0.053 (5.3 % of c) for ^{16}O. Therefore, no significant ICF contribution is expected at the given values of v_{rel}. However, the results presented in Fig.2(a) clearly demonstrate the onset of ICF at relatively lower value of v_{rel} i.e, ≈0.027 (F_{ICF} ≈7 %) in ^{12}C+^{169}Tm system, and at ≈0.014 (F_{ICF} ≈10 %) in ^{16}O+^{169}Tm system. In both cases, the observed value of F_{ICF} is significant at well below the proposed onset value of v_{rel} (i.e., 6 % of c). Further, as shown in Fig.2(a), the value of F_{ICF} for ^{12}C-projectile is lower than ^{16}O-projectile for the entire measured energy range. The difference in F_{ICF} for two systems (^{12}C,^{16}O+^{169}Tm) can be seen clearly which points towards projectile structure and/or μ_A effect on ICF.

3.2. Projectile dependence of ICF

The percentage fraction of ICF has been deduced and plotted as a function of μ_A for 11 projectile-target combinations (^{12}C+^{128}Te,^{165}Ho,^{169}Tm and ^{16}O+^{103}Rh,^{159}Tb,^{169}Tm) at a constant value of v_{rel} ≈0.053 in Fig.2(a) to understand projectile structure and/or μ_A dependence of ICF. It may be pointed out that both ^{12}C and ^{16}O are α-cluster nuclei, which may break up into several combinations of α-clusters. The probability of breakup increases with input ℓ-values imparted into system in peripheral interactions [4]. Some of the breakup combinations which have been observed in previous studies are; (*a*) $^{12}C \rightarrow {}^{8}Be + {}^{4}He(\alpha)$ and/or three α fragments, and (*b*) $^{16}O \rightarrow {}^{12}C + {}^{4}He$, ^{8}Be+^{8}Be and/or four α fragments. One or a group of fragments may fuse with target nucleus to form an incompletely fused composite system. The overlap between the interacting nuclei and the transformed mass depends on the number of nucleon occupying the overlapping volume. As such, ^{16}O may open up more ICF channels as compared to ^{12}C induced reactions. As can be seen from Fig.2(b), the Morgenstern's systematics does not explain the variation of F_{ICF} with μ_A for given systems. However, the value of F_{ICF} increases with μ_A individually for ^{16}O, ^{12}C and ^{14}N projectiles. It is interesting to note that the ^{12}C+^{169}Tm system is a more mass asymmetric (μ_A = 0.9337) system than ^{16}O+^{169}Tm system (μ_A = 0.9135), but the value of F_{ICF} is ≈ 18 % higher than that observed for ^{12}C+^{169}Tm system. The aforementioned observations based on 11 projectile-target combinations strongly contradict Morgenstern's mass-asymmetry systematics, and suggest strong projectile dependence of ICF.

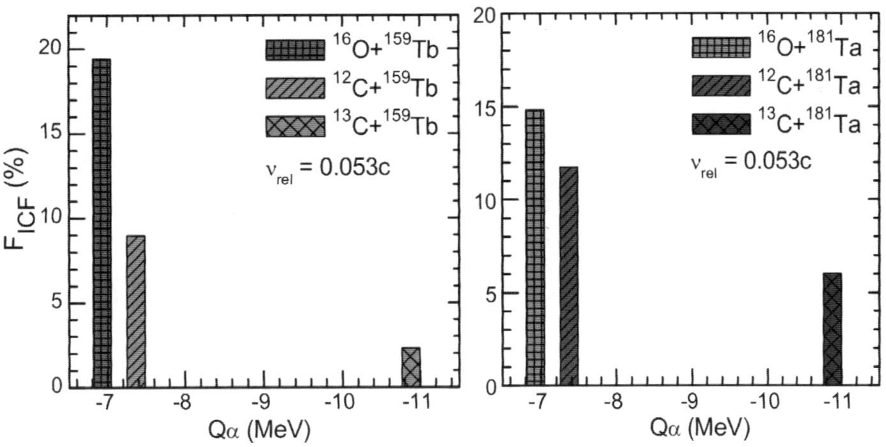

Figure 3. The values of F_{ICF} for ^{16}O,^{12,13}C+^{159}Tb and ^{16}O,^{12,13}C+^{181}Ta systems are plotted as a function of projectile α-Q-value at a constant ν_{rel} = 0.053c.

3.3. α-Q-value dependence of ICF

For better insight into the onset of ICF, the percentage fraction of ICF is analyzed in terms of projectile α-Q-value for 6 projectile-target combinations The value of F_{ICF} is plotted as a function of Q_α in Fig.3(a-b) at a constant $\nu_{rel} = 0.053c$. As shown in this figure, the probability of ICF for the ^{13}C projectile is noticeably smaller than that for the ^{12}C projectile. The strikingly different ICF fractions for ^{13}C and ^{12}C induced reactions point towards the projectile structure effect. It may be pointed out that ^{12}C is a well known α-cluster nucleus with Q_α=-7.37 MeV. However, ^{13}C has a larger Q_α-value (=-10.64 MeV) than ^{12}C. The higher Q_α-value for ^{13}C translates into the smaller breakup probability into constituent α-clusters, resulting in a smaller ICF fraction than that found in ^{12}C induced reactions. It is evident in Fig.3(a-b) that the probability of ICF is found to be less for larger negative Q_α-value projectiles. For example, the value of F_{ICF} for the $^{16}O(Q_\alpha$=7.16 MeV)+^{159}Tb system is found to be 19% which is reduced to only 3% for the $^{13}C(Q_\alpha$=-10.64 MeV)+^{159}Tb system. The same systematics was followed for the ^{181}Ta target. Hence, from the data presented in this figure, it can be inferred that the Q_α-value is an important entrance channel parameter which essentially dictates the probability of ICF.

3.4. Angular momentum dependence of ICF

The fact that the relative number of statistical and *yrast*-like transitions depend on entry state angular momenta (ℓ) and the excitation energy (E*). The CF residues are formed at high E* and at low ℓ leading to more statistical transitions, where 'yrast' states are expected to fed by statistical γ-transitions. However, the ICF residues achieve low E* (due to the involvement of partial degrees of excitations) and high ℓ (relatively higher values of impact parameters contribute to the high spin states) at a given projectile energy. In such a case, number of 'yrast'-like transitions are much larger than that of statistical ones, where less or no feeding is expected. As a result, the spin-distributions of CF and ICF residues are expected to be entirely different in nature and can be used as a sensitive tool to probe the mode of reaction by looking into the entry state spin population. In order to probe the role of high ℓ-values in the production of ICF residues, the spin-distributions of various CF and ICF channels have been measured [4].

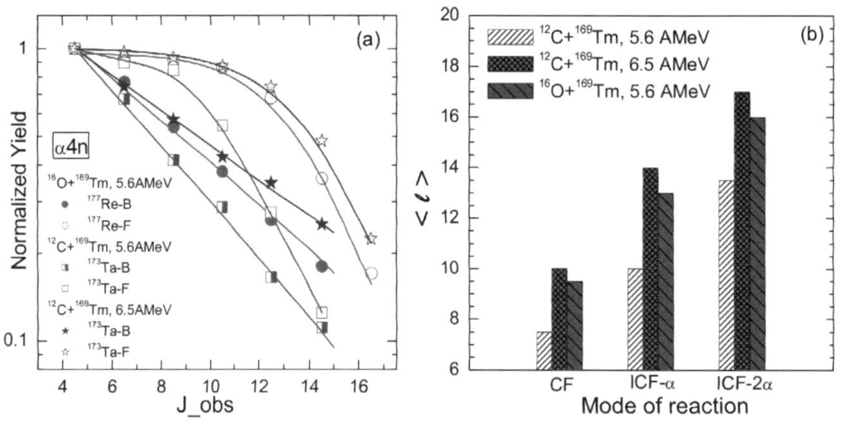

Figure 4. (*a*) the spin-distributions of xn/αxn/2αxn-channels populated via CF and ICF in ^{12}C,^{16}O+^{169}Tm systems. The nomenclature used in this figure is explained in the text, and the lines & curves are best fit to the data points. (*b*) mean 'ℓ-*values*' involved in different CF- and ICF-channels as a function of mode of reactions.

The spin-distributions of xn/αxn/2αxn-channels populated via CF and ICF in ^{12}C,^{16}O+^{169}Tm systems are plotted in Fig.4(a-b). The nomenclature used in this figure indicate the involved reaction dynamics, i.e., 'B' and 'F' respectively indicate CF-channel: identified from backward(B)-α-gated spectra, and ICF-channel: identified from forward(F)-α-gated spectra. As shown in Fig.4(a), there is a striking difference in the spin-distributions of different channels which indicate the involvement of entirely different mode of reactions in the production of these residues. The intensity of xn(singles)/αxn-B-channels (CF residues) falls off rather quickly with observed spin (J_{obs}), indicating strong feeding and/or broad spin population during the de-excitation of CN. However, for αxn-F-channels (ICF residues), the intensity appears to be almost constant up to a certain value of J_{obs}, and then decreases towards band head. This indicates the absence of feeding to the lowest members of the 'yrast' band and/or the population of low spin states are strongly hindered in ICF-channels.

In order to understand the multitude of mean ℓ-values, and to examine the possibility to populate high spin states via ICF, the value of ℓ involved in CF- and ICF-channels have been deduced from the analysis of experimental spin-distributions. The ℓ-values involved in various modes of reactions are plotted in Fig.4(b). As can be seen from this figure, the value of $< \ell >$ involved in the production of CF-xn/pxn-B/αxn-B, and ICF-αxn-F, and ICF-2αxn-F channels are found to be $\approx$ 7.5 $\hbar$, $\approx$ 10 $\hbar$ and $\approx$ 13.5 $\hbar$, respectively, at projectile energy $\approx$ 5.6 AMeV for ^{12}C+^{169}Tm system. However, at projectile energy $\approx$ 6.5 AMeV, the value of $< \ell >$ for CF-xn/pxn-B/αxn-B, and ICF-αxn-F, and ICF-2αxn-F channels are found to be $\approx$10$\hbar$, $\approx$14$\hbar$ and $\approx$17$\hbar$, respectively. The enhancement in the value of $< \ell >$ in case of direct-α-emitting channels (ICF residues) indicates their origin from high ℓ-values as compared to CF-channels. A very useful correlation between the value of $< \ell >$ and the successively opened ICF channels can be obtained from the data presented in Fig.4(b). The value of $< \ell >$ associated with ICF in contrast with CF can be represented as [4];

(a) for ^{12}C($E_{lab}$$\approx$5.6 AMeV)+^{169}Tm system,
$\ell_{(ICF-\alpha xn)} \approx 1.33 \ell_{(CF-xn/pxn/\alpha xn)}$,
$\ell_{(ICF-2\alpha xn)} \approx 1.35 \ell_{(ICF-\alpha xn)} \approx 1.8 \ell_{(CF-xn/\alpha xn)}$, and

(b) for ^{12}C($E_{lab}$$\approx$6.5 AMeV)+^{169}Tm system,
$\ell_{(ICF-\alpha xn)} \approx 1.4 \ell_{(CF-xn/pxn/\alpha xn)}$,
$\ell_{(ICF-2\alpha xn)} \approx 1.2 \ell_{(ICF-\alpha xn)} \approx 1.7 \ell_{(CF-xn/\alpha xn)}$.

It is interesting to note that the values of $< \ell >$ involved in ICF-αxn/2αxn-channels are found to be $\approx$ 30 to 70 % higher as compared to the CF-xn/pxn/αxn-channels. This clearly suggests the involvement of high ℓ-values in the production of ICF-products at a constant projectile energy, essentially due to peripheral interactions, where a significant amount of orbital angular momentum between projectile and target nucleus transformed into high spin states of final reaction products. It can also be observed from Fig.4(b), the involved ℓ-values in different reaction channels are found to increase linearly with the projectile energy. The CF is not able to approach the value of $< \ell >$ even at $\approx$ 6.5 AMeV which has been populated via ICF for the same residue at low projectile energy. The above striking features strongly support the possibility to populate high spin states via ICF in final reaction products even at low incident energy. The partial waves of lower ℓ-values do not contribute to the ICF significantly. Therefore, it may not be out of order to state that ICF events occur in the peripheral interactions, probably at finite values of impact parameters.

4. Summary and conclusions

In summary, the ICF fraction has been measured from the analysis of differential EFs within the framework of statistical model code PACE4. The value of F_{ICF} has been presented as a function of projectile energy (E_{lab}), mass-asymmetry (μ_A), Q_α, and input angular momenta (ℓ). Existence of ICF at energy as low as $\approx$ 7.5 % above the barrier is conclusively demonstrated. It has been found that the value of F_{ICF} increases from $\approx$7 % to $\approx$18 % with in the studied energy range (i.e., $1.02V_b$ to $1.64V_b$). The onset of ICF at relatively lower value of v_{rel} i.e, $\approx$0.027 (F_{ICF} $\approx$7 %) in ^{12}C+^{169}Tm system, and at $\approx$0.014 (F_{ICF} $\approx$10 %) in ^{16}O+^{169}Tm system has been presented. In both the cases, the observed value of F_{ICF} is significant at well below the proposed onset value of v_{rel} (i.e., 6 % of c).

A comparison three projectile-target combinations (i.e., ^{12}C+^{169}Tm and ^{16}O+^{169}Tm) displays higher ICF probability for ^{16}O+^{169}Tm system. This indicates strong projectile dependence on F_{ICF}. The existence of ICF at well below the proposed onset value of v_{rel} (i.e., 6 % of c) conclusively supplements Morgenstern's mass-asymmetry systematics. The effect of Q_α-value on ICF fraction has been observed for strongly bound projectiles. The fraction of ICF has been found to decrease for large negative Q_α-value projectiles. If confirmed for other projectile-target combinations, this may provide an important input to understand the complex ICF dynamics at low incident energies. More experiments are planned to cover this aspect.

The spin-distribution(s) associated with ICF are found to be originated from the narrow spin population, localized near and/or above to the critical angular momentum (ℓ_{crit}) for CF, where a given direct-α-fragment is emitted to release excess driving angular momenta. This reveals a competition from successively opened ICF channels for each value of ℓ above the ℓ_{crit} for normal fusion (CF). It has been found that the direct α-multiplicity increases with the value of $< \ell >$ at a particular projectile energy. For an example, at $\approx$ 5.6 AMeV the value of $< \ell >$ for CF-xn/pxn-B/αxn-B, and ICF-αxn-F, and ICF-2αxn-F channels are found to be $\approx$ 7.5 $\hbar$, $\approx$ 10 $\hbar$ and $\approx$ 13.5 $\hbar$, respectively. It can be safely inferred that the high ℓ-values associated with peripheral interactions contribute to open up direct-α-channels. The value of $< \ell >$ associated with 2α-emitting channel likely to be originated from higher impact parameters as that associated with the production of single α-channel.

Acknowledgments

The authors thank Prof. Amit Roy, Director of IUAC for extending experimental facilities to perform these experiments. One of the authors (P.P.S.) acknowledge the financial support from HIC for FAIR during his stay at GSI Helmholtz Center for Heavy-Ion Research GmbH, Germany, and Prof. Hans-Juergen Wollersheim and Prof. Juergen Gerl for their support and motivation during the analysis of this work. Authors R. P., B. P. S. & D. P. S. thank to DST and UGC for financial support.

References

[1] Kamal Kumar *et al.*, Phys. Rev. C 88, 064613 (2013)
[2] D. J. Hinde and M. Dasgupta, Phys. Rev. C **81**, 064611 (2010), and the references therein.
[3] Alexis Diaz-Torres, J. Phys. G: Nucl. Part. Phys. **37**, 075109 (2010), and the references therein.
[4] Pushpendra P. Singh *et al.*, Phys. Lett. **B671**, 20 (2009); Phys. Rev. C**80**, 064603 (2009), Phys. Rev. C**78**, 017602 (2008); and the references therein.
[5] L. R. Gasques *et al.*, Phys. Rev. C**79**, 034605 (2009); Phys. Rev. C**74**, 064615 (2006).
[6] Devendra P. Singh *et al.*, Phys. Rev. C**81**, 054607 (2010).
[7] P. R. S. Gomes *et al.*, Phys. Lett. **B601**, 20 (2004); Phys. Rev. C**73**, 064606 (2006); Phys. Rev. Lett. **B53**, 1630 (1984).
[8] A. Diaz-Torres *et al.*, Phys. Rev. Lett. **98**, 152701 (2007).
[9] E. Z. Buthelezi *et al.*, Nucl. Phys. **A734**, 553 (2004).
[10] Unnati Gupta *et al.*, Phys. Rev. C**80**, 024613 (2009).
[11] Pushpendra P. Singh *et al.*, Phys. Rev. C**77**, 014607 (2008), Euro. Phys. J. A **34**, 29 (2007).

[12] Unnati Gupta *et al.*, Nucl. Phys. **A811**, 77 (2008), and the references therein.
[13] Abhishek Yadav *et al.*, Phys. Rev. C **85**, 064617 (2012), Phys. Rev. C **86**, 014603 (2012), and the references therein.
[14] Pushpendra P. Singh *et al.*, Euro. Phys. Jour. WoC 21, 10009 (2012); Euro. Phys. Jour. WoC 17, 03007 (2011), and the references therein.
[15] J. Wilczynski *et al.*, Phys. Rev. Lett. **45**, 606 (1980); Nucl. Phys. **A373**, 109 (1982).
[16] K. Siwek-Wilczynska *et al.*, Phys. Rev. Lett. **42**, 1599 (1979).
[17] H. Morgenstern *et al.*, Z. Phys. **A313**, 39 (1983); Phys. Rev. Lett. **52**, 1104 (1984); Z. Phys. **A324**, 443 (1986).
[18] T. Inamura, *et al.*, Phys. Rev. C **32**, 1539 (1985).
[19] I. Tserruya *et al.*, Phys. Rev. Lett. **60**, 14 (1988).

Study of nuclear dynamical deformation in the synthesis of super-heavy nuclei

Nan Wang [1], **Enguang Zhao** [2,3], **and Shan-Gui Zhou** [2,3]

[1] College of Physics, Shenzhen University, Shenzhen 518060, China
[2] State Key Laboratory of Theoretical Physics, Institute of Theoretical Physics, Chinese Academy of Sciences, Beijing 100190, China
[3] Center of Theoretical Nuclear Physics, National Laboratory of Heavy Ion Accelerator, Lanzhou 730000, China

Abstract. The nuclear dynamical deformations in the fusion reactions for the production of superheavy nuclei (SHN) are studied within the framework of the di-nuclear system model with dynamical potential energy surface (DNS-DynPES model). The development of the dynamical deformation for the incident channel is investigated and the maximum dynamical deformations for different configurations are also calculated. The influences of the dynamical deformations on the potential energy surface (PES) and the fusion probability are also calculated. The results indicate that the structure of the PES changes significantly and the fusion for synthesizing SHN is also hindered for the inclusion of the nuclear dynamical deformation. The fusion probability as a function of angular momentum is also investigated.

1. Introduction

In recent years, the synthesis of superheavy nuclei (SHN) has already obtained much progress in both experimental and theoretical aspects. On the experimental aspect, the elements from 107 to 113 were synthesized in cold fusion reactions with Pb or Bi as target and the elements from 112 to 118 were synthesized in hot fusion reactions with the neutron-rich projectile ^{48}Ca [1-5]. Several theoretical models were put forward, such as the macroscopic dynamical model [6], the fusion-by-diffusion (FBD) models [7-12], the nuclear collectivization model [13,14], the di-nuclear system (DNS) model [15-25] and some others, e.g., Ref. [26], to understand the fusion process and predict the evaporation residue cross sections.

The theoretical investigation about the fusion of SHN is of vital importance for the experimental research. The evaporation residue (ER) cross sections are decreasing significantly as increasing the charge number of the SHN. After SHN with Z=116, the ER cross sections drop exponentially. Because of the very low ER cross sections and very narrow excitation functions for synthesizing SHN, the SHN could be synthesized only at the optimal projectile-target combination and the optimal bombarding energy. The predicted optimal reaction condition and the expected maximum ER cross section are anticipated by experimentalists. However, the predictions of ER cross sections for SHN are very difficult because the fusion for producing SHN is a very complicated process where the relaxations of the nuclear deformations, angular momentum, and radial kinetic energy are coupled with the mass transfer. Although much progress has been made in understanding the fusion reaction, there are still many open problems to be explored. As mentioned above, the nuclear deformation relaxation plays an important role in the heavy-ion reactions. This kind of dynamical deformations [27-29] developed in the fusion process should be taken into consideration for the investigations in the synthesis of SHN. In this paper, the development of the dynamical deformation is investigated and its effects on the potential energy surface and fusion probability will be investigated with the di-nuclear system model with dynamical potential energy surface (DNS-DynPES model) [30]. The paper is organized as follows. In section 2, an introduction to the theoretical model is given. In section 3, some results about the dynamical deformation and fusion probability are studied. Finally, a summary is given.

2. DNS-DynPES model

In Ref. [30], we developed a DNS-DynPES model. In this section we briefly mention the formalism for the theoretical model. Generally, the production of SHN can be separated into three stages: (1) the capture process in which the system overcomes the Coulomb barrier, (2) the fusion process in which an excited compound nucleus (CN) is formed, (3) the survival process in which the emission of one or more neutron(s) leads the excited CN to an SHN with expected Z. The evaporation residue cross section in a heavy-ion fusion reaction can be written as the summation over all partial waves J as

$$\sigma_{\mathrm{ER}}(E_{\mathrm{c.m.}}) = \frac{\pi\hbar^2}{2\mu E_{\mathrm{c.m.}}} \sum_J (2J+1) T(E_{\mathrm{c.m.}}, J) P(E_{\mathrm{c.m.}}, J) W_{\mathrm{sur}}(E_{\mathrm{c.m.}}, J) \qquad (1)$$

with $E_{\mathrm{c.m.}}$ being the incident energy in the center of mass (c.m.) frame. Here $T(E_{\mathrm{c.m.}}, J)$ is the transmission probability and the survival probability $W_{\mathrm{sur}}(E_{\mathrm{c.m.}}, J)$ is calculated with a statistical model.

The DNS concept was proposed by Volkov [31] and later used to study the competition between complete fusion and quasi-fission and to calculate the fusion probability in fusion reactions [32-34]. The basic idea of the DNS model is that after the capture process, a DNS in the entrance channel is formed. Then the DNS evolves via nucleon transfer along the mass asymmetry coordinate $\eta = (A_1 - A_2)/(A_1 + A_2)$ instead of the direction of the relative distance R between the projectile and target. If the nucleon transfer occurs from the light nucleus to the heavy one, the mass asymmetry increases. Otherwise, if the nucleon transfers from the heavy nucleus to the light one, the mass asymmetry decreases and the quasi-fission rate increases, and that hinders the fusion. Therefore, when η is close to 1, a CN is formed with a high probability, while the decrease of η away from 1, the CN formation is hindered and it may be formed with much lower probability.

We calculate the formation probability P_{CN} of a CN based on the DNS concept. During the nucleon transfer process, any configuration of DNS (A_1, A_2) with $A_1 = 0, 1, \ldots, A_P + A_T$ and $A_2 = A_P + A_T - A_1$ can be formed. The evolution of the distribution function of each DNS with time can be described by a master equation

$$\frac{dP(A_1,t)}{dt}=\sum W_{A_1A_1'}(t)[d_{A_1}P(A_1',t)-d_{A_1'}P(A_1,t)]-\Lambda_{A_1}^{\rm qf}(t)P(A_1,t) \quad (2)$$

Since $A_1 + A_2 = A_P + A_T$, only A_1 is explicitly included in the above equation. d_{A_1} is the microscopic dimension for a DNS with a local excitation energy $E^*_{\rm DNS}$ defined in Eq. (3) which is shared by the two nuclei in this DNS.

For each nucleus, a valence space is opened due to the excitation and those nucleons in the states within the valence space are active for the transfer between the two nuclei. $d_{A_1} = C_{m_1}^{N_1}\cdot C_{m_2}^{N_2}$ where N_k is the number of valence states and m_k is that of valence nucleons. $\Lambda_{A_1}^{\rm qf}(t)$ is the quasi-fission rate of the DNS(A_1, A_2) and $W_{A_1A_1'}=W_{A_1'A_1}$ is the mean transition probability between the DNSs (A_1, A_2) and (A_1', A_2').

In Eq. (2), the microscopic dimension, the quasi-fission rate and the mean transition probability are all related to the local excitation energy of the DNS which reads

$$E^*_{\rm DNS}(A_1,t)=E_{\rm total}-E^0_{\rm DNS}(A_1,t)-E^{\rm rot}_{\rm DNS}(A_1,t) \quad (3)$$

where

$$E_{\rm total}=E_{\rm c.m.}+(M_{\rm T}+M_{\rm P})c^2\ , \quad (4)$$

$$E^0_{\rm DNS}(A_1,t)=V_{\rm DNS}(A_1,t)+(M_1+M_2)c^2\ , \quad (5)$$

$$E^{\rm rot}_{\rm DNS}(A_1,t)=\frac{J(J+1)}{2I_{\rm DNS}(A_1,t)}\ , \quad (6)$$

$$V_{\rm DNS}(A_1,t)=V_{\rm N}(A_1,t)+V_{\rm C}(A_1,t)\ . \quad (7)$$

Here $V_{\rm N}(A_1,t)$ is the nuclear potential calculated with a double-folding method [35] and $V_{\rm C}(A_1,t)$ is the Coulomb interactions potential from the Wong's formula [36]. In this work a tip-tip orientation of the two deformed nuclei is assumed. The potential energy in the mass asymmetry degree of freedom (the driving potential) at t=0, is defined as

$$V_{\rm PES}(A_1,t)\equiv V_{\rm N}(A_1,t)+V_{\rm C}(A_1,t)+(M_1+M_2-M_{\rm P}-M_{\rm T})c^2 \quad (8)$$

The interaction potential between the two nuclei is related to the distance between two centers and in this paper the interaction potential takes the minimum value in the pocket. Because of the attractive nuclear force and the repulsive Coulomb force, both nuclei in a DNS are distorted and dynamical deformations develop during the process of nuclear reactions [27-29]. This results in a time-dependence of the potential energy surface (PES). The dynamical deformations of the two nuclei satisfy $\delta\beta_1^2 C_1/A_1=\delta\beta_2^2 C_2/A_2$ [13] with the stiffness parameter C_i (i=1 and 2) calculated from a liquid drop model [37]. The total dynamical deformation is the average of the dynamical deformations of the two nuclei $\delta\beta=(\delta\beta_1+\delta\beta_2)/2$. Following Refs. [27-29], we assume that the dynamical deformation evolves in an over-damped motion,

$$\delta\beta_t(t)=\delta\beta_{\rm max}(1-e^{-t/\tau_{\rm def}})\ . \quad (9)$$

In Eq. (9), the relaxation time $\tau_{\rm def}$ =40×10^{-22} s and the maximal dynamical deformation is found when the total "intrinsic" energy reaches the minimum,

$$E_{\rm int}(A_1,\delta\beta)=V_{\rm N}(A_1;\beta_1,\beta_2)+V_{\rm C}(A_1;\beta_1,\beta_2)+\sum_{i=1,2}\frac{1}{2}C_i\delta\beta_i^2 \quad (10)$$

where the quadrupole deformation $\beta_i=\beta_i^0+\delta\beta_i$ (i = 1 or 2) with the static deformation parameter β_i^0 taken from Ref. [38].

3. Results and discussion

The intrinsic energy of the di-nuclear system $^{48}Ca+^{244}Pu$ as a function of nuclear dynamical deformation is shown in Fig.1. For the configuration above, the maximum dynamical deformation for the system is about 0.31, where the minimum value for the intrinsic energy could be found and the maximum dynamical deformation $\delta\beta_{max}$ is indicated as dashed line in the figure. The intrinsic energy increases as increasing the dynamical deformation when it is larger than $\delta\beta_{max}$. As explained in Ref. [39], the phenomenon can be attributed to the rise of the nuclear energies due to their distortions. The interaction potential for the di-nuclear system $^{48}Ca+^{244}Pu$ as a function of dynamical deformation is also indicated in Fig. 1 and it is found that the interaction potential decreases as increasing the dynamical deformation. The interaction potential at the maximum dynamical deformation is about 145 MeV.

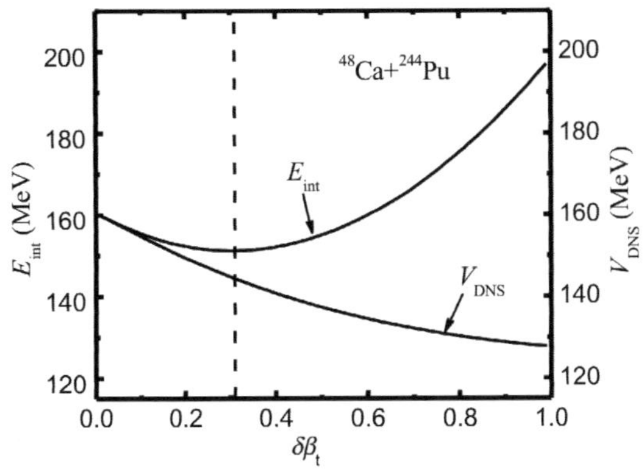

Fig. 1 Intrinsic energy and potential energy as functions of the dynamical deformation for the reaction . The maximal dynamical deformation is indicated with the vertical dashed line.

The dynamical deformations for the two fragments are shown in Fig.2. The solid and dashed lines are for the dynamical deformations of ^{48}Ca and ^{244}Pu, respectively. The dynamical deformations are found to be increasing in the beginning 120×10^{-22} seconds and then gradually get saturated. The heavier fragment may get more significant dynamical deformation. The saturated dynamical deformation parameters for ^{244}Pu and ^{48}Ca are 0.42 and 0.18, respectively.

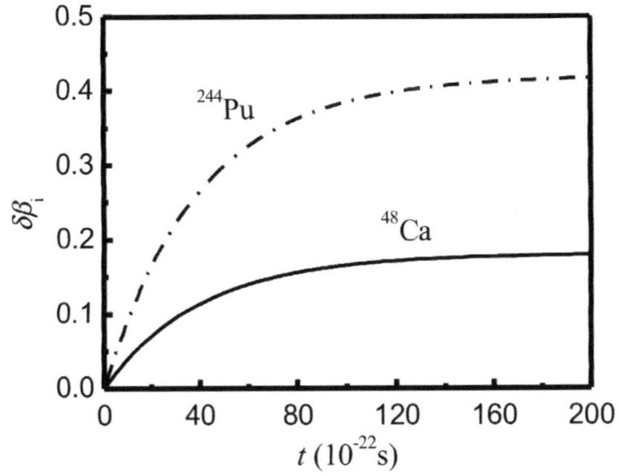

Fig. 2 Dynamical deformations for ^{48}Ca and ^{244}Pu as a function of the reaction time. Taken from Ref. [39]

.The maximum dynamical deformations for the incident channel $^{48}Ca+^{244}Pu$ as a function of mass asymmetry are shown in Fig. 3. It can be seen that the maximum dynamical deformations are relatively large for the symmetric configurations with the mass asymmetry ranging between –0.5 and 0.5 and are decreasing as increasing the mass asymmetry of the di-nuclear configuration. The maximum dynamical deformation for the incident channel is about 0.3 as the dashed line indicates. For some very asymmetric cases, the maximum dynamical deformations are nearly zero.

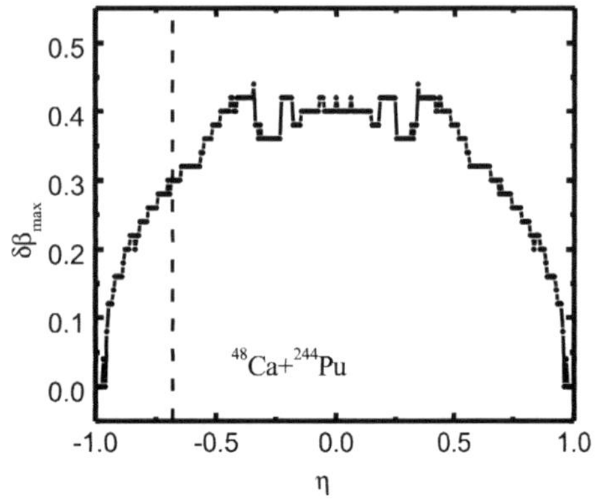

Fig. 3 Maximum dynamical deformations as a function of the mass asymmetry for the $^{48}Ca+^{244}Pu$ reaction.

The potential energy surface, namely the driving potential, for the reaction $^{48}Ca+^{244}Pu$ is shown in Fig.4. The di-nuclear configurations from the incident one evolve along the mass asymmetry degree of freedom. To form a CN, the di-nuclear configuration must overcome the BG point on the driving potential and the barrier between the BG point and the incident channel is defined as inner fusion barrier. The dashed line is for the case when the dynamical deformation is fully developed and the solid line is for the case without dynamical deformation. It is found that the potential energy surface changes significantly due to the nuclear dynamical deformations. The potential energies for the incident channel are −0.808 MeV and −16.392 MeV for the cases when t=0 (solid line) and the dynamical deformations are fully developed (dashed line), respectively. In Fig. 4, the inner fusion barrier for the two cases are found to be 15.4 MeV and 26.8 MeV, respectively. It can be seen that the inner fusion barrier increases significantly due to the inclusion of the dynamical deformation and the fusion probability may be hindered for the increase of the inner fusion barrier.

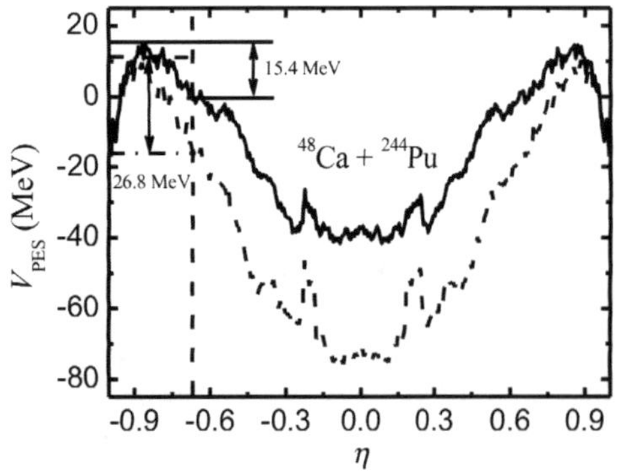

Fig. 4 Potential energy surface for the reaction $^{48}Ca+^{244}Pu$. Taken from Ref. [39].

The fusion probability for the reaction ^{48}Ca + ^{244}Pu at zero angular momentum as a function of the incident energy is depicted in Fig. 5(a). The solid line and dashed line are for the cases with and without dynamical deformation, respectively. It is found that the fusion probabilities increase with the incident energy for both the cases. When the incident energy is below 210 MeV, the fusion probabilities increase rapidly. After that, the fusion probabilities increase slowly with increasing the incident energy. It is also found that the fusion probability decreases significantly due to the dynamical deformation. For the case of ^{48}Ca+^{244}Pu, the fusion probability with dynamical deformation is about one order of magnitude smaller than that for the case without dynamical deformation. The inclusion of the dynamical deformation leads to the decrease of the fusion probability.

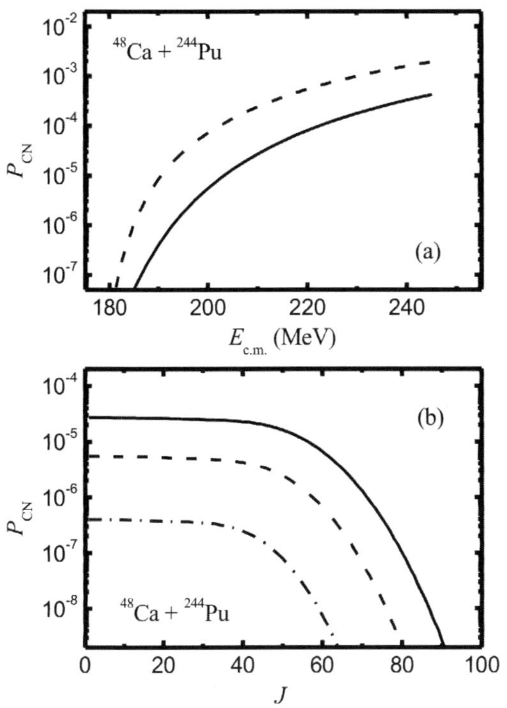

Fig. 5 **(a)** Fusion probabilities for the reaction ^{48}Ca + ^{244}Pu at J=0 as a function of the incident energy in c.m. frame. The solid and dashed lines are for with and without dynamical deformations, respectively. Taken from Ref. [39]. (b) Fusion probabilities for the reaction ^{48}Ca + ^{244}Pu as a function of angular momentum in unit of for $E_{c.m.}$ being 210 (solid line), 200 (dashed line), and 190 MeV (dash dotted line), respectively.

The fusion probabilities for the reaction ^{48}Ca + ^{244}Pu as a function of the angular momentum (in unit of $\hbar$) at the incident energies $E_{c.m.}$ = 210 MeV, 200 MeV, and 190 MeV are also depicted in Fig.5(b). The solid, dashed, and dash dotted lines are for the cases with incident energy being 210 MeV, 200 MeV, and 190 MeV, respectively. The Q value for the reaction is about 161 MeV. It is found that the fusion probability decreases as increasing the angular momentum. In the region with small angular momentum, the fusion probability decreases slowly while it drops rapidly when the angular momentum is larger than about 40 $\hbar$.

4. Results and discussion

The nuclear dynamical deformation effects in the fusion reactions for the production of superheavy nuclei are studied within the framework of the di-nuclear system model with dynamical potential energy surface (DNS-DynPES model). The development of the dynamical deformation for the incident channel is investigated and the maximum dynamical deformations for different configurations are also calculated. The influences of the dynamical deformations on the potential energy surface (PES) and fusion probability are studied. The calculation results indicate that the structure of the PES changes significantly because of the dynamical deformations. The fusion for synthesizing SHN is also hindered and it is found the fusion probabilities decrease by nearly one order of magnitude because of the inclusion of the dynamical deformations. The fusion probability as a function of angular momentum is found to decrease with increasing the angular momentum.

Acknowledgement

This work is supported by the National Natural Science Foundation of China (NSFC) projects (No.10975100, 11275098, 1121112015211275248, 11175252, 11120101005), MOST of China (Project No. 2013CB834400) and the Knowledge Innovation Project of CAS (Projects No. KJCX2-EW-N01). Some of the numerical results were obtained on the ScGrid of the Supercomputing Center, CNIC of CAS.

[1] Hofmann S and Munzenberg G 2000 *Rev. Mod. Phys.* **72** 733
[2] Morita K et al. 2004 *J. Phys. Soc. Jpn.* **73** 2593
[3] Oganessian Yu T 2007 *J. Phys. G.* **34** R165
[4] Oganessian Yu T et al. 2010 *Phys. Rev. Lett.* **104** 142502
[5] Zhang Z Y et al. 2012 *Chin. Phys. Lett.* **29** 012502
[6] Bjørnholm S and Swiatecki W 1982 *Nucl. Phys. A* **391** 471
[7] Liu Z H and Bao J D 2009 *Phys. Rev. C* **80** 054608
[8] Liu Z H and Bao J D 2009 *Phys. Rev. C* **83** 044613
[9] Siwek-Wilczynska K, Cap T and Wilczynski J 2010 *Int. J. Mod. Phys. E 19* 500
[10] Siwek-Wilczynska K, Cap T and Kowal M 2012 *Phys. Rev. C* **86** 014611
[11] Liang Y J et al 2012 *Phys. Rev. C* **83** 037602
[12] Shen C et al 2002 *Phys. Rev. C* **66** 061602(R)
[13] Zagrebaev V I 2001 *Phys. Rev. C* **64** 034606
[14] Zagrebaev V I and Greiner W 2008 *Phys. Rev. C* **78** 034610
[15] Adamian G G, Antonenko N V and Scheid W 1997 *Nucl. Phys. A* **618** 176
[16] Adamian G G, Antonenko N V and Scheid W 2004 *Phys. Rev. C* **69** *011601*
[17] Diaz-Torres A, Adamian G G, Antonenko N V and Scheid W 2001 *Phys. Rev. C* **64** 024604
[18] Giardina G, Hofmann S, Muminov A I and Nasirov AK 2000 *Eur. Phys. J. A* **8** *205*
[19] Mandaglio G, Giardina G, Nasirov AK and Sobiczewski A 2012 *Phys. Rev. C* **86** 064607
Nasirov AK et al 2013 *Eur. Phys. J. A* **49** 147
[20] Li W et al 2003 *Eurphys. Lett.* **64** 750
[21] Li W et al 2006 *J. Phys. G.* **32** 1143
[22] Wang N, Li J Q and Zhao E G 2008 *Phys. Rev. C* **78** 054607
[23] Wang N, Zhou S G and Zhao E G 2010 *Chin. Phys. Lett.* **34** 1615
[24] Wang N, Zhao E G and Schied W 2010 *Chin. Phys. Lett.* **27** 061502

[25] Gan Z G et al. 2011 *Sci. Chin. Phys. Mech. Astron.* **54** s61
Huang M. et al 2011 *Phys. Rev. C* **84** 064619
[26] Wang N, Tian J and Schied W 2011 *Phys. Rev. C* **84** 061601(R)
[27] Riedel C, Wolschin G and Noerenberg W 1979 *Z. Phys. A* **290** 47
[28] Riedel and Noerenberg W 1979 *Z. Phys. A* **290** 385
[29] Wolschin G 1979 *Phys. Lett. B* **88** 35
[30] Wang N, Zhao E G, Schied W and Zhou S G 2012 *Phys. Rev. C* **85** 041601(R)
[31] Volkov V V 1978 *Phys. Rep.* **44** 93
[32] Antonenko N V, Cherepanov E A and Nasirov A K1993 *Phys. Lett. B* **319** 425
[33] Antonenko N V, Cherepanov E A and Nasirov A K1995 *Phys. Rev. C* **51** 2635
[34] Adamian G G, Antonenko N V and Scheid W1997 *Nucl. Phys. A* **627** 361
[35] Adamian G G et al. 1996 *Int*. J. Mod. *Phys. E* **5** 191
[36] Wong C Y1973 *Phys. Rev. Lett.* **31** 766
[37] Myers W D and Swiatecki W J1966 *Nucl. Phys.* **81** 1
[38] Moeller P et al. 1995 *Data Nucl. Tables* **59** 185
[39] Wang N, Peng J S, Zhao E G and Zhou S G 2013 *Nucl. Phys. Rev.* **30** 294 (in Chinese)

Role of wave packet width in quantum molecular dynamics in fusion reactions near barrier

X. G. Cao[1,a], Y. G. Ma[1,b], G. Q. Zhang[1], H. W. Wang[1], A. Anastasi[2], F. Curciarello[2], V. De Leo[2]

E-mail: [a]caoxiguang@sinap.ac.cn; [b]ygma@sinap.ac.cn
[1]Shanghai Institute of Applied Physics, Chinese Academy of Sciences, Shanghai 201800, China
[2]Istituto Nazionale di Fisica Nucleare, Sezione di Catania, and Dipartimento di Fisica e di Scienze della Terra dell'Università di Messina, I-98166 Messina, Italy

Abstract. The dynamical fusion process of ^{48}Ca + ^{144}Sm with different impact parameters near barrier is studied by an extended quantum molecular dynamics (EQMD) model, where width of wavepacket is dynamically treated based on variational principle. The time evolution of different energy components such as potential energy, kinetic energy, Coulomb energy and Pauli potential are analyzed when dynamical or fixed width is assumed in calculation. It is found that the dynamical wavepacket width can enhance the dissipation of incident energy and the fluctuations, which are important to form compound nuclei. Moreover, we compare the fusion barrier dependence on the incident energy when it is determined by both dynamical and fixed wavepacket width.

1. Introduction

The use of dynamical process is important in order to understand the formation of complete fusion in a heavy ion reaction. In experiment, it is hard to correctly determine the compound nucleus cross section in massive nucleus reactions since it is difficult to unambiguously identify the complete fusion products from quasifission fragments [1–5]. The determination of superheavy nuclei (SHN) cross section is overestimated if the fusion-fission products from compound nuclei strongly overlap with the one of quasifission process. Therefore it is highly needed to study the quasifission and fusion-fission dynamics and how the nuclear system evolves from the capture forming the dinuclear system to statistically equilibrated compound nucleus (CN).

In the last decades, in some almost mass symmetric reaction (namely "cold fusion reaction") were used to synthesize some superheavy elements (SHEs) with atomic numbers from 107 to 112, by using double magic target nuclei such as ^{208}Pb, ^{209}Bi, and projectile nuclei heavier than ^{40}Ar [6]. For heavier SHEs, mass asymmetric reactions (namely "hot fusion reaction") were used to synthesize elements with Z up to 118, by using ^{48}Ca as beam and actinide nuclei as targets [7].

However, to go beyond Z = 118, there are not results for Z = 119 and Z = 120 elements by using reactions such as like ^{50}Ti, ^{54}Cr, ^{58}Fe and ^{64}Ni (as beams) + Actinides (as targets) [8]. Furthermore, all synthesized transfermium nuclei are neutron-deficient, lying above the β stability line. Therefore, new reaction mechanisms and methods are mandatory for exploring neutron rich superheavy nuclei in north-east direction. Alternative reaction mechanisms –such as fusion with radioactive neutron-rich projectile, transfer reactions in damped collisions with very heavy projectile and target nuclei– are very promising to access neutron-rich side and superheavy nuclei beyond Z = 118.

The macroscopic liquid and dinuclear models are widely used in fusion reactions near barrier. However, for damped collisions with massive nuclei, a large number of degrees of freedom related to the excitation and deformed projectile and target nuclei, the neck formation, nucleon transfer, different types of re-separation of the composite system and nucleon emission, will simultaneously play relevant roles. Therefore, it is difficult to have a realistic description of the dynamical process.

The microscopic dynamic approaches, such as time-dependent Hartree-Fock (TDHF) [9] and quantum molecular dynamics (QMD) [10, 11] have been used to consider the full dynamical degrees of freedom highly needed to determine the fusion reaction with massive nuclei. For TDHF and QMD, the calculations are fully microscopic, without assuming any reaction mechanism and cluster formation in advance. It is found that the dynamical conditions produce apparent effects on a series of processes such as barrier distribution, pre-equilibrium collective motion, nucleons transfer, dynamical deformation and equilibration of the shape etc..

TDHF is essentially one-body theory, where the fluctuation and correlation are not treated sufficiently and explicitly. QMD is based on many-body theory and it can link the simulation to experimental observables. In comparison with TDHF, the QMD model can also be used at higher energies besides around barrier. The recent Langevin dynamics + quantum molecular dynamics calculations show that the non-Gaussian random forces play a very important role in the dissipation dynamics of the relative motion between target and projectile nuclei [12]. However, the wavepacket width is treated as fixed and uniform for every nucleon in QMD-like models [13–15] except that the extended quantum molecular dynamics (EQMD) [16, 17], fermionic molecular dynamics (FMD) [18] and one generalization of Gaussian trial wave functions in quantum molecular dynamics [20]. The dynamical width provides the nonclassical degree of freedom and it is expected to compensate some important quantum effects, which is insufficiently considered in the traditional QMD model. In this paper, the fusion process of ^{48}Ca + ^{144}Sm near barrier is studied by the EQMD model. The role of how dynamical width of wavepacket affects the energy dissipation of incident energy is carefully checked by comparing with the case of fixed and uniform wavepacket.

2. Model and approach

In the present paper, we use a microscopic transport dynamics model, EQMD, to simulate the time evolution of nuclear phase-space. This model was presented by Maruyama [16], and also used in [17]. In addition to advantage of dynamical wavepacket width, this model has a much better conversion of energy and enough long stability for initialized ground nucleus. It can also well describe the ground properties of nuclei through the periodic table. Therefore, it is very suitable to study low energy fusion reactions. A brief introduction of EQMD model is shown as following.

In the EQMD model, the total wave function of the nuclear system was assumed as a product of Gaussian wave packets of all nucleons

$$\Psi = \prod_i \phi_i(\mathbf{r}_i), \tag{1}$$

where

$$\phi_i(\mathbf{r}_i) = \left(\frac{\nu_i + \nu_i^*}{2\pi}\right)^{3/4} \exp\left[-\frac{\nu_i}{2}(\mathbf{r}_i - \mathbf{R}_i)^2 + \frac{i}{\hbar}\mathbf{P}_i \cdot \mathbf{r}_i\right]. \tag{2}$$

Here $\mathbf{R}_i$ and $\mathbf{P}_i$ are the centers of position and momentum of the ith wavepacket. ν_i is the Gaussian width written as

$$\nu_i \equiv \frac{1}{\lambda_i} + i\delta_i, \tag{3}$$

where λ_i and δ_i are its real and imaginary parts, respectively. The Hamiltonian is written as

$$H = \left\langle \Psi \middle| \sum_i -\frac{\hbar^2}{2m}\nabla_i^2 - \hat{T}_{c.m.} + \hat{H}_{int} \middle| \Psi \right\rangle \tag{4}$$

$$= \sum_i \left[\frac{\mathbf{P}_i^2}{2m} + \frac{3\hbar^2(1 + \lambda_i^2\delta_i^2)}{4m\lambda_i}\right] - T_{c.m.} + H_{int}, \tag{5}$$

where $T_{c.m.}$ and H_{int} denote the spurious zero-point center-of-mass (CM) kinetic energy and the potential energy term, respectively. For the effective interaction, this model uses Skyrme, Coulomb, symmetry, and the Pauli potential

$$H_{int} = H_{Skyrme} + H_{Coulomb} + H_{symmetry} + H_{Pauli}. \tag{6}$$

Specifically, the Pauli potential is written as

$$H_{Pauli} = \frac{c_P}{2}\sum_i (f_i - f_0)^{\mu}\theta(f_i - f_0), \tag{7}$$

$$f_i \equiv \sum_j \delta(S_i, S_j)\delta(T_i, T_j)| < \phi_i|\phi_j > |^2, \tag{8}$$

where f_i is the overlap of a nucleon i with the same kind of nucleons (including itself), c_P is the strength of the potential and f_0 is the threshold parameter ~1. S and T denote spin and isospin of a nucleon, respectively.

The phase-space is obtained firstly from a random configuration. To get the energy-minimum states as initial ground state, one can solve the damped equations of motion as

$$\dot{\mathbf{R}}_i = \frac{\partial H}{\partial \mathbf{P}_i} + \mu_R \frac{\partial H}{\partial \mathbf{R}_i}, \quad \dot{\mathbf{P}}_i = \frac{\partial H}{\partial \mathbf{R}_i} + \mu_P \frac{\partial H}{\partial \mathbf{P}_i}, \tag{9}$$
$$\frac{3\hbar}{4}\dot{\lambda}_i = -\frac{\partial H}{\partial \delta_i} + \mu_\lambda \frac{\partial H}{\partial \lambda_i}, \quad \frac{3\hbar}{4}\dot{\delta}_i = \frac{\partial H}{\partial \lambda_i} + \mu_\delta \frac{\partial H}{\partial \delta_i}.$$

Here μ_R, μ_P, μ_λ, and μ_δ are damping coefficients. After this cooling process, the system goes to its energy minimum point. For free evolution of coordinate, momentum and corresponding widths in reaction, μ_R, μ_P, μ_λ, and μ_δ are set to zero.

The proper initialization of target and projectile nuclei are necessary to describe reasonably the reactions. Especially for fusion near barrier, one of the necessary conditions, which must be satisfied, is that the stability of initialized nuclei has to be taken for enough time and at meantime no spurious nucleon and cluster emission occur, in comparison with the time scale of fusion dynamical process. The stability checks: binding energy and root mean square (RMS) radius vs evolution time are shown in Fig. 1 (a) and (b), respectively. The binding energy of initialized ^{48}Ca and ^{144}Sm is 8.305 MeV and 7.659 MeV, respectively, which are near the experimental binding energy: 8.667 MeV and 8.304 MeV [19], respectively, for ^{48}Ca and ^{144}Sm. These good initialized samples are stored for fusion simulations. The fluctuation of binding energy and RMS radius are more than one order smaller compared with many other version QMD models, where fluctuation of binding energy and RMS radius can be as large as 0.15 MeV/u and 0.3 fm, respectively, for the best initialized nucleus such as ^{144}Sm. The small enough and negligible energy fluctuation is very important for fusion with massive nuclei at very low energy since a fraction fluctuation of binding energy per nucleon will bring a big enough total energy fluctuation into the fusion system, compared with the incident energy.

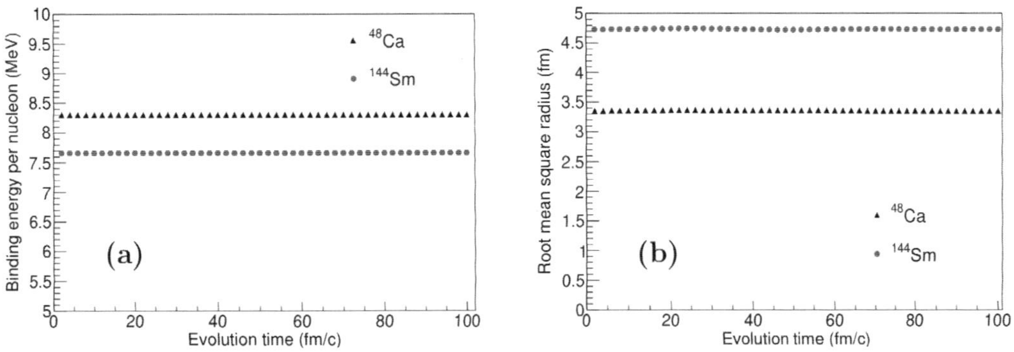

Figure 1. (Color online). (a) The examination of the average binding energy for initialized ^{48}Ca and ^{144}Sm versus the evolution time; (b) The examination of the root mean square radius for initialized ^{48}Ca and ^{144}Sm vs the evolution time.

3. Results and discussion

One of the distinct feature in our fusion calculations is the dynamical evolution of the wavepacket width. The dynamical evolution of one wavepacket width in ^{48}Ca + ^{144}Sm with CM frame energy 180 MeV is shown in Figure 2. Large fluctuations can be seen clearly, which has not be considered in fusion reaction before. The average width increases till 1400 fm/c and then saturates even if large fluctuations are present. The rise of average width corresponds to the early stage of projectile-target touching. In later stage, the larger width is due to the compound system formation, and to the nucleon diffusion between the projectile and target contact region. The momentum width largely fluctuates. However, it is perfectly anti correlated with the coordinate width. The product of them well satisfies the uncertainty principle shown by the green dotted line in Figure 2. The lines in Figure 2 show that the width evolutions are reasonable and reliable in our calculations.

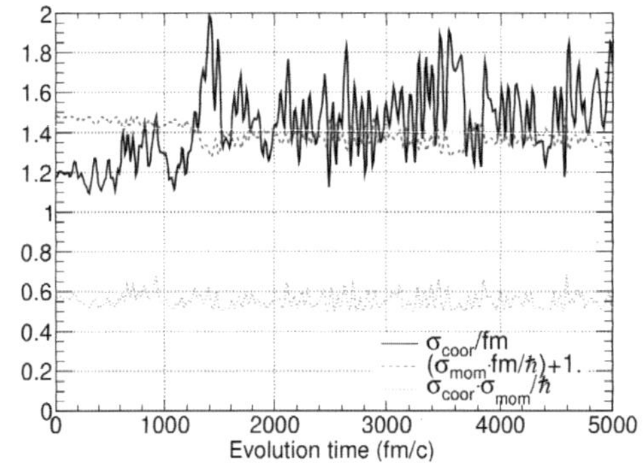

Figure 2. (Color online). Coordinate width (black solid line), momentum width (red dashed line) and the product of coordinate width and momentum width (green dotted line) of one wavepacket as functions of the evolution time for a ^{48}Ca + ^{144}Sm event at 180 MeV CM energy.

In Figure 3, the functions of dynamical width of wavepacket can be clearly seen through the time evolution of the dynamical nucleus-nucleus interaction potential in comparison with fixed wavepacket calculations. The nucleus-nucleus interaction potential is defined as

$$V(R) = E_{tot} - E_{tar}(R) - E_{proj}(R), \tag{10}$$

where R is relative distance between centers of mass of target and projectile, and E_{tot}, $E_{tar}(R)$, and $E_{proj}(R)$ represent the total energy of the system, target, and projectile, respectively; they can be calculated from the energy density function. Each of them includes kinetic energy as well as all potential energy components as reported in equation (6). From panel (a) we can see that the dynamical width leads to a more faster fusion process and a more deeper binding between target and projectile, in comparison with the fixed width. However, there is no much difference for these two cases when the distance between target and projectile is larger than 10 fm, as shown in panel (b) of Figure 3. After the touching point, a large difference emerges between the two approaches. The nucleus-nucleus potential becomes more soft due to the fact that the dynamical width plays an important role of additional degree of freedom. Therefore, it should be easier and quicker to reach equilibrium for energy, shape, isospin, etc. when the dynamical width is considered.

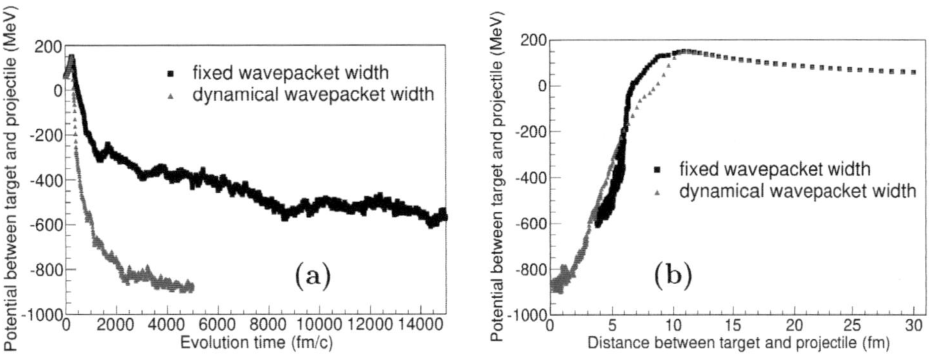

Figure 3. (Color online). (a) Potential between target and projectile for fixed and dynamical wavepacket width in the process of approaching as a function of the evolution time in the ^{48}Ca + ^{144}Sm head-on fusion with 180 MeV CM energy; (b) The same as (a) but for dependence of distance between target and projectile.

In order to clarify the energy dissipation process before equilibrium, the total potential energy, total kinetic energy, Coulomb energy of projectile and Pauli potential of target and projectile as functions of the distance between target and projectile, are shown in panels (a), (b), (c), (d), and (e) of Figure 4, respectively. The anti correlation between potential energy and kinetic energy is described well since EQMD keeps energy conservation with a very good accuracy. Projectile with dynamical width has larger Coulomb energy than the one with fixed width at short distance, since the former has more compact shape when the target and projectile merge together. At the path of approaching, the Coulomb energy does not have apparent difference. However, for Pauli potential, the case is different. Pauli potential from target anti correlates strongly with projectile for dynamical width, which is based on the fact that the Pauli potential is calculated from overlap between wave packets shown by equation (7) and (8). From Liouville theorem, we know that the total size of phase space of system is constant. Therefore, more larger overlap between nucleons in target, more smaller overlap in projectile, and vice versa. The Coulomb and Pauli potential play dominant roles on the dynamical nucleus-nucleus potential at large distance. They both push the target and projectile to adjust their density distributions in the approaching process.

The dependence of the potential evolution on different impact parameter is calculated, as shown in Figure 5. From panel (a) we can see that the speed of energy dissipation in head-on collisions is not the fastest compared with all other cases. Fusion with impact parameter around 2 fm evolves fastest towards equilibrium.

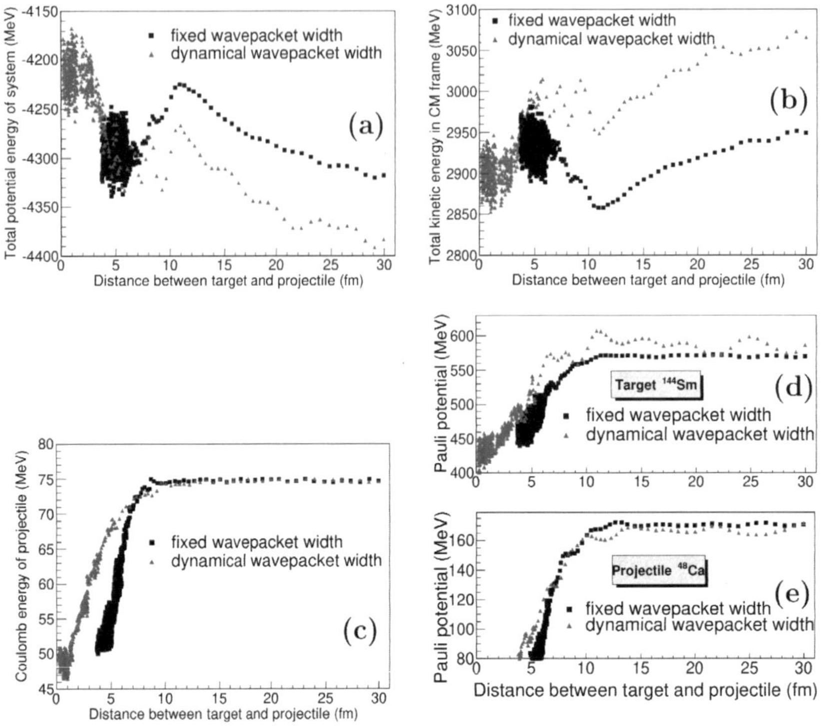

Figure 4. (Color online). Total potential energy, total kinetic energy in CM frame, Coulomb energy of projectile, Pauli potential of target and projectile in ^{48}Ca + ^{144}Sm fusion with 180MeV CM energy head-on collision as functions of distance between target and projectile are shown by (a), (b), (c), (d) and (e), respectively, where the calculations with fixed and dynamical wavepacket width are compared in each figures.

For impact parameters larger than 4 fm, obviously a longer time is needed to reach equilibrium. Panel (b) shows that the evolution path with larger impact parameters is more complicated with walking back.

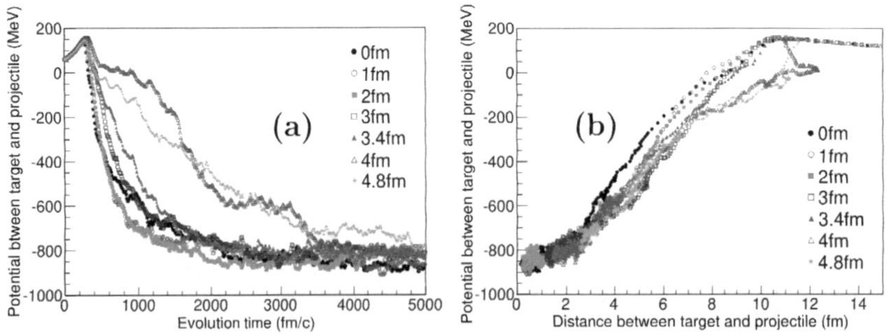

Figure 5. (Color online). Potential between target and projectile with dynamical width in ^{48}Ca + ^{144}Sm fusion with 180MeV CM energy as a function of evolution time and distance between target and projectile are shown by (a) and (b), respectively, where the calculations with different impact parameters are compared in each figures.

The energy dependence of the fusion barrier is extracted from head-on collisions for both dynamic and fixed wavepacket width shown in Figure 6 . The barrier decreases with the decrease of incident energy in the two different calculations. However, the dynamical wavepacket width reduces the barrier height a lot, which is based on the fact that dynamical wavepacket width, acting as another additional degree of freedom, allows the approaching target and projectile to adjust their densities quickly. This should be very different from the nucleus-nucleus barrier extracted from frozen approximation.

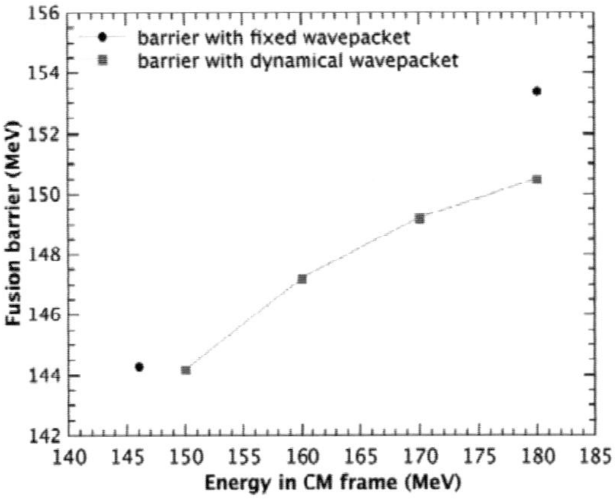

Figure 6. (Color online). Fusion barrier from ^{48}Ca + ^{144}Sm head-on fusion as a function of incident energy in CM frame, where the fixed wavepacket width is calculated with two incident energies.

4. Summary

The role of the dynamical width of wavepacket is studied in the ^{48}Ca + ^{144}Sm fusion reaction near barrier energy by the EQMD model. In comparison with the fixed width (widely used before), the dynamical wavepacket width, as additional degree of freedom which before was scarcely considered in the QMD model, can enhance the fluctuation and accelerate the incident energy dissipation in the target-projectile approaching process. As a result, the fusion barrier is strongly reduced and it is easier to run towards equilibrium in calculations with dynamical width.

5. Acknowledgements

This work is partially supported by Major State Basic Research Development Program in China under Contract Nos. 2014CB845401 and 2013CB834405, National Natural Science Foundation of China under Contract Nos. 11305239, 11075195 and 11205230. X. G. Cao is grateful to G. Giardina, A. K. Nasirov, and G. Mandaglio for the interesting and fruitful discussions.

References

[1] Giardina G, Nasirov A K, Mandaglio G, Curciarello F, De Leo V, Fazio G, Manganaro M, Romaniuk M and Saccá C 2011 *Journal of Physics: Conference Series* **282** 012006
[2] Mandaglio G, Nasirov A K, Curciarello F, De Leo V, Romaniuk M, Fazio G and Giardina G 2013 *Journal of Physics: Conference Series* **420** 012008
[3] Nasirov A K, Giardina G, Mandaglio G, Manganaro M, Hanappe F, Heinz S, Hofmann S, Muminov A I and Scheid W 2009 *Phys. Rev. C* **79**(2) 024606
[4] Nasirov A K, Mandaglio G, Giardina G, Sobiczewski A and Muminov A I 2011 *Phys. Rev. C* **84** 044612
[5] Giardina G, Nasirov A K, Mandaglio G, Curciarello F, De Leo V, Fazio G and Romaniuk M 2013 *Nuclear Science and Techniques* **24** 050519
[6] Hofmann S 2011 *Radiochim. Acta* **99** 405–428
[7] Hofmann S 2010 *Int. J. Mod. Phys. E* **19** 483–492
[8] Oganessian Y 2013 *Nucl. Phys. News* **23** 15–21
[9] Simenel C 2012 *Euro. Phys. J. A* **48** 152
[10] Feng Z Q, Jin G M and Li J Q 2009 *Phys. Rev. C* **80** 067601
[11] Zhu L, Su J, Xie W J and Zhang F S 2013 *Nucl. Phys. A* **915** 90–105
[12] Wen K, Sakata F, Li Z X, Wu X Z, Zhang Y X and Zhou S G 2013 *Phys. Rev. Lett.* **111**(1) 012501
[13] Cao X G, Zhang G Q, Cai X Z, Ma Y G, Guo W, Chen J G, Tian W D, Fang D Q and Wang H W 2010 *Phys. Rev. C* **81**(6) 061603
[14] Cao X G, Cai X Z, Ma Y G, Fang D Q, Zhang G Q, Guo W, Chen J G and Wang J S 2012 *Phys. Rev. C* **86**(4) 044620
[15] Zhang G Q, Ma Y G, Cao X G, Zhou C L, Cai X Z, Fang D Q, Tian W D and Wang H W 2011 *Phys. Rev. C* **84**(3) 034612
[16] Maruyama T, Niita K and Iwamoto A 1996 *Phys. Rev. C* **53** 297
[17] Wada R, Hagel K, Cibor J, Li J, Marie N, Shen W Q, Zhao Y, Natowitz J B and Ono A 1998 *Phys. Lett. B* **422** 6–12
[18] Feldmeier H and Schnack J 2000 *Rev. Mod. Phys.* **72** 655–688
[19] Kiderlen D and Danielewicz P 1997 *Nucl. Phys. A* **620** 346–362 ISSN 0375-9474
[20] Wang M, Audi G, Wapstra A, Kondev F, MacCormick M, Xu X and Pfeiffer B 2012 *Chin. Phys. C* **36** 1603

AUTHOR INDEX

AUTHOR INDEX

Institute of Physics
Dirac House, Temple Back
Bristol BS1 6BE UK

CURRAN ASSOCIATES INC
proceedings
.com